DIMENSIONS V

D1511669

90 16.40

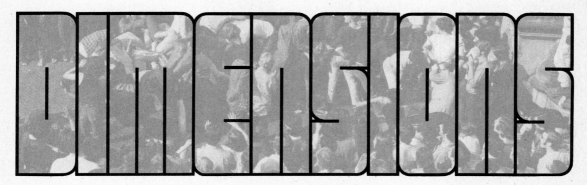

A CHANGING CONCEPT OF HEALTH

FIFTH EDITION

KENNETH L. JONES
LOUIS W. SHAINBERG
CURTIS O. BYER
Mt. San Antonio College

HARPER & ROW, PUBLISHERS, New York
Cambridge, Philadelphia, San Francisco,
London, Mexico City, São Paulo, Sydney

1817

Sponsoring Editor: Claudia Wilson
Project Editor: Jon Dash
Text and cover designed by Michel Craig
Production Manager: Willie Lane
Photo Researcher: Mira Schachne
Compositor: Ruttle, Shaw & Wetherill, Inc.
Printer and Binder: The Murray Printing Company
Outline type for display designed by Michel Craig
Art Studio: Fine Line, Inc.
Cover Photo: Four by Five, Inc.

Photo credits: i, iii, v, ix Michel Craig; 1, 17 Ritscher, Stock, Boston; 6 Brandeis University; 7 Stone, Peter Arnold; 8 Weldon, de Wys; 10 de Wys; 11 Southwick, Stock, Boston; 12 Hamlin, Stock, Boston; 16 Forsyth, Monkmeyer; 20 Skytta, Jeroboam; 28 Reno, Jeroboam; 29 Stone, Peter Arnold; 29 O'Neil, Stock, Boston; 30 Diakopoulos, Stock, Boston; 31 Bodin, Stock, Boston; 32 Webb, Magnum; 33 Kalver, Magnum; 35 Conrad, de Wys; 36 Kaye, de Wys; 37, 41 Mercado, Jeroboam; 47 Carrol, Nancy Palmer; 48 Chih, Peter Arnold; 49 The Menninger Foundation; 51 Rieger, de Wys; 52 McNamara, Nancy Palmer; 54 Coit, Jeroboam; 59 Hrynewych, Stock, Boston; 60 Michael Fenster; 61 Beckwith Studios; 64 Fusco, Magnum; 68, 69 Conklin, Monkmeyer; 70 Weldon, de Wys; 71 Berndt, Stock, Boston; 72 Stone, Peter Arnold; 73 Beckwith Studios; 75 Wolinsky, Stock, Boston; 76 Bartlett, 1980, Woodfin Camp; 79 de Wys; 82, 108 Wide World; 86 Footho rap, Jeroboam; 91 Addiction Research Foundation of Ontario; 95 Shames, Photon West; 107 Smolan, Stock, Boston; 110 Albertson, Stock, Boston; 111 Weldon, de Wys; 117 Franken, Stock, Boston; 121 de Wys; 122 Weldon, de Wys; 123 National Clearing House for Alcohol Information; 126 Kliewe, Jeroboam; 140 Addiction Research Foundation of Ontario; 141 Siteman, Stock, Boston; 142 Malloch, Magnum; 146 Wide World; 153, 226 de Wys; 171 Anderson, Photo Trends; 178 Hopker, Woodfin Camp; 182 Caraballo, Monkmeyer; 183 Strickler, Monkmeyer; 186 Herwig, Stock, Boston; 187 Vilms, 1979, Jeroboam; 195 Siteman, Stock, Boston; 202 Naval Medical Research Institute; 206 Kliewe, Jeroboam; 209 Smolan, Stock, Boston; 211 Urtis, Peter Arnold; 213 Anderson, 1978, Woodfin Camp; 216 Hulstein, de Wys; 216 Peter Arnold; 217 Liebman, de Wys; 218 Weisbrot, Stock, Boston; 225 Joel Gordon, 1981; 228 Christelow, Jeroboam; 231 Grace, Stock, Boston; 236 Beckwith Studios; 240 Glenn, Magnum; 244 Weldon, de Wys; 248 FDA; 252 Beckwith Studios; 266 Kaplan, DPI; 267 Miller, Magnum; 268 Wilks, Stock, Boston; 271 Bodin, Stock, Boston; 272 Wilks, Stock, Boston; 276, 282 Franken, Stock, Boston; 280 Lejeune, Stock, Boston; 285 Charles Gatewood; 286 Grace, Stock, Boston; 289 Charles Gatewood; 290 Wide World; 293 Heron, 1981, Woodfin Camp; 307 Johnson; 312 Alper, Stock, Boston; 317 Fortin, 1981, Stock, Boston; 326 Ylvisaker, Jeroboam; 328 Forsyth, Monkmeyer; 330 Lejeune, Stock, Boston; 337 Park, Monkmeyer; 340 Carlson, Stock, Boston; 346 Stewart, Jeroboam; 357 Charles Gatewood; 353 Beckwith Studios; 355 Leinwand, Monkmeyer; 354 Charles Gatewood; 363 Vine, de Wys; 372 de Wys; 374 Forsyth, Monkmeyer; 376, 379 Vandermark, Stock, Boston; 384 Traendly, Jeroboam; 388 Druskis, Taurus; 396 WHO; 398 Pfizer; 408 Culver; 408 Mathers, Peter Arnold; 412 Morrow, Stock, Boston; 414 Hall, Stock, Boston; 417 Shelton, Peter Arnold; 427 Rotker, Taurus, 436 Chih, Peter Arnold, 446 Reno, Jeroboam; 450 Potker, Taurus; 456 WHO; 457 Bodin, Stock, Boston; 462 Dain, Magnum; 463 de Wys; 466, 486 Caraballo, Monkmeyer; 470 Reno, Jeroboam; 476 de Wys; 478 Brody, Stock, Boston; 478 Alper, 1978, Stock, Boston; 481 Perry Studio, DPI, 482 Sobol, Stock, Boston, 483 Cott, Jeroboam; 485 Vandermark, Stock, Boston

DIMENSIONS: A Changing Concept of Health
Fifth Edition

Copyright © 1982 by Kenneth L. Jones, Louis W. Shainberg, and Curtis O. Byer

All rights reserved. Printed in the United States of America. No part of this book may be used or reproduced in any manner whatsoever without written permission, except in the case of brief quotations embodied in critical articles and reviews. For information address Harper & Row, Publishers, Inc., 10 East 53d Street, New York, NY 10022.

Library of Congress Cataloging in Publication Data

Jones, Kenneth Lamar, 1931–
    Dimensions: a changing concept of health.
    Bibliography: p.
    Includes index.
    1. Health   I. Shainberg, Louis W.   II. Byer,
Curtis O.   III. Title.
RA776.J6918   1982   613   81-20250
ISBN 0-06-043442-2

CONTENTS

PREFACE   ix

SECTION ONE EMOTIONAL HEALTH   1

Chapter 1: Health: The Personal Basis   3
Emotional Health 3 | A Model of Emotional Health 3 | Personality 4 | The Self and Identity 5 | Human Needs 6 | Sound Emotional Health 10 | Stress 13 | Stress Management 18 | What Is Holistic Health? 23 | In Review 25

Chapter 2: Emotional Development   27
Early Infancy 28 | Later Infancy 28 | Early Childhood 29 | Middle Childhood 30 | Adolescence 30 | Adulthood 32 | Career Burnout 33 | The College Years 36 | Developing Assertiveness 39 | Developing Communication and Intimacy 40 | In Review 43

Chapter 3: Emotional Problems   45
Common Emotional Problems 45 | Emotional Dysorganization 48 | Types of Treatment 59 | Types of Specialists 62 | Types of Facilities 63 | In Review 65

Chapter 4: Aging and Death   67
A New View of Aging 67 | Coming to Terms with Death 72 | In Review 80

SECTION TWO DRUGS, ALCOHOL, AND TOBACCO 83

Chapter 5: Drugs 85
Medical Use of Drugs 85 | The Drug-Abuse Problem 90 | Psychoactive Drugs 92 | Classifying the Drug-Abuse Problem 105 | Drug-Abuse Behavior 106 | Drug-Abuse Treatment 108 | The Return to Society 111 | Control and Enforcement 111 | In Review 114

Chapter 6: Alcohol 117
Alcoholic Beverages 118 | Fetal Alcohol Syndrome (FAS) 123 | Alcoholism as Drug Abuse 126 | The Treatment of Alcoholism 132 | The Acceptable Use of Alcoholic Beverages 135 | In Review 136

Chapter 7: Smoking 137
The Extent of Smoking 138 | The Effects of Smoking 138 | The Smoking Habit 140 | Breaking the Smoking Habit 143 | Benefits to Ex-smokers 147 | Rights of the Nonsmoker 149 | In Review 150

SECTION THREE GOOD HEALTH AND THE MARKETPLACE 153

Chapter 8: Nutrients and Nutrition 155
What Is Food? 155 | The Chemistry of Nutrition 159 | Dietary Guidelines 163 | U.S. Four Food-Group Plan 168 | Food Facts and Fallacies 168 | Nutrition Labeling 173 | What's In Your Food? 173 | Nutrition and Health 177 | In Review 179

Chapter 9: Diet and Weight Control 181
A Healthy Diet 182 | Determining Desirable Weight 190 | Obese or Overweight? 192 | Reducing Your Weight 194 | In Review 204

Chapter 10: Total Fitness 205
The Measure of Total Fitness 205 | Sleep 206 | The Meaning of Total Fitness 209 | Activity Programs 211 | Fitness Programs 219 | Sports 229 | Use of Leisure Time 230 | In Review 231

Chapter 11: Consumer Affairs 235
Drug Products—Good and Bad 236 | Cosmetics 241 | Quackery 247 | In Review 253

Chapter 12: Health Services 255
The Health of Americans 255 | The High Cost of Health Care 257 | Paying Medical Bills 258 | Public Health Care 259 | Private Health Insurance 260 | Purchasing Health Insurance 263 | The People in Medicine 264 | Facilities for Patient Care 272 | In Review 274

SECTION FOUR HUMAN SEXUALITY AND REPRODUCTION 277

Chapter 13: Human Sexual Behavior 279
The Concept of Sexuality 279 | Sexuality in the Life Cycle 280 | Nonmarital Sexual Adjustment 283 | Sexual Orientation 288 | Homosexual Orientation 288 | Transsexuality 289 | Variant Sexual Behavior 290 | In Review 294

Chapter 14: Sexual Physiology and Response 297
Sexual Anatomy and Physiology 297 | Sexual Stimuli 306 | Problems in Sexual Response 316 | In Review 318

Chapter 15: Sexual Partnership 321
Forming Sexual Partnerships 321 | Being Ready for Marriage 322 | Society and Marriage 329 | Adjusting to Marriage 331 | Extramarital Sexual Relations 332 | Divorce 334 | The Decision to Have a Child 338 | Alternative Lifestyles 339 | In Review 341

Chapter 16: Fertility Management 343
Selecting a Method 343 | Contraception 348 | Sterilization 354 | Abortion 355 | Infertility 358 | In Review 361

Chapter 17: Pregnancy and Birth 363
Pregnancy 364 | Sex Education 377 | Heredity 380 | Congenital Defects 384 | In Review 386 |

SECTION FIVE DISEASES 389

Chapter 18: Communicable Diseases 391
Pathogens 391 | Stages of Communicable Diseases 392 | Body Defenses Against Disease 394 | Stress and Communicable Disease 397 | Public Health Efforts 398 | Treating Communicable Diseases 398 | Some Major Communicable Diseases 401 | In Review 406

Chapter 19: Sexually Transmitted Diseases   407
A New Name for an Old Problem 407 | The History of STD 407 | Today's Incidence of STD 408 | Gonorrhea 410 | Syphilis 412 | Genital Herpes Virus Infections 415 | Candida 417 | Trichomonas 418 | Nonspecific Urethritis 419 | Pubic Lice 420 Genital Warts *(Condyloma acuminata)* 420 | Scabies ("The Itch") 420 | Preventing STDs 420 | In Review 422

Chapter 20: Cardiovascular Health   425
Anemias 426 | Congenital Defects 427 | Rheumatic Heart Disease 427 | Atherosclerosis 428 | Hypertension 429 | Cerebrovascular Accidents (Strokes) 430 | Coronary Heart Disease 431 | Symptoms of Heart Disease 433 | Treatment of Heart Disease 433 | Prevention of Cardiovascular Disorders 434 | In Review 438

Chapter 21: Cancer   439
Nature of Cancer 439 | Causes of Cancer 443 | Stress and Cancer 447 | Symptoms of Cancer 447 | Diagnosis of Cancer 447 | Recommended Cancer Checkups 450 | Treatment of Cancer 451 | In Review 453

Chapter 22: Other Major Health Problems   455
Arthritis 455 | Diabetes 456 | Noncommunicable Respiratory Disorders 459 | Immune Disorders 461 | Epilepsy 462 | In Review 464

SECTION SIX ENVIRONMENTAL HEALTH   467

Chapter 23: A Healthful Environment   469
What Is a Healthful Environment? 469 | Pollution 470 | Air Quality 470 | The Environmental Protection Agency 475 | Water Quality 476 | Chemical Wastes 478 | Pesticides in the Environment 479 | Radiation 481 | Noise 483 | Housing 484 | In Review 487

Chapter 24: Population Dynamics   489
Population Worldwide 489 | United States Population 494 | The Boundaries of Population Growth 499 | In Review 507 |

Glossary   509
Bibliography   519
Index   523

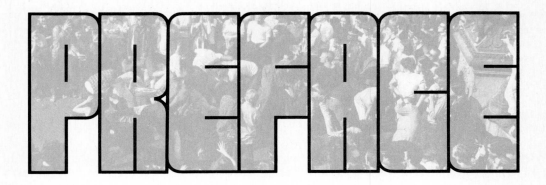

PREFACE

In recent years the college health course has taken on a new vitality. Many people now believe that the subject of health is one of the most important aspects of humanity and the quality of life. Many major challenges facing the world today are seen to be rooted in peoples' physical, emotional, and social health. But the definitions of health and the responsibilities for maintaining it have changed. No longer is health defined as merely the absence of physical disease. *Health* now refers to an individual's entire physical, emotional, and social well-being. Also, we now believe that each of us must assume the responsibility for his or her own health.

A variety of interdependent factors are now generally acknowledged as determining a healthful way of life. As health information continues to expand, each person must make more of an effort to remain well informed. Valid health decisions must be based upon accurate information. The great amount of health information being disseminated requires each of us to keep abreast of the new developments in order to take appropriate health actions.

We are constantly bombarded with information about health from government agencies, private hospital groups, our casual and close friends, our families, and television and radio advertisements. It is essential to our development as human beings and to our relationships with others that we find a way of sorting out this information and advice. Today's health courses are designed to help accomplish just that. *Dimensions V* is intended as a basic text for a college health course. This new fifth edition provides a broad background and, we hope, a stimulus for enlightened classroom discussion. We do not expect every reader to agree with everything contained here, since some of the material is, by its very nature, controversial.

A Holistic View of Health

Health science today is a "holistic," "multidimensional" subject, and we have therefore made some major departures from the traditional approaches. For example, we have shifted away from the anatomical and physiological toward the behavioral and sociological aspects of health. Of course, biological material must be introduced for a proper understanding of many contemporary health issues. *Dimensions V* embodies our concept of health science as a dynamic, rapidly changing field, reflecting not only advances in knowledge but changes in the political and cultural environment as well.

Organization

The organization of *Dimensions V* is based on the suggestions of users of previous editions. The book has several natural divisions. We have started with personality and emotional development because sound emotional health profoundly influences an individual's physical and social health. Health habits have lifelong effects. Individuals who plan for their later years and understand the naturalness of aging and death are able to live a more complete and full life. The study of emotional problems leads directly to the exploration of the use and abuse of substances — alcohol, tobacco, other drugs, and finally food. The overuse of any of these substances is often rooted in problems in emotional health. Continuing with substances, we discuss the proper kinds and amounts of nutrients needed for health. Our discussion of food and caloric balance leads into the importance of a fitness program in maintaining weight and health. Consumer information — from the buying of food to the selection of health services — is presented to help people deal with today's harsh economic realities. Then follows human sexual behavior, the many types of sexual partnerships, and reproduction. The communicable and noncommunicable diseases which affect the individual, the family, society, and the world are presented. These lead to the concluding section, which explains the ultimate implications for health of environmental quality and population control.

New in *Dimensions V*

Dimensions V represents the most extensive revision of all the previous editions of this text. Every chapter has been completely updated. The number and sequence of chapters have been changed. The chapter on emotional problems now precedes aging and death. The material on drugs and drug abuse has been reorganized and integrated into a single chapter. The discussion of human sexuality and reproduction has undergone major revision and has been expanded from three chapters to five. The material on cardiovascular health, cancer, and other noncommunicable diseases has been separated into three shorter, more manageable chapters.

All chapters have been revised in light of the world's changing social, political, and economic conditions as well as rapidly expanding scientific knowledge. We have attempted to present a balanced approach and to integrate emotional, social, physical, and environmental concepts. We have tried to keep *Dimensions V* from being a one-issue book. We feel that today's issues require as broad a perspective as possible. We have not proposed any simple solutions to complex problems.

Many new topics have been added with this revision. There is a new emphasis on stress throughout the text, with special sections on stress management and the relationship of stress to cardiovascular health, cancer, and communicable disease. New sections on sleep, women's alcoholism, and career burnout have been added.

The concept of holistic health is introduced and applied. The chapter on aging and death has been expanded.

Aids to Understanding   Several major student aids have been added to *Dimensions V*. Thought-provoking self-assessment questions ("Ask Yourself") have been distributed throughout each chapter. These questions should stimulate readers to apply the information to their own lives. Features presenting topics of special interest have been added to each chapter. End-of-chapter reviews provide quick summaries of the major points of each subject. As in previous editions, careful attention has been given to reading level throughout the text.

Artwork   Much attention has been given to the visual aspects of *Dimensions V*. Much of the artwork has been redone for this edition. Photos are more numerous and are carefully selected for their instructional value. Increased use has been made of tables and graphs for emphasis and explanation.

Glossary   The extensive glossary presents hundreds of important health terms designed to increase the students health vocabulary. Knowing these terms is essential for understanding the health sciences.

Instructor's manual   The instructor's resource manual provides health instructors with factual, accurate, and current material. Besides a complete outline of the text material, supplemental information is presented which the instructor may want to use to expand upon the text. Sources of current statistics and other information needed for a dynamic health course are included. Many of the tables, charts, and figures from the text are reproduced in a form that can be used for overhead transparencies or duplication masters.

Acknowledgments

Many individuals have helped in the preparation of this fifth edition. Throughout its history numerous people contributed time and ideas. For their contributions to this and previous editions we would like to thank Herb Booth, Mt. Hood Community College; Don L. Calitri, Eastern Kentucky University; Barbara Combs, City College of San Francisco; Janet Faurot, Long Beach City College; Dicky Hill, Abilene Christian University; Kenneth Hurst, Merritt College; Allure Jefcoat, Diablo Valley College; Tim Knickelbein, Normandale State Junior College; George M. Larsen, Sacramento City College; Alfred Mathews, California State University at Hayward; Michael Perrine, Portland Community College; Victor Petreshene and James Webster, both of the College of Marin; James Pryde, American River College; Jesse Thomas, Morgan State University; Marian S. Weiser; and H. Mark Whittleton, San Diego City College. We are particularly grateful to Abbey Stitt of Bronx Community College, New York, for her guidance in establishing the continuing philosophy of *Dimensions*.

KENNETH L. JONES
LOUIS W. SHAINBERG
CURTIS O. BYER

ONE

EMOTIONAL HEALTH

Central to the quality of life is emotional health. It largely determines our life-long happiness and productivity. Our own behavior has great impact on our physical well-being, and that behavior is largely dependent on the view we hold of ourselves. The higher we esteem ourselves the better care we tend to give ourselves.

How completely we maximize our genetic potential for physical and mental development relates directly to our success in adjusting to constantly changing life situations. The complexity of living, family and vocational mobility, and social isolation are a continued threat to this adjustment. To fulfill our basic needs we must come to terms with the world in transition around ourselves.

This on-going process culminates with the spector of aging and, ultimately, death. Rather than being a subject for denial, we can face aging and death confidently. With insight, there can be a sense of fulfillment for ourselves, our families, and intimate circles of friends.

HEALTH: THE PERSONAL BASIS

Each of us largely determines the quality of his or her own health. Rather than being "victims" of poor health or illness, we often cause or contribute to them. Obviously, we have little control over the genes we carry and some social and environmental factors. But for most of us, the collective health significance of such uncontrollable factors is minor compared to the impact of our own behavior. Our emotions and behavior play a role in virtually every illness we develop, from colds to cancers. We alone choose what we eat, whether or not to use alcohol, tobacco, or other drugs, whether to get adequate exercise and rest, and how to deal with stress. No physician or health-care system can guarantee the good health of anyone who lives in an unhealthy way. At best, a physician can patch up the damage. But most health problems are preventable, and prevention is far superior to treatment. Throughout this book, we shall emphasize everyone's personal responsibility for maintaining good health.

EMOTIONAL HEALTH

Emotional health is basic to all health. Our emotional state so strongly influences the body's physical functioning that emotional conflicts can lead directly to serious physical problems. It has also been clearly shown that the establishment of healthful behavior patterns (good diet, adequate exercise, avoidance of harmful substances) depends on a sense of self-esteem. The more highly you think of yourself, the better care you will give yourself.

More specifically, such problems as alcoholism and other drug dependencies are an indication of poor emotional health. Today, the impact of emotional problems on human welfare far exceeds that of all other health problems.

A MODEL OF EMOTIONAL HEALTH

There are many valid approaches to emotional health. In this and the next two chapters, we will explore emotional health from the point of view of need fulfillment.

Every one of us has certain basic needs. A first step toward emotional health is to recognize these needs and accept them as such. We should not confuse superficial goals such as wealth, status, or power with underlying real needs such as a sense of self-esteem. Nor should we deny the needs that we so strongly feel. For example, we may deny that we need the love and companionship of other people. But in reality we do feel this need, although we may fear possible rejection by those we need (and so we reject them first). In such ways we often prevent our own needs from being fulfilled.

Once we recognize and accept our needs, the next step toward emotional health is finding effective and appropriate ways of fulfilling those needs. This is apparently not an inherent ability, but a learned one that is gained with increasing experience and maturity. One method we will explore is the development of *assertiveness*—letting others know what our needs are and standing up for our right to have them fulfilled. Another method is the development of emotional *intimacy* with others — establishing relationships that fulfill the needs of both persons.

Emotionally healthy persons also accept the fact that inevitably some of their needs will not be met. No person, regardless of wealth or fame, can have all his or her needs fulfilled at all times. There are just too many uncertainties in life. Too many limitations are imposed by the unpredictability of weather, the unreliability of machines, and the frailty of human beings.

Thus, we must develop effective, appropriate *coping* methods of dealing with such frustration. One important way of dealing with frustration is to maintain a positive outlook on life. We can focus on the fortunate circumstances of each life situation instead of the unfortunate ones. Learning to talk out frustrations instead of bottling them up is another effective device. Still another is to "work off" frustration through activity of some kind, instead of just sitting and brooding over it and attempting to blot it out with alcohol or other drugs. We often marvel at the strength of some individuals who weather difficult situations. This same strength is attainable by anyone who approaches life in a positive way.

PERSONALITY

All living things attempt to adjust to their environment. We human beings try to establish and maintain an equilibrium with it. Since the environment is constantly changing, maintaining an equilibrium demands our constant attention.

The task of maintaining an equilibrium with our environment is further complicated by our basic inner needs which also must be satisfied. Very often we find that environmental situations make the satisfaction of inner needs difficult or even impossible. All of us are frustrated to some extent in our attempts to satisfy our needs. The ways in which we react to this frustration are indications of our emotional health and vary with each personality.

Personality is made up of a number of behavior patterns, traits, attitudes, and all the other interactions we have with our environment. Everything that has ever happened to us has left some mark on our personality. We are born with a certain genetic

makeup, and the personality we eventually develop depends both on our original genetic endowment and how the environment acts on that endowment. Our personality also includes the ways in which we interpret our environment and how we learn to deal with the conflicts between our basic inner needs and the obstacles to their fulfillment presented by the outer world.

Many attempts have been made to describe the makeup of personality. For example, the pioneer psychoanalyst Sigmund Freud conceived of personality as being divided into three "processes": the *id,* which is unconscious; the *ego,* which is mostly conscious; and the *superego,* which has both conscious and unconscious elements. According to this scheme, the mind is like an iceberg. The conscious mind is like the visible tip that projects above the surface of the water, and the unconscious mind is like the much greater bulk that lies below.

The Id

The id, which is unconscious, is the most primitive of the three portions of the personality. It contains all the basic instinctive drives. These instincts, which are present at birth, represent unconscious urges to *survive* and *enjoy life.* Throughout life, the id remains as the storage place for the primitive instincts.

The Ego

The ego represents the conscious mind, and acts as a moderator between the id and the outer world. It is in contact with the environment (reality) and with the id and the superego (which imposes judgments of goodness and worthiness). The ego must integrate all factors and *determine appropriate behavior* for any situation. The ego must decide which urges can be allowed satisfaction and which urges must be suppressed.

The Superego

The superego contains our judgment mechanisms regarding right and wrong, good and evil. It advises and threatens the ego. As very young children, we have little concept of what is right and wrong, but we gradually develop one as codes and values of society are impressed upon us.

The word "conscience" is sometimes used to describe the superego, but actually the conscience is only part of the superego. It is this portion of the well-developed superego that makes cheating or stealing difficult or impossible for many persons, because they have learned that these actions are wrong and they would feel guilty if they did them.

It is possible for a person with a weakly developed superego to live according to the principle of personal pleasure alone, without regard for the rights and privileges of others, and yet suffer little or no guilt. On the other hand, it is possible for the superego to be so rigid and severe that it prevents the person from having any pleasure, regardless of the situation. The most healthy superego is one that fits *within the demands of society* and yet *allows adequate individual fulfillment.*

THE SELF AND IDENTITY

The *self* is our individual awareness or perception of our own personality. It is a composite of how we feel about ourselves and how we imagine other people feel about us. Much of what we "know" about ourselves is based on our experiences with other people. Our chief sources of information about ourselves are the reactions of others to us. Of course, we evaluate these reactions subjectively. It is possible for us to see ourselves somewhat differently from the way others see us.

Many factors contribute to our self-image. Our experiences with others are very important in determining what our self-perception will be. We find that our appearance and personality elicit responses of acceptance or rejection, kindness or hostility, attention or indifference. We hear (or overhear) ourselves described by other people in terms of various personality traits. When these traits are consistently applied to us, we usually adopt them as descriptions of ourselves. Praise and attention help us to form a picture of ourselves as desirable persons. Rejection, indifference, criticism, or hostility lead to a poor self-image, with resulting feelings of inferiority.

The treatment that we receive from others may or may not reflect accurately our own traits and abilities. For example, we tend to react most positively to those whose traits are most like our own. Thus, people may meet with indifference or rejection merely on the basis of their racial, religious, or cultural background. As a result, many effective and pleasant people may come to perceive themselves as inadequate, inferior, or undesirable because of the prejudices or misconceptions of others.

Knowing how people see themselves helps us to understand their behavior, especially when they see themselves much differently from the way others see them. Behavior is largely determined by how people perceive a situation with reference to themselves.

People whose self-images are too different from their true or objective personalities may experience serious adjustment problems. They must constantly explain or ignore circumstances that are inconsistent with their view of themselves. They rely on *ego-defense mechanisms,* discussed later in this chapter. Consider, for example, a capable student whose self-esteem is so low that academic success would actually create anxiety by conflicting with his or her self-image. Such a student may purposefully, though unconsciously, create barriers to success.

Most of us tend to underestimate our capabilities. We fall into the trap of feeling that our lives are completely structured for us by circumstances beyond our control and that we are powerless to make any significant change in their course. In reality, all of us have amazing abilities to accomplish things if we just accept and believe in them. This is the concept of *self-determination.*

Self-determination depends on two basic beliefs. First is a belief that human beings in general are self-determining and have considerable control over the courses of their lives. Second is a belief or confidence in their own personal ability to accomplish what they want. This belief in self is based on faith and experience. When people learn that they are successful in most of the things they attempt, they develop faith in themselves. Even their failures do not bother them excessively, because they have confidence in their ability to recover from failure. Through their experiences in living, they have learned what they can and cannot do and they have a basis for self-determination.

When people do not believe in themselves, they act in ways that are limiting, or even self-destructive. When they "know" they are going to fail, they almost always do. This concept is valid in every realm of life. It is true for the student in course work, for the athlete in competition, for anyone in his or her relations with people—especially in sexual relationships, and for a critically ill patient balanced between life and death.

● ASK YOURSELF

1. What do psychologists mean by the "self?"
2. Have you spent any time considering your own self-image?
3. How are you feeling about yourself *at this moment*?
4. Is your self-image productive? Is it helping you to achieve your life goals? Might your self-concept have caused you to set your goals too low?

A. H. Maslow (1908–1970). Often referred to as the "father of humanistic psychology," Maslow emphasized the importance of human feelings, emotions, and experiences.

An integral part of our self-concept is our sense of *identity*. This is our appraisal of who we are, what we mean to ourselves and others, and where we fit into the general scheme of things, especially in our relations with other people. The process of establishing an identity is a source of much internal conflict for many people. By the time we reach our high school years we are deeply involved in the task of separating our identities from those of our parents and finding our own place in the world. But changes in identity (and identity crises resulting from conflicts in identity) can occur at any age throughout life.

Many people never do really discover their own identities. They may go through life hiding in a career, a marriage, organizations, or a succession of "causes." Or they base their identities on what they have rather than on what they are. Either of these approaches is just a way of dodging the difficult task of defining one's identity. But for a truly fulfilling life, each of us must find his or her own identity as an individual.

HUMAN NEEDS

One of the first requirements for good emotional health is to know ourselves. Why do we feel the way we do? Why do we do the things we do? The answers to these questions will come largely from our understanding of our basic needs, the extent to which these needs are fulfilled in us, and how we react to the frustration of these needs.

Abraham H. Maslow, in his classic book *Motivation and Personality* (1970), made an interesting interpretation of human needs. He arranged these needs in sequence from the most basic to the highest. Maslow believed that the higher needs develop only when the basic needs are fulfilled. For example, since food is a very basic need, people

who are starving may have little interest in pursuing knowledge or reaching their full artistic potential.

In the following discussion we shall examine some of the basic human needs Maslow identified.

The Physiological Needs

The most basic human needs are the physical drives. Among them are hunger, thirst, sleep, sexuality, and many others. When physiological needs are unsatisfied, all other needs are pushed into the background. Not only is interest in higher needs lost, but plans for the future are modified. For example, extremely hungry people may think of paradise merely as a place where there is plenty of food. They think that if only they were guaranteed enough to eat for the rest of their lives, they would never want anything more.

Children have a strong need to feel secure.

Viktor Frankl's observations of behavior in a concentration camp suggest that foremost among the physiological needs are those related to survival (1963). Survival so dominated individuals' thoughts and actions that overt sexual activity disappeared. Even dreams were reported to be free of sexual content typical of a more normal environment.

But what happens when the physiological needs are satisfied? Other and higher needs emerge and begin to dominate people's thinking and behavior. As long as a need is satisfied, it has little effect on behavior. Behavior is governed much more by unsatisfied needs.

The Safety (Security) Needs

Maslow rated the safety needs as second only to the physiological needs. After the physiological needs are fairly well satisfied, the safety needs emerge, and they make everything else appear less important. The safety needs are most easily seen in the fright reactions of infants and children, because adults in our society have been taught to hide or inhibit these reactions. Thus, when most adults feel their security threatened, they are likely to mask their anxiety; they show their fear in physical changes such as increased heartbeat.

Parents must provide children with a feeling of security in the home. The relationship between parents is especially important to children. Quarreling, physical abuse, separation, or divorce may be particularly frightening to them. When the parents'

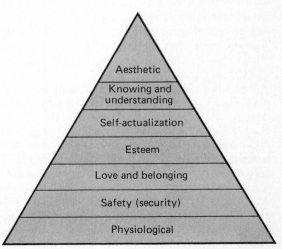

According to Maslow, human needs arise in a progression, physiological needs being most basic.

People of all ages feel the needs for love, acceptance, and companionship.

anger is directed toward them, children often react with such total terror that it is clear that their fear involves more than the pain of physical punishment. They fear that their parents' love will be taken from them.

Another indication of children's need for security is their preference for a certain amount of routine and ritual in the daily schedule of the family. They like a predictable, orderly world. Young children seem to thrive under a structured system that contains basic rules of conduct. They prefer and need freedom *within limits* rather than total permissiveness. Consistency is more important, however, than the degree of rigidity or permissiveness. Children raised under a consistent system are happier and emotionally healthier than those who are not because they feel more secure. They know what their parents expect of them and how their parents will react to their behavior.

The Needs for Love and Belonging

Maslow placed the needs for love and belonging (companionship) right after the physiological and safety needs. People whose physiological and safety needs are fulfilled then experience needs for love and belonging. They feel a strong need for friends or a mate. They have a need for companionship with people in general, for a feeling of belonging and group acceptance, and a need to love and be loved.

Belonging to a group and being accepted by it is central to a person's social life. One of the greatest causes of unhappiness is the feeling of not being accepted by a group. Along with social satisfactions, we also derive other benefits from belonging to a group. We develop the sense of security that we need to become more independent of our families. We are exposed to values different from those of our parents. We gain experience in reacting as equals with our peers. School, church, play, and work provide many opportunities for group involvement.

Our society is inhibited and cautious in its attitude toward love and affection, especially the sexual expression of them. This attitude has made it difficult for many people to establish close personal relationships, including those in which sex plays no role. Love and sex are not synonymous. There may be strong love relationships in which sex is not a part, just as there may be loveless sexual relationships. Love is characterized by a genuine concern for the welfare of the loved one.

It is of the utmost importance that parents be responsive to a child's love needs. Children who feel unloved and rejected by their parents often develop a poor self-image that may endure and cause much trouble at home and at school. They may suffer continued emotional problems as adolescents and adults. Parental rejection can leave lifelong emotional scars. However, parents are not the only source of love and affection. Children often receive love from relatives, neighbors, teachers, and leaders of community organizations. Growing children have many opportunities to learn that life is essentially what they make of it and to discover the rich possibilities of self-determination. Children who have been totally rejected, however, may later be reluctant to establish close personal relationships for fear of further rejection; they may never be able to give and receive love and thus go through life as "loners." People who have "loved and lost" may also withdraw because of fear of further rejection. But to reasonably secure persons, the fulfilling of love needs is worth the risk involved.

Self-esteem

Every one of us has a need for *self-esteem*—a feeling of personal value or worth, of success, achievement, self-respect, and confidence in the face of the world. In order to attain this feeling, most of us depend on the respect or esteem of other people; we show this excessively in our striving for status, dominance, attention, and appreciation.

Self-esteem, however, seems to work in a vicious circle. A well-developed sense of self-esteem includes feelings of self-confidence, value, strength, adequacy, and usefulness. Such feelings lead us to accomplishments that further reinforce our self-esteem. However, a lack of self-esteem includes feelings of dependency, inferiority, weakness, helplessness, and despair. Such feelings, of course, effectively prevent the kinds of achievements that could build our self-esteem.

The most obvious way to build self-respect is to do something of value well—to excel. The specific area in which to excel should be determined by one's interests and ability. Anyone who makes the effort can find some constructive activity in which to excel. It is especially fortunate when people receive a feeling of satisfaction from their work, since they must devote a great amount of time to it. But a sport or hobby can be equally rewarding in building self-respect, as can social and love experiences.

Most people also feel a desire for respect from *other* people, in the form of reputation or prestige. As a person's self-respect is strengthened, the need for the esteem of others diminishes. People who feel confident of their own worth are less dependent on the praise of others. It is certainly better to base self-respect on actual achievements rather than on the opinions of others.

The constant seeking of praise may suggest insecurity on the part of the seeker. Still, it is important in our dealings with other people to recognize that almost everyone has a need for some recognition from others. Sincere praise is one of the foundations of good personal relationships. It is important to give praise when praise is due. Too often we tell people when we are annoyed with them but neglect to tell them when we are pleased.

● ASK YOURSELF

1. What is the idea behind Maslow's sequence of human needs?
2. What needs are most strongly motivating your own behavior now?
3. In what ways do you rely on other people to help you fulfill your need for self-esteem? Has this ever created any problems for you?
4. To what extent do you feel that you are self-actualized? What efforts are you making toward that goal?

The Need for Self-actualization

The need for self-actualization is the need to do what one is capable of doing to achieve self-fulfillment. To be at peace with themselves, artists must paint, poets must write, and musicians must make music. In order to feel fulfilled, people must not only do what they are able to do, but they must also do it as well as they are capable of doing. Most of us (Maslow said 99 percent) are not operating at this level. We are stuck at a lower level of need-satisfaction, sometimes through our own fault and sometimes through circumstances outside our control. According to Maslow, we can best increase our level of self-actualization by discovering what we are really like inside.

The Need to Know and Understand

Among the higher needs is the desire to satisfy curiosity, to know, to explain, and to understand. Curiosity is a natural rather than a learned characteristic. Children are naturally curious. They may lose this curiosity, however, through overly strict discipline or lack of encouragement at home or too much regimentation at school. Parents and teachers must be careful not to destroy this valuable characteristic. When curiosity has already been suppressed, every effort should be made to restore it, for it leads to creativity and invention.

Although many people live fairly satisfying lives without much intellectual stimulation, these lives are not often highly fulfilling. People whose thoughts are confined to the mundane are often bored, feel constantly tired, and have little zest for life. In contrast, the satisfaction of the need to know and understand often gives people a bright, happy feeling of fulfillment in their emotional lives.

Beauty is a subjective quality that defies any scientific measurement.

The Aesthetic Needs

Beauty is a subjective quality that defies scientific measurement. But it is a basic need for many adults and probably for all children. Like curiosity, an appreciation of beauty may or may not be retained into adulthood.

Although beauty is often perceived through the senses (visual, auditory, tactile, and so on), this is not necessarily always the case. Beauty can also be a purely internal experience; it is possible to feel the beauty of ideas or philosophical concepts.

SOUND EMOTIONAL HEALTH

It would be very difficult to set up an exact standard by which to judge an individual's level of emotional health. No line neatly divides the emotionally healthy from the emotionally ill. There are many degrees of emotional health. The following characteristics are all signs of sound emotional health. But the lack of one or more of them in a person does not indicate emotional illness. Actually, no one has all the traits of emotional health at all times. At the very least, emotionally healthy, fully functioning, or self-actualizing people "feel good" and enjoy themselves and their personal existence.

Dealing Constructively with Reality

Healthy people accept reality, whether it is pleasant or not. They do not generally attempt to escape from reality through fantasies or by excessive use of alcohol or other drugs. Nor do they take the opposite path by becoming preoccupied with their own problems and spending much time worrying or brooding.

Dealing constructively with reality means acknowledging and accepting our capabilities and limitations. Then, when a problem arises, we can do something about it (if the solution is within our capabilities) or else realistically accept the problem as beyond our ability to solve. When someone else might be able to solve a problem, mature people do not hesitate to ask for help. When it is apparent that they are dealing with a problem that is impossible to solve even with help, they then adjust to the situation.

Healthy people set realistic goals for themselves. They try to make the fullest use of their natural abilities, but they also realize their natural limitations and do not frustrate themselves trying to do something that is obviously beyond their natural ability. People usually are happiest working at or near their full ability. They usually are not very happy working far below their ability or when they are trying to work far above it.

Another indication of emotional health is taking responsibility for ourselves as well as for others. Responsibility for ourselves is of primary importance. Unless our own needs are met, we can do very little for others. Except for the very young, the very old, or the severely handicapped, each person is individually responsible for his or her own actions and life.

Less healthy people may work harder at evading responsibilities than they would if they accepted them and did whatever was necessary to fulfill them. More healthy people not only accept responsibilities, but also enjoy and need a certain number of them.

Adapting to Change

We all have a natural tendency to resist change and even to fear the future. This results from our basic need for safety and security, as discussed previously. But healthy people are confident of their ability to adapt to change. They realize that the world constantly changes, and they expect to change with it. They plan ahead and do not live in fear of the future.

Many characteristics of maturity improve as we grow older, but one characteristic that we often possess when young and then gradually lose is the ability to adapt readily to change. When we are young, most of us welcome new experiences and new ideas. We should try to keep this open-minded approach to life as time passes. In past generations this adaptive ability was less important than it is nowadays, because conditions in the past changed much more slowly than they do today.

Autonomy (Independence)

People in sound emotional health can function autonomously, think for themselves, and make the most of their own decisions. They can plan their lives and follow through with their plans. The inability to make decisions is very common in immature, insecure people. They may spend much time in confusion, not knowing what to do. They are afraid to face the consequences of whatever decisions they make, so they make as few as possible. Growth involves mistakes as well as successes.

Yet a certain amount of dependence is also desirable. People should not try to divorce them-

People in sound emotional health plan their lives and follow through with their plans.

selves completely from the society in which they live and of which they are a part. For example, the enjoyment of being loved by another person is a normal kind of dependency on another person. In our complex society we are dependent on other people in many ways. Healthy people enjoy this interdependency.

Managing Stress

All of us are subject to many stresses and tensions both in our everyday lives and in occasional crises. It is perfectly normal to experience emotional reactions to these stresses and tensions. It is the type and degree of emotional response that indicates the state of our emotional health.

Healthy people are able to control their emotions. They are not overpowered by fears, anger, hatred, jealousy, guilt, or worries. They can take life's disappointments in stride, minimizing the unpleasant aspects. They have a tolerant attitude toward themselves as well as toward others. They are inclined to look at the brighter side of life rather than to concentrate on their troubles. They can laugh at themselves. This attitude tends to minimize problems and may make them more solvable. Stress-management techniques will be discussed later in the chapter.

The ability to work productively is a prime indicator of emotional health.

● ASK YOURSELF

1. According to the eight characteristics just described, how do you rate your own current level of emotional health?
2. In what ways are you currently dependent upon other people? Are you satisfied with the degree of independence you have attained?
3. How could you increase your independence?

Concern for Others

The healthy personality combines self-respect with a concern for the rights and happiness of other people. Healthy people truly find as much satisfaction in giving as in receiving. They respect the differences they find among other people, accepting them as they are. They do not bully or use people unreasonably in achieving their own goals. They are careful in their words and actions to avoid offending others. They avoid saying things to build their own egos at the expense of others.

Satisfactory Relationships with Others

People who enjoy emotional health are able to relate to other people in a consistent manner with mutual satisfaction and happiness. They like and trust most people and expect that people will like and trust them. Their relationships with others are satisfying and lasting. They have enough self-confidence to be able to feel a part of a group. They feel accepted and, conversely, make others feel accepted.

People who seem to have trouble dealing with others should examine their own attitudes and actions. They may be using ego-defense devices in ways which make it difficult for other people to relate to them. They may feel that they are being rejected by others when they have actually been hostile or inconsiderate themselves.

The Ability to Love

Mature people are able to feel and express affection for other people. Even a rather immature person can receive love, but it takes a mature person to give love unselfishly. Although we refer to this type of love as "unselfish," it is actually very emotionally rewarding. It is one of the most fulfilling human activities.

It must be stressed that the love to which we are referring is not just sexual love, but includes a love for children (not just our own, but all children) and for all people, regardless of their racial, religious, or cultural backgrounds. It is a love for humanity. Only the person with a well-developed sense of inner security can freely express such acceptance of others. Before people can love others, they must learn to accept and really love and respect themselves.

Working Productively

It is widely agreed among authorities today that one of the prime indicators of good emotional health is the ability to work effectively and productively. The inability to be productive may be a symptom of emotional illness, especially since inner conflict requires a great amount of energy that would otherwise be available for other things.

There are many examples. In school, students who have trouble completing their work often have emotional conflicts. Chronically unemployed adults sometimes suffer from emotional problems. Their

A Mature Value System

A well-developed value system is an important aid to effective living. Decision-making is simplified because you know what is important to you. Much stress is eliminated from your life because you also know that certain things are not important to you.

There is no cut-and-dried approach to developing a satisfying value system. Much depends on personal factors such as the philosophy of life that you accept and the many experiences that are exclusively your own. But a few general suggestions may be helpful.

1. *Draw on the experiences of others.* Spend some time reading the works of the great philosophers of the past as well as those of the present. Talk to older people whose lives you admire. Study their value systems and how they express their values in their daily actions.

2. *Accept only that which is meaningful.* Evaluate what you read and hear with respect to your own experiences. From all the conflicting ideas presented, accept only the ideas that seem personally meaningful and significant to you. It is not that some value systems are basically right and others wrong, but that ideas often conflict because a philosophy or value system is highly personal and individual. What is right for one person may be very wrong for another. You should not automatically accept another's philosophy, regardless of how famous he or she may be. If your philosophy is not really your own, it is inadequate.

3. *Respect the rights of others.* A value system can work successfully only if it recognizes that an individual's freedom is limited by the rights and freedoms of others. A value system is unrealistic and will be an unsuccessful guide through life if it demands rights and freedoms for yourself, while denying these same rights to others. For example, people who demand the right to freely express their own ideas must extend this same right to all others.

4. *Reappraise your values periodically.* From time to time your philosophy should be reevaluated in light of the kind of life it is producing for you. Are you truly satisfied with the way things are going? Perhaps your values are inadequate. As you gain in wisdom and experience, it is normal to alter your value system accordingly. Values that are typical and acceptable for a fifteen-year-old could seem immature and inappropriate for a thirty-year-old.

problems can keep them from actively looking for a job, work against them during job interviews, or lower their productivity to the point where they repeatedly lose jobs. Of course, much unemployment is the result of economic conditions, a lack of marketable skills, and ethnic or sexual discrimination. But emotionally healthy individuals normally have enough energy available to adapt to the situation, whatever it may be, so that their productivity is not reduced.

The emotional problems of people at home may appear in their inability to keep up with housework. On the other hand, routine household chores can consume so much energy that they are able to do little else.

STRESS

Stress is now recognized as a major factor in the development of almost every human disorder, from the most trivial to the most threatening. The word

TABLE 1.1   PROBLEMS THAT MAY BE STRESS-RELATED

Allergies
Arthritis
Asthma
Cancers
Colds
Headaches
Heart disease
High blood pressure
Impaired sexual response
Inability to concentrate
Indigestion
Menstrual disorders
Muscle spasms
Skin disorders
Insomnia
Ulcers

This is a sampling from among thousands of stress-related conditions. Of course, these problems may have causes other than stress. Stress related problems are sometimes referred to as *psychosomatic* (from the Greek words *psyche,* meaning "mind," and *soma,* meaning "body") disorders.

"stress" appears in almost every discussion of emotional health, though it has various meanings. To avoid any misunderstanding, we shall follow the noted Canadian physiologist, Hans Selye, and define *stress* as a group of bodywide, nonspecific defense responses induced by any of a number of stressors. A *stressor* is any force that produces stress. Obvious stressors include: emotional conflict, fear, fatigue, physical injury, disease organisms, and noise. Less obvious stressors can include very pleasant experiences such as a passionate kiss or lovemaking.

Not all stress is harmful. Quite to the contrary, the human body functions at its best when it is in a reasonable state of stress. When we speak of a team being "up" for a game or a student being "up" for an examination, we are referring to the benefits of the stress response. Selye uses the terms *eustress* and *distress* to distinguish between helpful and harmful stress. Eustress (*eu* is the Greek word for "good") is stress that is appropriate in degree and duration to prepare the body for optimum functioning. Distress is stress that is inappropriate because

of its source, degree, or duration and is harmful in its effects.

The General Adaptation Syndrome

The body's stress response appears to be regulated by a part of the brain called the *hypothalamus.* The hypothalamus constantly adjusts the body's physical functions in response to a person's emotional state. It does this in at least two ways. By governing the release of hormones by the pituitary gland, the hypothalamus influences the levels of most of the body's hormones. Also, the hypothalamus influences the actions of the autonomic portion of the nervous system, which controls the glands and involuntary muscles.

Almost every body function is modified by stress. For example, when under stress the heartbeat speeds up and the blood pressure rises. Sweating increases. The digestive process slows as blood is diverted from the digestive organs to the body muscles. The pupils of the eyes dilate. And, numerous hormone levels change. Such effects are very useful when we are responding to an emergency, competing athletically, or performing any activity that requires peak body functioning. But when they continue over long periods of time, the same effects may be very damaging; they can lead to ulcers, high blood pressure, heart disease, and a host of other disorders (*see* Table 1.1).

Many aspects of contemporary life can produce prolonged stress. Typical stressors include personal problems, family tensions, career-related problems, financial problems, noise, the threat of crime, and, for students, grades and career decisions. Few, if any, lives are free of stress; and uncontrolled stress can have disastrous physical consequences.

Selye called the entire stress response the *General Adaptation Syndrome* (GAS). (A *syndrome* is a group of related symptoms.) The GAS includes three successive stages. The first is the *alarm reaction,* consisting of the immediate mobilization of the body mechanisms just mentioned. For example, an individual's alarm reaction could be triggered by the anticipation of taking an exam.

If the stress continues for some time, however, the person enters the second stage, called *resistance to stress.* This is the stage of maximum ability to withstand the stressor. It may continue for days, weeks, or even months, depending on the vitality of the person and the amount of rest the individual is able to obtain during this period. Sustained endurance, however, puts a considerable strain on the body's resources and often results in psychosomatic disorders.

One of the causes of depression is related to stress. Emotional and physical stressors alike initially induce a state of psychological and somatic excitement that is followed by a secondary phase of depression ("letdown").

If the stress continues long enough, the third stage, *exhaustion,* may be reached. The person becomes progressively devitalized and loses his or her ability to resist stress. The internal resources for dealing with the continued assault have been exhausted. Every function of the body is weakened. If prolonged, this stage inevitably results in death.

Consider the person preparing for an exam. Let us suppose that he or she has also taken on a very heavy course load and has set high standards of achievement. In order to reduce the anxiety over possible failure to meet such self-imposed demands, the person devotes a great amount of time and energy to study. For a certain period of time he or she adapts and therefore gets good grades because of diligent study habits—the resistance to stress phase. However, now let us assume this person has additional responsibilities such as a job and perhaps a family to support. The amount of time given over to meeting these demands is likely to reduce the time available for study, and grades begin to drop. Thus the person is no longer able to meet expectations of success. In order to raise his or her exam scores, the person tries to allow more time for study by sleeping less and also by eliminating previous recreational diversions. It is only a matter of time until the person reaches the third stage—exhaustion. At this point certain symptoms are likely to appear. Common emotional symptoms include depression and guilt (because of failure to live up to one's standards). Physical symptoms may include colds, ulcers, high blood pressure, backaches, headaches, and many others. If stress is not brought under control, both body and mind can break down.

Since all types of stressors produce the same stress response, excess exposure to any one stressor can exhaust our resistance to all other stressors. For example, prolonged emotional stress interferes with our ability to fight off infectious diseases. Thus we may become physically ill more readily while under emotional stress. Or, conversely, physical illness lowers our ability to resist emotional stress. Thus, the health of the body and of the mind are interrelated, interdependent, and inseparable.

Defense Mechanisms

Stress situations are a part of life, and everyone develops methods for coping with them. These coping methods help us avoid a conscious feeling of stress. These stress-preventing devices have been called *ego-defense mechanisms* or just *defense mechanisms.*

Defense mechanisms should not be thought of as abnormal. Everyone makes use of them. They are recognized by most authorities today as necessary and valuable in dealing with the stress situations that we all face throughout life. Hardly anyone could get through life without using them.

In examining a few common defense mechanisms, we will make no attempt to classify them as good or bad. In most cases, the value of a particular mechanism depends on how and when it is used. Used in moderation and in the proper circumstances, a mechanism might be of great value. Yet the same mechanism, used in excess or in inappropriate situations, might be definitely undesirable.

Avoidance One of the simplest and most common methods of defense against anxiety is to avoid situations that produce it. We all use this defense to some extent. People who fear airplanes travel by car or train. People who are bothered by speaking in front of groups try to avoid situations in which they would be obliged to do so. People who feel threatened by an intimate, caring relationship with another person often avoid this anxiety by keeping other people at "arm's length" emotionally.

Certainly, none of these examples of avoidance could be considered extreme or even unusual. But the avoidance can become so intense that it indicates a serious emotional conflict. Such would be the case if someone became so fearful that he or she refused to leave the house for any reason or lacked the self-confidence to perform any kind of job at all. Thus it can be seen that the same defense mechanism can be normal and harmless or seriously disabling, depending on its degree of use.

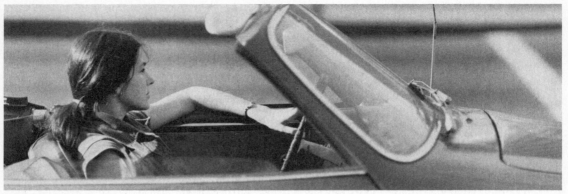

Avoidance. People who fear flying often travel by car, even though auto travel is statistically much more hazardous than flying.

Denial of reality In using this mechanism, people protect their egos from stressful situations by refusing to perceive them. Denial of reality is usually an unconscious process, or at least partly so. Individuals using this mechanism usually are not aware that they are denying anything.

Denial appears in several forms. One is the refusal to see or hear certain things that might lead to stress reactions, such as when children "tune out" their mother's calling them home. Another form is to deny the existence of a reality. Many people deny the risk of pregnancy, for example, and fail to take adequate precautions. Others deny the danger inherent in the use of hard drugs or insist that only "stupid" people are victims of drug abuse.

Still another variation is the denial of inward feelings. We often deny our feelings of sexual desire for persons if we know we have no chance of fulfilling that desire. We may even find something about them to criticize or ridicule. We also tend to react in the same way to material things that we believe are entirely out of our reach. We usually avoid anxiety by restricting our serious desires to what is attainable, or nearly so.

Like avoidance, denial can be either a helpful or harmful defense, depending on the situation and the extent to which it is used. For example, we must admit that the world is, to some extent, a dangerous place. Every day, each one of us faces the possibility of being cut down by a fatal disease, a war, a murderer, an automobile accident, a fire, a flood, a tornado, lightning, or a building collapse — to name but a few of life's hazards. But the probability of any of these happening to a given person on a given day is extremely remote. If we continually considered all the dangers we face, we would be in a constant state of anxiety. Through denial, we can ignore the remote dangers and live normal lives.

But denial can be harmful if it causes us to ignore more immediate dangers. For example, some people fail to get immunizations or physical examinations because they deny the possibility of illness. Others will not use automobile seat belts because they deny the possibility of an accident. The denial of any clear and present danger is obviously a misuse of the mechanism.

Repression Repression is the process of forgetting (or more accurately, restricting to the unconscious mind) an event, feeling, or memory that would produce anxiety. The repressed material, though blocked from entering the conscious mind, continues to have an influence. It constantly seeks expression in some indirect way. It may cause tension in certain situations, or it may influence preferences, decisions, attitudes, or beliefs. Repression may enlist other defense mechanisms to keep anxiety-producing material in the unconscious.

Material that is repressed is pushed back into the unconscious mind much more rapidly than material that is ordinarily forgotten. Repression may occur almost immediately in order to defend the ego against anxiety, as in the case of frightening childhood experiences such as a bad fall or an attack by an animal. Adults similarly tend to repress memories of accidents, attacks, and other frightening events or socially unacceptable thoughts and

Regression. Sometimes we try to return to the less stressful days of childhood.

feelings. Interestingly, repression underlies most other defense mechanisms.

In the school of psychology called psychoanalysis, repressed conflicts are seen as the cause of much vague anxiety. In order to relieve their anxiety, patients in psychoanalysis make elaborate efforts to bring these conflicts from the unconscious into the conscious mind so that they may discuss them rationally.

Projection Projection is a mechanism by which people bolster their own self-images by unconsciously attributing their own undesirable feelings or characteristics to other people. For example, people who feel guilty about lying, cheating, or stealing will want to believe that everyone else lies, cheats, or steals. Thus, they can excuse their own behavior because "everybody does it." Very often the things that people accuse others of doing are exactly the things that they themselves are doing or would like to be doing.

Rationalization In rationalization, which is closely related to projection, people explain and excuse their behavior by giving it socially acceptable motives and disguising its unacceptable motives. Rationalization is among the most commonly used of the defense mechanisms and we all probably use it to some extent. Students who fail exams or courses often blame the instructor, the curriculum, their best friends—in fact, anything but their own lack of effort or ability. People who get fired from their jobs

or get traffic tickets similarly find someone or something else to blame. People using rationalization really believe what they are saying or thinking. They use this mechanism to protect their own opinion of themselves, not just the opinion that others have of them.

Regression Regression occurs when, in times of acute stress, people unconsciously try to return to an earlier stage of life in order to escape their current anxiety. As we gain in age and maturity, we find that our responsibilities are equally increased. We become more responsible for the consequences of our own behavior and therefore must maintain a greater degree of control over our emotions and impulses. This naturally tends to produce some stress.

Most people think of the earlier stages of their lives as being less stressful than their present stage. While this may be true, it is equally possible that they have forgotten or repressed the less pleasant aspects of their earlier ages. But when stress occurs, a person may behave in ways characteristic of earlier periods—periods in which they were dependent on someone and had few responsibilities.

Many examples of regression can be cited. The wife or husband who deserts a marriage and "goes home to mama" is regressing. The satisfaction received from sucking on a cigarette has been attributed to regression. A most extreme example of regression is the emotionally ill person who curls up, mute and withdrawn, into the fetal position (knees tucked under the chin).

● ASK YOURSELF

1. We all make use of ego-defense mechanisms, so it's no disgrace. What mechanisms do you seem to use most often?
2. Does your use of defense mechanisms sometimes prevent you from correcting basic conflicts in your life?

Fixation Fixation is closely related to regression. In fixation, however, people do not regress from an advanced stage of development; they simply never reach it. They remain emotionally immature, either in all phases of personality or only in certain phases of it. Such people may never gain emotional maturity, or may gain it at a later than average age.

Sublimation The word *sublimation* originally indicated the process of satisfying frustrated sexual desires in nonsexual substitute activities. The term is now used more loosely by many authorities to mean satisfying higher needs at the expense of—or in substitution for—lower ones.

Through sublimation, we repress undesirable thoughts and actions by developing more socially acceptable forms of behavior. Instead of expressing our aggressive impulses toward others in the form of destructive acts, we can act them out in various forms of competition, as in sports, politics, and business.

Sublimation thus means converting basic emotional drives, sexual and otherwise, into acceptable and useful activities. Sublimation is considered by many authorities to be among the most constructive of the defense mechanisms.

Need for Stress Management

The defense mechanisms just described, as well as many similar mechanisms, help us reduce or avoid anxiety (the conscious awareness of stress). Yet, at the same time, the use of these mechanisms does not resolve any of the basic sources of stress and may even contribute to them. Thus, we must consider stress again and the methods of dealing with it.

STRESS MANAGEMENT

We know the results of uncontrolled stress: it produces countless physical disorders ranging from itches, rashes, and colds all the way to heart attacks and cancers. But are all these problems inevitable? What can be done about them?

Stress management involves a new way of looking at life. Its philosophy is that you, as an individual, are basically responsible for your own emotional and physical well-being. You can no longer allow other people to determine whether or not you are happy. No other person can make you happy and secure. You must do this for yourself. Nor should you allow other people's behavior to make you miserable. Your perception of events, not the events themselves, is what causes stress. The same life situation may be perceived as either stressful or nonstressful, depending on your individual interpretation of it.

General Guidelines

Woolfolk and Richardson (1979) have presented some general guidelines for a low-stress life style:

1. Relief from stress and the beginning of wisdom come when we start accepting the basic uncertainty of life and our lack of control over many of its outcomes. Much stress originates in our efforts to maintain control over events which are basically beyond our control.
2. It is not what happens to us, but our perceptions, beliefs, and what we tell ourselves about what happens to us that cause almost all emotional distress.
3. Happiness cannot be achieved when pursued as a goal. It is always a byproduct of other activities.
4. Find activities from which you derive intrinsic satisfaction. Stress can be reduced by focusing on the *process* of your activities, rather than on their results or outcomes. Whatever you are doing, focus on and enjoy the activity itself, rather than focusing on how well you are performing or what the activity will bring you.
5. Find something other than yourself and your achievements to care about and believe in. A sense of purpose in life comes from dedication to people, relationships, ideas, and values.
6. Learn to recognize and accept both your

Warning Signs of Excess Stress

If you are experiencing one or more of these symptoms, your stress level is probably too high. If you do not reduce your stress level, you can expect to develop physical or emotional health problems.

Difficulty in thinking clearly, making decisions, or solving problems
Nervousness
Vague anxieties
Impulsive behavior
Strong urge to hide or cry
Loss of joy of living
Tendency to tire easily
Sighing
Irritability
Depression
Trouble sitting still or maintaining a physically relaxed position
Difficulty in sleeping
Feelings of weakness or dizziness
Increased use of alcohol, tobacco, or other drugs
Loss of appetite or eating too much
Intestinal disturbances such as indigestion, diarrhea, or belching
Menstrual problems
Loss of sex drive or impaired sexual response
Tendency to be easily startled
Pounding heart
Dryness of mouth
Nervous tics or twitches
Trembling
Tooth grinding
Sweating

personal shortcomings and your lack of control over much of what will ultimately happen to you. Acceptance of the uncertainty and ambiguity of life relieves you of the need to always be "right," another common source of stress.

7. Develop an "unhostile" sense of humor. Although much humor has a hostile quality (for example, ethnic and sexist jokes), there is a tension-relieving type of laughter that springs from a sense of the comedy and absurdity in life and the ability to laugh at your own self.

8. Learn to tolerate and forgive both yourself and others. Intolerance of your own failures leads to stress and low self-esteem. Intoler-

ance of others leads to blame and anger, and thus to stress.

9. Learn to see the world and yourself through the eyes of others. Interpersonal relationships are less stressful when we understand the viewpoints of others.

10. A low-stress life style is usually reasonably efficient and well managed. Laziness, procrastination, and sloppiness usually create more stress than they relieve.

11. Don't live in the past. To focus on the past is to rob the present of its joy and vitality.

12. The struggles of life never end, they only change. Stop waiting for the day when your problems will be over; it will never come. Every day offers us a wealth of pleasures.

Physical activity is one of the best tension relievers.

Most good things in life are fleeting and transitory (temporary). Enjoy them while they last. Live in the Golden Now!

"Instant" Stress Relievers

Many "instant" stress relievers have been proposed by Litvak (1979) and others. While these techniques may not eliminate the basic sources of stress, they very effectively enable a person to quickly relax and reestablish normal body function. Some, like meditation, may also improve your outlook on life. Space does not permit a full explanation of each of these methods, but many widely available books do so.

1. Physical activity One of the best tension relievers is physical activity: walking, running, dancing, calisthenics, swimming, or almost any sport. The many benefits of physical activity are described in Chapter 10.

2. Relaxation techniques One effective relaxation technique is to lie or sit down, close your eyes, and starting with your toes and working to the top of your head, concentrate on relaxing each body part, one by one. When you are completely relaxed, lie or sit quietly, enjoying the moment and learning how your body feels in the complete absence of tension. Before getting up, tell yourself, "I am completely relaxed, refreshed, and confident."

In order to fall asleep when stress is keeping you awake, first tense up each body part (contract the muscles) for about seven to twenty seconds, then let that part relax and feel the difference. This technique effectively relaxes tense shoulder and neck muscles.

3. Breathing exercises Deep breathing has several benefits. As your body takes in more oxygen,

---

Deep Breathing for Relaxation

1. Lie or sit comfortably and close your eyes. If sitting, sit upright so you can breathe freely.
2. Inhale slowly and deeply through your nose. Breathe with both diaphragm and ribs (your stomach should push out, not pull in).
3. Exhale slowly and fully through your mouth.
4. While breathing, either:
   a. Count slowly to about six or eight while inhaling and to about eight or ten while exhaling (a longer count for exhalation); *or*
   b. During each inhalation say slowly to yourself, "I am," and during each exhalation say very slowly, "relaxed."
5. Repeat about eight times or as long as is comfortable.
6. Breathe normally and rest quietly for a few minutes, enjoying the feeling or total relaxation.

---

you feel better and experience less stress. Also, just the act of concentrating on the process of breathing removes the source of stress from your mind temporarily and may help put the stressor back into its proper perspective. One breathing exercise is explained in the accompanying feature.

4. Meditation There are many varieties of meditation. We will describe several simple types that many people find very relaxing. Practiced regularly, meditation can help bring about feelings of well-being and inner peace. The purpose of meditation is to give the mind a brief vacation from the hassles of daily life. This short vacation from reality rests and revitalizes your coping abilities. Interrupting a fruitless, emotion-generating train of thought by interjecting a relaxing activity such as meditation or a breathing exercise diminishes stress even after the activity is over.

Virtually all forms of meditation involve three common features: assuming a comfortable body position, maintaining physical immobility, and continuously focusing attention on some object, sound, or bodily process. Simply do these things and, by definition, you are meditating.

When you practice meditation be sure you are in a private, quiet place. Wear comfortable, loose clothing. Get into a comfortable position, sitting upright in a chair or in the cross-legged "lotus" position, if you can. (You might fall asleep if you are lying down.)

Don't be concerned about success or failure or how well you are performing the technique; this defeats the purpose of meditation. One does not meditate "well" or "poorly." If your attention is focused on the object of meditation, you are meditating; if not, you are not meditating.

Many people meditate by concentrating on a *mantra*—a word or several words. Close your eyes, relax, clear your mind of all worldly thoughts and concerns. Concentrate on your breathing. Silently repeat your words to yourself with each cycle of inhalation and exhalation: "I . . . am" or "I am . . . relaxed" or "inner . . . peace" or "in . . . out" or any other words or sounds.

Or instead of meditating with eyes closed, open your eyes and gaze at a visually soothing object such as a flower or a picture. Here the aim is to attend to the visual perception of the object instead of the bodily sensations arising from breathing. This type of meditation can be performed with or without a mantra.

Many people like to meditate twice a day for about ten to twenty minutes each time. Others prefer one longer session or three or more shorter sessions.

5. Imagery The imagination can be used as a means of relaxation or as a method of "rehearsing" more effective ways of coping with stressful situations. To use imagery for relaxation, sit or lie comfortably, close your eyes, and imagine a pleasant, peaceful scene. It might be a lake, meadow, beach, forest, stream, desert, or any place you really enjoy. Place yourself in the scene. Concentrate on all of the details—the colors, sounds, smells, feelings. Some people like to add a companion—a person or an animal—to the scene. Continue the imagery for a few minutes or as long as desired. It will relax your mind and body and, like all of these stress-management techniques, will have a residual relaxing action for some time after its completion.

6. Massage Massage is among the most relaxing and stress-reducing activities. Massage by a good friend or mate is extremely fulfilling and, if desired, can serve as a means of sharing affection. Self-massage of many parts of the body such as the head, neck, face, shoulders is also easy to do and provides instant relief from tension.

7. Yoga There are many forms of yoga. Usually, yoga exercises comprise only a portion of the yoga discipline, which encompasses an entire way of life. Many good books on yoga are available, and classes are given in most cities and colleges.

8. Slowing down One of the simplest ways to reduce stress is to slow down. Much of our stress is the result of trying to do too much, too quickly. You can instantly reduce stress just by slowing down whatever you are doing: working, driving, making love, bathing, or any other activity. A good exercise is to take some familiar activity that you tend to rush (such as shaving or dishwashing) and do it in slow motion.

9. Cutting back Another common cause of stress is forcing yourself to do too many things. Rushing is in itself stressful, as is the feeling that you are doing

During periods of stress, a few minutes of imagery can relax the mind and the body.

● **ASK YOURSELF**

1. Are you presently (within the past week) experiencing any stress symptoms?
2. If so, what basic changes in your attitudes or life style might lower your stress level?
3. What might happen if you neglect your stress symptoms?
4. What "instant" stress relievers seem appropriate to your situation?

things less well than you are capable of doing because of constant rushing. Further, no time remains for relaxation activities like we have just described. If you tend to be rushing much of the time, slow down, take inventory of what you are doing, deter-mine which activities are least rewarding, and eliminate them or approach them in a new way.

Breaking Free of Traps

Successful stress reduction often requires breaking free of some of the traps that society imposes upon us or we impose upon ourselves. For example, society approves, rewards, and reinforces many stressful behavior patterns. We are encouraged to be ever more productive—to do more work and do it faster—and are rewarded accordingly. Many people suspect that the nine stress-reducing activities just discussed are a "waste of time," if not worse. Society is very suspicious of people who in any way take a break from reality, which is what many of these activities amount to. Some unfortunate people have incorporated such activities so firmly in their value systems that the time they spend on stress-reducing activities causes guilt and actually produces stress!

Many other traps are self-imposed. Often our behavior is just a matter of habit; we engage in activities that are neither productive nor stress relieving. Watching too much television or spending a lot of time with people we don't really enjoy are just two examples. Our prejudices and stereotypes produce stress because they pattern our thinking in ways that provoke predictable, emotionally intense responses to various circumstances. We may feel needless stress when we encounter unfamiliar foods, people engaging in nontraditional sex roles, people who look "different" or live differently and so forth. The key to overcoming your prejudices is to identify and challenge their underlying faulty assumptions.

Finally, successful stress management absolutely requires a positive approach to life. Many people are unaware of how their lives are bogged down in negativity: pessimism, defeatism, faultfinding, self-pity, jealousy, envy, greed, and arrogance. A major cause of stress is the feeling of being "trapped," of having no options in life. But we all have *many* options. Sometimes we just fail to see them because of negative thinking. Become a positive thinker! Be aware of the many options life presents to you! Make the most out of every situation! Develop the confidence and decisiveness that will enable you to break free of the bonds you have placed upon yourself!

WHAT IS HOLISTIC HEALTH?

Holistic health is a positive approach to health, based on the interaction of the mind, body, and spirit—the whole person. The mind, body, and spirit are interdependent; none can function optimally unless all do. Physical disorders are prevented or treated by maintaining the health of all three.

Holistic health strongly emphasizes nutrition, exercise, relaxation, and meditation and holds the

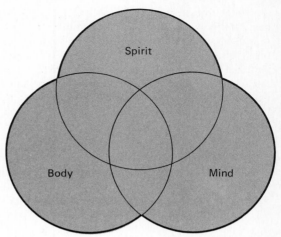

Holistic health recognizes the interdependence of mind, body, and spirit. None can function well unless all do.

use of medications to a minimum. It teaches freedom of body movement, stretching and body limbering, deep breathing, and massage techniques. "Orthodox" medicine has been slow to accept some holistic teachings, but many health professionals are now applying holistic health methods.

Levels of Health

The levels of health lie along a continuum that ranges from death at one extreme to positive wellness at the other. The health of most people lies somewhere near the center; they are not seriously ill, but they are not really healthy either. Positive wellness is a state of radiant physical and mental well-being. Life is a pleasure. Every day is approached with enthusiasm and lived joyfully. This seemingly impossible goal is attainable by anyone who is willing to apply holistic health methods.

Individual Responsibility for Health

Holistic health makes each of us responsibile for the state of his or her own health. Holistic health is a matter of individual practice. No one else can ex-

Positive wellness · Good health · Average health · Minor illness · Life-threatening disorders ← → Death

Levels of health may fall anywhere along a continuum ranging from death at one extreme to positive wellness at the other. Note that the left side of the continuum is open-ended—there is no limit to the level of wellness we can achieve.

ercise for you, relax for you, or see that you eat healthful food. In fact, according to the principles of holistic health, most illness is at least partly the result of an individual's behavior and is thus preventable by a change in behavior. This concept applies to thousands of conditions ranging from the trivial to the life threatening. Of course, adults are largely responsible for the health of their dependent children.

The Function of Illness

Another holistic concept is that much (some people say all) illness serves a function for the sick individual. For example, it may provide an excuse for getting out of doing something we really don't want to do. It may allow us to put off dealing with an unpleasant problem. It may help us gain attention or sympathy. It may be a means of coping with an unresolvable life situation. Some people adopt illness as a life style. Year after year they avoid decisions, responsibilities, or sexual or social interaction by being sick. Perfectionists, having set impossible goals for themselves, may use illness as an excuse for their inevitable failures.

Whenever you experience illness, you may find it highly productive to engage in some self-examination. Think for a moment about your most recent illness. First, consider what behavioral factors might have contributed to your being ill. Had you been careless about your nutrition? Had you failed to exercise regularly? Had you pushed yourself to the point of exhaustion? Had you been living with a lot of unresolved emotional stress or conflict? Next, consider whether your illness might have served some function for you. Can you relate it to anything that was happening in your life? If you consider illness as a message from your body, you may be able to manage your life in ways that minimize future illness. Remember that the same forces that lead to minor illnesses such as colds can, in time, contribute to major problems like heart disease and cancers.

The Benefits of Holistic Health

The time for a holistic approach to health has clearly arrived. In most industrialized countries, science and technology have done much to reduce the threat of many infectious diseases such as ma-

laria, typhoid fever, polio, and tuberculosis. Most of the major causes of death and disability today appear to be beyond the reach of current technology. Though we can certainly expect continued medical advances, a point of diminishing returns is being reached where great expenditures of money and effort produce only moderate practical benefits. Yet many people could immediately reduce their risk of early death from cancers and heart disorders by applying well-known holistic health methods.

Today's important diseases are symptoms of an entire cultural life style that has become increasingly unhealthful. Our best hope for overcoming these diseases and enjoying more rewarding lives is through basic life-style changes. By applying the holistic concept of individual responsibility for health, we can adopt a healthful life style even though we live in an unhealthful cultural milieu.

Kenneth R. Pelletier, in *Mind as Healer, Mind as Slayer* (1977), summarizes the characteristics of holistic medicine:

1. All states of health and all disorders are considered to be psychosomatic to some degree.

2. Each individual is unique and represents a complex interaction of body, mind, and spirit; disease results when stress disrupts this complex balance.

3. The patient and the health practitioner share the responsibility for the healing process. The patient must be an active participant in adopting a healthful life style.

4. Health care is not the exclusive domain or responsibility of orthodox medicine. Instructors in physical exercise, nutrition, meditation, relaxation, stress management, and other life-enhancing skills all have something to contribute to health care.

5. Illness is a creative opportunity to learn more about yourself and your fundamental values. Conversely, exceptionally healthy persons should be the subject of research to identify healthful factors in their life styles.

6. Effective health practitioners must know themselves as total persons in order to treat their patients as total persons. Too often patients are referred to as "the appendectomy in room 223"; while their physical needs are

looked after, their emotional and spiritual needs are largely ignored. (We shall return to this subject in Chapter 2.)

We have seen that holistic health integrates techniques that range from ancient systems of meditation to the latest biomedical technology. By applying its teachings to our own lives, we can maintain a healthful balance of mind, body, and spirit within ourselves and live in harmony with the world around us.

● ASK YOURSELF

1. What is holistic health?
2. Where does your current level of health fall on the health continuum?
3. Are you satisfied with your level of health?
4. What is meant by the statement that illness usually has a function? What examples can you think of in your own life?

IN REVIEW

1. Each of us largely determines the quality of our own health. Our emotions and behavior play a role in virtually every illness we develop.
2. Emotional health may be viewed in terms of need fulfillment — how well we learn how to fulfill our needs and how we learn to deal with the frustration of our needs.
3. Personality is the total reaction of a person to his or her environment.
4. In Freudian theory, the id represents the primitive survival instincts; the ego represents the conscious mind; the superego represents the judgment mechanisms regarding right and wrong.
5. The self is our individual perception of our own personality. Our identity is our perception of who we are relative to other people and the general scheme of things. Many people suffer from a poor self-concept or a lack of identity.
6. Human needs are felt sequentially: physiological needs, safety or security needs, needs for love and belonging, self-esteem, self-actualization, needs to know and understand, and aesthetic needs. Behavior is motivated by unfulfilled needs.
7. Some characteristics of sound emotional health include:
   a. The ability to deal constructively with reality
   b. The ability to adapt to change
   c. A reasonable degree of autonomy (independence)
   d. The ability to manage stress
   e. A concern for other people
   f. Satisfactory relationships with other people
   g. The ability to love
   h. The ability to work productively
8. Excessive stress is a major health problem which, if uncontrolled, can have disastrous physical results.
9. One way we deal with stress is by using ego-defense mechanisms: avoidance, denial of reality, repression, projection, rationalization, regression, fixation, and sublimation.
10. Effective stress-management techniques can reduce the impact of excessive stress:
    a. Stress management starts with a new way of looking at life.
    b. "Instant" stress relievers include physical activity, relaxation techniques,

breathing exercises, meditation, imagery, massage, yoga, slowing down, and cutting back.

11. Holistic health is a positive approach to health, based on the interaction of the mind, body, and spirit.

a. Each of us is responsible for our own health.

b. Illness often results from our own unhealthful behavior and often serves some function for the sick individual

CHAPTER 2

EMOTIONAL DEVELOPMENT

Each of us is conceived with a certain genetic potential for our physical and mental development. The extent to which this potential is realized depends heavily on environmental, psychological, and sociological factors. Few people are lucky enough to come even close to achieving their full potential.

Environmental factors begin their influence even before conception. Exposure of the ovaries or testes to radiation or to certain chemicals, for example, may result in genetic damage that severely limits the potential of offspring. During pregnancy, diet, drugs, diseases, and hormonal factors exert lifelong influences on development.

The experiences of early childhood are extremely important in our development. Many basic lifetime personality characteristics are formed before the age of 4 years. Important influences continue throughout the school years and, in fact, until death. As young adults, we are faced with choices concerning life styles, careers, and our emerging adult identities.

Continued emotional health requires an ongoing process of adjustment to our ever-changing life situation. The world around us changes constantly, frequently presenting new opportunities for the fulfillment of our basic needs. Our own physical and mental selves are also in a state of perpetual transition. Hopefully, we continue to gain insight and understanding throughout life, though some of our new insights may in themselves threaten our sense of emotional security. Finally, every one of us faces the unavoidable reality of the physical aging pro-

cess and the ultimate challenge of death itself. Unfortunately, we live in a society that has been largely unsuccessful in dealing with the psychological aspects of aging and death, even though the number of elderly people has greatly increased and will continue to do so.

In this chapter, we will examine the stages of human personality development as presented by Erik Erikson in his classic *Childhood and Society* (1950, 1963). We will relate Erikson's concepts to Maslow's thoughts on human needs and to our own discussion of intimacy and need fulfillment. We will then explore in greater depth some emotional factors that affect college students and ways of dealing with these factors. We will investigate the development of assertiveness, effective communication, and emotional intimacy. All of these can be important contributions to the fulfillment of our needs.

In Erikson's concept of development, each phase of the human life cycle is characterized by a specific developmental task that must be resolved in some way before the individual proceeds to the next level. The success with which each task is resolved may have lasting effects on the future personality of the individual. It should be noted that these tasks are never completely resolved, but are worked out further during successive stages. Again, we would emphasize the dynamic nature of personality, with the ever-present possibility for improvement (that is, the substitution of more appropriate or effective responses). Erikson has formulated each stage in terms of psychological contrasts that

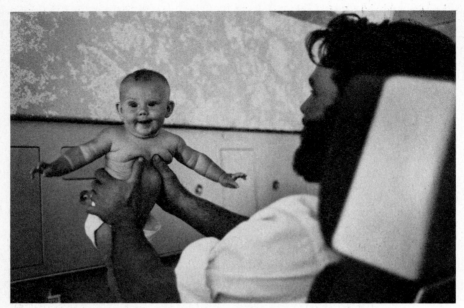

Babies who receive plenty of sensory stimuli develop more rapidly.

represent the two extremes of successful and un-successful task resolution (personality development). The actual outcome is always somewhere between the two polar extremes for each of the eight developmental tasks.

EARLY INFANCY

This period is the first year or year and a half following birth. The lifelong importance of early childhood environment and experiences would be difficult to overstate. Right from birth, many basic human needs are strongly felt. The needs for food, love, and security are overwhelmingly strong and demanding at this time. Deprivation during the first few years, whether physical (such as diet) or emotional, will likely have effects spanning one's entire life.

Studies have shown that an infant is capable of learning new responses from the first few days following birth. The experiences of the first year are very significant. For example, babies who receive plenty of sensory stimulation in the form of holding and other loving attention show more rapid mental (and often physical) development than babies who are deprived of such stimulation.

Since the child is so totally dependent on others at this time, Erikson rates its emotional health in terms of either *trust* or *mistrust*. Trust (sound emotional health in the infant) results from affection and adequate gratification of the infant's needs. Mistrust (developmental failure) results from inadequate need fulfillment—neglect, isolation, deprivation of love, too early or harsh weaning, or physical abuse. For many, this is the first step toward a lifetime of isolation and inability to develop intimacy with others. Intimacy is based on mutual trust and can never develop in an atmosphere of mistrust.

LATER INFANCY

This period lasts from about the first to the third year of age. Much of one's lifelong self-concept, gender identity, and sexual orientation is formed during this period. In addition to the more basic needs—love, security, and the physiological needs—the need for self-esteem is strongly felt during this period. In fact, Erikson rates developmental success at this age in terms of the child's self-image. Successful development is seen as a sense of *autonomy*. While still recognizing its dependence on its parents, the child feels itself to be an independent person and will often exert its independence. Developmental failure at this age is characterized by feelings of inadequacy, doubts about one's individuality, and a desire to hide one's inadequacies (a

Successful development in late infancy results in a sense of independence.

Children have a strong need to know and understand.

sense of shame). Development of basic skills like walking and talking is unusually slow. Many of these same characteristics carry over into our adult lives. Those of us who believe in ourselves tend to be much more successful in anything we try than are those of us who doubt our abilities. Again, the development of intimacy may be blocked by shame and the desire to hide imagined inadequacies.

Parents can make an important contribution to their child's lifelong emotional health by building his or her self-esteem at this time. A feeling of individuality, ability, and value gained at this age may stick with the child for life.

EARLY CHILDHOOD

A child goes through this developmental period between the ages of 4 and 5. In addition to the previously mentioned needs, the need to know and

to understand becomes pressing at this time. For many children at this age, the most commonly used word is *why*. During this period, children thrive on an abundance of sensory input. Erikson sees emotional health at this age as characterized by either *initiative* or *guilt*. The child who is developing well has a lively imagination, vigorously tests the limits of reality, imitates adults, and anticipates adult roles. In contrast, developmental failure is characterized by guilt over goals contemplated and acts initiated. This guilt results in an inhibition of normal activity. The child lacks spontaneity and is jealous, possessive, suspicious, evasive, and has difficulty in assuming a social role. Once again, we see that these symptoms of developmental failure carry over in some adults. These unfortunate people gain little pleasure from life because anything that they might do to produce pleasure will also produce guilt. Their relationships with other people tend to be mutually unrewarding and difficult.

● ASK YOURSELF

1. What did your parents do, or fail to do, to help you develop feelings of adequacy and self-esteem?
2. In raising children of your own, what things would you try to do differently from the way your own parents did them?

MIDDLE CHILDHOOD

This period lasts from about 6 to 11 years of age. During this prepubertal stage, most children exhibit fairly stable personality traits. There is seldom a sudden upheaval of established patterns, but rather a gradual evolution of more mature responses. Erikson sees the basic inner conflict at this time as being between *industry* and feelings of *inferiority*. Again, the need for a sense of self-esteem seems dominant in the successful development of children of this age.

The child who is developing successfully holds a positive self-image as someone who has value and is capable of doing worthwhile things. Such a child has a sense of duty and accomplishment, functions up to his or her capability in school, and relates well with other children and adults. He or she undertakes and successfully completes tasks at home and at school, and can clearly distinguish fantasy from reality.

Children who are held back by their feelings of inferiority tend to have poor work habits because they have little expectation of success. Their atti-

The ability to undertake and complete tasks is essential to successful development.

tude may be "What's the use? Why try? I'm going to fail anyway." They may expend little effort at schoolwork and contribute little to chores at home.

An important contributing factor in the development of feelings of inferiority in many children is that they are seldom if ever complimented on things they have done, or told that they are "OK." Instead, they hear only negative things about themselves. They are constantly reminded of the "bad" things they have done or the "good" things they have failed to do. Many parents could make a great contribution to the lifelong happiness of their children if they would merely emphasize their positive values and minimize their shortcomings.

School failures can often be traced to prejudice, preconceived notions, stereotyped thinking, and self-fulfilling prophecies. For example, teachers' expectations for a child are sometimes influenced by such factors as the child's sex, ethnic group, appearance, or the opinions of previous teachers. If a teacher expects little from a child, the child will quickly become aware of this feeling. The expectation of failure will be incorporated into the child's self-image. Once a person of any age adopts a self-image of incompetence in any area, his or her failure in that area is pretty well assured. It is a self-fulfilling prophecy. Teachers, parents, and anyone interacting with children can, by taking a positive view of the potential of each child, contribute to a self-image that will lead to success, not failure.

ADOLESCENCE

In most Western or industrial societies, adolescence (the period between puberty and full maturity) is recognized as a period of special difficulty in adjustment. It is seen as a critical period in the individual's development. This should not be surprising if we consider the special stresses to which the adolescent is typically subjected, in comparison to people at most other age levels. Adjustment is less difficult when a well-stabilized individual is confronted with familiar environmental demands for which appropriate responses have already been developed. This condition is seldom true of adolescence in our culture. (Similar periods of stress are created by entering or leaving school, divorce, career adjustments such as changing occupations or retirement, the death of a spouse, and declining health.)

Successful adolescent development depends on a firm sense of identity and a feeling of self-assurance.

The onset of puberty brings with it a host of physiological changes, including increases in hormone levels and changes in body structure and function. These changes present special adjustment problems in themselves. They also challenge the individual's basic sense of self and identity. In adolescence, one's previous concept of self and sense of continuity is challenged by the rapidity of body growth and the entirely new addition of sexual maturity, with its multitude of physical and emotional changes. Thus, Erikson sees adolescent development in terms of *identity* versus *role confusion* or *diffusion*.

Adolescents who have an adequately developed sense of identity are characterized by a feeling of self-assurance. They have a definite sense of who they are, where they are, and in what direction they are moving. They are certain of their gender identity and their sexual orientation. They have been able to integrate their varied partial identifications. This includes their identification with each of their parents or other significant adults, their sexual identifications, their intellectual and emotional endowment, and their social roles. All of this is integrated into a single identity. The consequence of this integrative process is a sense of rightness about who

● ASK YOURSELF

1. How did the difficulties you experienced in your own adolescence seem to compare with the problems of other people you knew at that time?
2. What might have contributed to these differences?

they are and what they are doing. Another result of this integration is a well-defined sense of values.

Identity diffusion (poor self-integration or role confusion) occurs when people cannot adequately integrate their varied identifications. As a result, they may be unable to make definite choices of career, roles, values, or sexual objects. There may be a sense of total confusion — identity confusion, role confusion, authority confusion, value confusion, and even time confusion. All of this uncertainty leads to self-consciousness, anxiety, depression, and a sense of alienation or loneliness.

Many adolescents are forced by these feelings into inadequate and ultimately self-defeating defense mechanisms. Some lean heavily upon alcohol or other drugs as coping devices. Some get caught up in compulsive or irresponsible sexuality. Some seek shelter in a premature marriage. Some withdraw into themselves. In some of the most tragic and severe cases, suicide seems the only way out.

Adolescents with emotional problems are usually identified first by parents or school authorities. Less often, they come to the attention of police, youth leaders, or the clergy. Unfortunately, many adolescent problems are not identified at all until the individuals are having serious emotional problems as adults. Persons outside the home who feel that an adolescent has serious problems may take the matter up with the parents. In many instances, parents are reluctant to take any action unless they are firmly convinced of the need for evaluation and treatment. Perhaps they fear that their child's problems will be seen by others as a poor reflection of the job that they themselves have done as parents. This, of course, is not necessarily the case. The problems of adolescents should be treated promptly before more serious trouble results. A therapist should be chosen who has specific training and experience in working with this age group. Various individual, family, and group therapy methods are used with adolescents.

The ability to form intimate relationships indicates successful development in early adulthood.

ADULTHOOD

Erikson divides adulthood into three stages, each with its own characteristic developmental task. These are early adulthood, from about ages 18 to 25; middle adulthood; and late adulthood, corresponding to the retirement years.

Early Adulthood

If the more basic human needs are being more or less adequately fulfilled, the full range of Maslow's higher needs should be manifest by this time. With adequate need fulfillment, the young adult should now be in a position to make a full commitment of him- or herself to others. In fact, Erikson judges the success of one's development at this age on the basis of *intimacy* versus *isolation*.

Some of the characteristics of satisfactory personality development include the ability to establish intimacy, as discussed later in this chapter; the ability to work productively; and adequate sexual adjustment. All of these traits have the prerequisite of an adequate sense of personal worth, or self-esteem. Lacking this feeling, many people avoid the development of intimacy. Or they may exhibit various "personality problems" that work against them, especially in their relations with other people. Sexual dysfunction is quite common among these people, often simply because they lack confidence in their sexual ability. There may be a feeling of "something missing" in sexual relationships. That missing element is intimacy.

As a substitute for intimacy, many people resort to "game playing" in their interpersonal relationships. A "false front" is presented to others, so every relationship involves a certain element of deception or dishonesty. People who fail to develop intimacy often experience the failure of one or more marriages during their early adult years. They may be unable to establish the kind of mature relationship necessary to provide mutual need fulfillment over a longer period of time.

People experiencing problems of this type need to explore (perhaps with a therapist) the true nature of their relationships with others. What is the true content of what they are verbally or nonverbally expressing to others? People are often quite amazed to learn how they "come across" to others.

Middle Adulthood

In the middle years of life, successful development depends on having one's needs adequately fulfilled so that one can help others in the fulfillment of their needs. Erikson uses the terms *generativity* (productivity) as contrasted to *stagnation* for the possible development of the middle adult. Others have applied the phrase *generosity* as contrasted to *self-absorption*.

Successful emotional development in the middle years is characterized by productivity and creativity for one's self and for others. If one has children, they are the source of pride and pleasure. The life of the middle adult should be rich with new

Successful development in the middle years is characterized by productivity and creativity.

experiences. The individual should gain satisfaction in establishing and guiding the next generation.

Poor personality development in the middle adult might be characterized by stagnation and/or self-absorption. Such people might be self-centered, self-indulgent, and threatened by new ideas or experiences ("in a rut"). They are less productive than they could be, and may even be forced into premature invalidism by psychogenic or psychosomatic disorders. Either term means a physical condition that is caused or aggravated by the state of the mind. Highly self-centered people tend to dwell upon every minor ache or pain. A slight pain that most people would ignore can seem so serious that it is disabling. No amount of physical treatment can cure such a problem, but appropriate psychotherapy can be rapidly effective in treating psychosomatic disorders.

Career burnout is a common experience in middle adulthood. We will examine this in more detail.

CAREER BURNOUT

Career burnout is an incapacitating psychological condition that occurs in response to continuous work-related stresses and frustrations. It is an important health consideration because it not only reduces individuals' job effectiveness but impairs their emotional, spiritual, and physical well-being.

Who Is at Risk?

Though most characteristic of people-oriented occupations, burnout can occur in any type of career. Health professionals, educators, police officers, paramedics, and social caseworkers have burnout rates that rank among the highest of all occupations. Students also commonly experience burnout. Individuals who tend to become "burned out" are usually idealistic and enthusiastic, with a strong commitment and dedication to their profession. They often assume too much responsibility and work too intensely. They may view their job as their major reason for being and thus must succeed at any cost. They typically rise to demanding, challenging, and responsible positions and are seen by others as experts at the job they do.

Those experiencing burnout lose their concern for the people (patients, students, clients, customers) with whom they were working (*see* Table 2.1).

TABLE 2.1 SYMPTOMS OF CAREER BURNOUT

Psychological	Behavioral	Physical
Depression	Negative attitudes with increasing expressions of job dissatisfaction	Fatigue
Feelings of isolation		Excessive sleeping or inability to sleep
Withdrawal from decision-making situations and responsibility	Inability to tolerate change	Digestive disturbances
Decreased self-esteem	Faultfinding	Weight gain or loss
Apathy toward work, colleagues, and patients, clients, students, or customers	Increased use of denial and projection mechanisms	Ulcers
	Uncharacteristic behavior	Shortness of breath
		Frequent colds
		Eventual total exhaustion

In addition to physical exhaustion (and often physical illness), burnout is characterized by an emotional exhaustion. All positive feelings, sympathy, and respect for one's patients, students, or clients is lost. A very cynical and dehumanized perception of these people often develops. They are labeled in derogatory ways and treated accordingly. Burnout is a major factor in low worker morale, impaired job performance, absenteeism, and high job turnover.

People experiencing burnout often increase their use of alcohol and other drugs as a way of reducing tension and blotting out strong feelings. They may see themselves as failures. Job-related stresses often result in family problems, such as increased fighting and feelings of isolation. Sometimes job frustrations are vented as anger or hostility toward those at home. After an emotionally trying day of work, a person often feels a strong need to get away from all people for a while. But this solitude may come at the expense of family and friends. And it may actually isolate the worker from the emotional support that could help reduce the stress.

Personal factors and bad working conditions can both lead to career burnout. Personal factors can include unattainable career goals; financial, marital, or parenting problems; poor health; and poor outlook on life. These factors will cause some individuals to burnout, no matter what kind of work they do or where they do it.

Many aspects of the job environment can contribute to burnout. Lack of management (administrative) support or understanding is high on the list. The feeling of having no options—of being "trapped" in a job—is another common factor. Long hours, overtime, high noise levels, and crowded work areas contribute to burnout. Under-staffing and "tight" scheduling that requires one to keep working constantly at maximum capacity are major causes of burnout.

Preventing Burnout

Burnout is preventable. Management can help combat it by making the workplace more physically attractive, giving greater recognition to employees for their efforts, and allowing more flexibility in working hours and the scheduling of breaks and duties.

As individuals we can do much to prevent our own burning out. A regular exercise program is important in stress control. The fatigue experienced in burnout is emotional and mental, seldom physical. The stress relief provided by physical exercise can sharply reduce feelings of fatigue and contribute to restful sleep. Use of alcohol or other drugs such as tranquilizers usually decreases when a worker begins an active program of physical exercise.

It is critically important to reserve a significant portion of your life for activities that are not job related. Everyone needs regular free time to spend with family or friends, engage in sports, read or pursue hobbies, or simply rest and relax. Regular vacations and weekends away from your work environment are essential in preventing burnout. People who burnout are often those who fail to take full advantage of weekends, holidays, and vacations. Burnout can also be prevented by maintaining a clear separation between work and home. By leaving your work at work, you confine its emotional stresses to a smaller part of your life (this includes not only doing work at home but worrying about it as well). If you socialize with people you work with, there should be a rule forbidding "shoptalk" away from the work place.

Regular weekends and vacations away from the work environment are essential in preventing career burnout.

Finally, each of us needs to maintain a support system. This is typically a group of people who are "in the same boat" as you are. As a student, your support system would probably consist mostly of other students, though it might include nonstudent friends, family members, professors, and others as well. On the job, your support system would typically include your co-workers. In organizing your support system, it is important to include people who approach life in a positive way. A support system can provide advice and comfort, and ease the stress and pain of difficult situations. It can give a fresh perspective on how to deal with troublesome situations and a sense of shared responsibility for difficult decisions. It always helps to be able to talk

● ASK YOURSELF

1. By what process did you arrive at your current career goal?
2. Was it a rational choice, in terms of your true interests and abilities?
3. Is your chosen career among those with a high incidence of burnout? If so, how do you plan to keep it from happening to you?

about your problems, and the feedback from your support group can help you see a situation in its proper perspective. When we don't talk about our problems, we tend to blow them all out of proportion in our minds. But the mere act of talking about a problem with a supportive person or persons can eliminate much of the stress it is causing.

Late Adulthood

The way in which younger people look ahead to maturity and the way in which mature people experience it reveals much about their successful personality development. Emotionally healthy younger people, while perhaps not eager to grow older, accept the aging process as part of life. They try to make the most of every phase of their lives.

Erikson sees the emotional health of the elderly in terms of *integrity*, as contrasted to *despair*. Emotional integrity includes a love and appreciation of the spirit of humanity that goes beyond the love of the self. It is an acceptance of the continuity of past, present, and future. Most important, it is an acceptance of the life one has lived as having been valid and worthwhile. People holding such views do not find their mature years clouded by a morbid fear of death.

Conversely, if one lacks this feeling of integrity, one experiences despair. There is a fear of death. There is often the feeling that one's life has been wasted, that the things one has done have not been worthwhile. There is the feeling that given another chance, one's life would be lived differently. But it is now too late to start anew, to try out alternate roads to integrity. Older people who lack a sense of integrity feel disgust with themselves and with humanity in general. They frequently feel that human existence has no meaning. It is a striking study in contrasts to compare the inner peace and happiness of a well-adjusted elderly person to the bitterness and depression of someone in a state of despair.

Emotionally healthy mature people consider their lives to be valid and worthwhile.

The next chapter will explore in more detail the psychological factors in aging and death.

THE COLLEGE YEARS

The so-called college-age group in reality includes a broad spectrum of the population. It ranges from those in late adolescence to an increasing number of people of more advanced ages. The majority of college students, though, are in their late adolescence or early adulthood. They are subject to the common problems that Erikson associates with those age groups: identity problems and isolation.

In colleges that provide adequate mental health services, about 10 percent of the students seek such help each year, either voluntarily or by referral. But most colleges do not have adequate mental health services. Thus many students are left to work out their problems on their own or seek help elsewhere.

Most of the problems of college students are easily treated—only 2 percent of students seeking help are hospitalized. However, of the 40 percent of all students who leave college before completing their studies, it has been estimated that half do so for reasons of emotional health.

In the following paragraphs we will briefly examine some of the problems experienced by college students:

Identity Formation Versus Identity Diffusion

During the college years, various identities and activities may be tried, such as different types of love affairs, living arrangements, and participation in organizations. If a student lacks integrative ability (as discussed in the section on adolescence), if identifications are too ambivalent, or if the pressures from the college environment are too severe, diffusion occurs.

Independence Versus Dependence

Whether students go away to college or commute from home, they are certain to be involved in new forms of independence (and sometimes dependency). They are exposed to new value systems and are put in the position of having to choose between the old and new. They may find their established value system strengthened, or they may identify with a totally new value system. Most will undergo an evolutionary process in which they integrate new values gradually into their existing value systems. Students must learn to trust and respect both the person they currently are and the person with new knowledge and values that they will become.

If the need to be dependent is too great or if accepting new values or developing a new identity seems to imply the rejection of the safe and familiar family structure, separation anxiety or depression may result. Such feelings may lead to compulsive behavior patterns such as compulsive sexuality, compulsive use of substances (food, alcohol, or other drugs), compulsive activity, or the inability to concentrate on studies.

Intimacy Versus Isolation

College students must learn to establish new relationships at a level of intimacy possibly not experienced before. A strong sense of self-esteem and

Emotional health is a major factor in the college students' academic and social success.

identity enables a young person to be emotionally close to another or to accept a new idea without fear of losing his or her identity. For someone lacking this confidence, the prospect of an intimate relationship produces anxiety, perhaps a very intense anxiety. But the failure to establish intimacy leads to feelings of isolation, rejection, and loneliness.

Sexual Decisions

Despite a general liberalization in sexual attitudes and practices, decisions on sexual behavior remain a source of anxiety and emotional conflict for many students. Many students come from families or religious groups where nonmarital sexual activity is still sternly limited. Elements of identity, dependency, and intimacy influence the sexual behavior of most students. These students also face possible internal conflicts and guilt feelings if they are sexually active. Yet a sexual decision entered into more out of concern for performance than for relationship may also cause concern. Anxiety may be caused either way. Sexual problems are best avoided by having a value system of one's own, not imposed by parents or friends, and with which one can be comfortable at both the intellectual and the emotional, or "gut," level.

Problems of the New Student

The student entering college may encounter for the first time a wide variety of values and attitudes that conflict with those of his or her family. The conflict may be interpreted by the student as one of individuality versus family loyalty. Students who pursue individuality, which is natural, may feel guilty about rejecting and devaluing their families.

A student's transition from high school to college is often assumed by parents to mean the acceptance of adult responsibilities—financial and otherwise. Not all college freshmen are prepared or willing to give up their dependent status. They may feel that their family has abandoned them. At the same time, the parents may not be tolerant of the new values and ideas that the student brings home. The student who lives with parents and commutes to school is in a particularly difficult position. He or she may have to function at two conflicting levels in two quite different environments.

A common problem among college freshmen is the depression brought about by the first failures to maintain the level of grades received in high school. A student who was outstanding in high school may be competing in college with many others of equal ability. A student in this position must rebuild the basis of his or her identity and self-esteem before

● ASK YOURSELF

1. Do you sometimes have trouble concentrating when you are trying to study? If so, can you identify the emotional conflicts that are contributing to this problem?
2. Can you identify with any of the other college-related problems just discussed? If so, can you formulate a plan for dealing with those problems?
3. What counseling and vocational guidance services are available to you through your college?

feelings of inadequacy take hold and make success in college impossible. Another frequent problem is adapting to other students. Conflicts often develop between roommates who may have differing value systems and living habits.

A supportive family can be a tremendous asset to a student away from home for the first time as he or she adjusts to independent living. Unhealthy family relationships may manifest themselves when a student leaves home to attend college. Previously suppressed negative feelings about parents may enter the consciousness, causing guilt and anxiety. If a student has been an ego-extension of one parent, separation from that parent may create anxiety or guilt at having "abandoned" that parent. If the parents of a student divorce soon after his or her entering college, there may be a feeling of responsibility for the breakup of the family. On the other hand, the parents may get along better after the student leaves home. This is not unusual. In this case, the student may feel rejected because of the apparent ease with which his or her absence is tolerated.

Transferring Students

A transfer from one college to another may be based on a real need such as a change of career choice, financial problems, marriage, or divorce. On the other hand, many students with social or academic problems try to solve them by meaningless transfers from one college to another. Students who find their problems transferring right along with them should recognize that those problems stem from their own personal difficulties. They should seek appropriate counsel in resolving them.

Problems of Academic Performance

Poor grades or anxiety over grades are among the most common college problems. For some students, the problem may be one of inadequate preparation or unrealistic goals. For these students, the answer may be a reduced course load or perhaps a change to a more realistic goal. More commonly, students have the basic ability to achieve their goals but have difficulty in channeling their efforts to that end. They may sit at their desks for hours with little to show for their time because they are unable to direct their thoughts to the task before them. Identifying the source of the distractions and dealing with the underlying problems, perhaps with the aid of a therapist, is the only way to break such a study block. The problem may involve social adjustment to college life, dependency conflicts, sexual problems, or even an unconscious effort to fail as revenge against parents or others. Or feelings of guilt or inadequacy may make a student feel unworthy of success, and the unconscious mind acts to insure lack of success.

Many students have trouble because they have chosen a major in which they are just not very interested. Often they have done so in an effort to please family members. Sometimes they simply are not aware of the many options that life offers them. In any event, it is almost impossible to do well at something you don't enjoy. A change of major is frequently an appropriate decision. Students need to be able to adapt to this change and to accept guidance from their academic advisors.

The Alienation Syndrome

Some students react to college environment with attitudes of apathy, boredom, lack of involvement or commitment, and general unhappiness. This has been called the *alienation syndrome*. This syndrome often arises when students consciously or intellectually reject their value system while at the same time they unconsciously hold on to it. Or students may reject their existing value system without formulating an adequate replacement for it. There may also be elements of repressed guilt if the system being rejected is that of their parents or former religion.

Rejection Reactions

Some students (actually, people in general) suffer extreme reactions in situations that they perceive

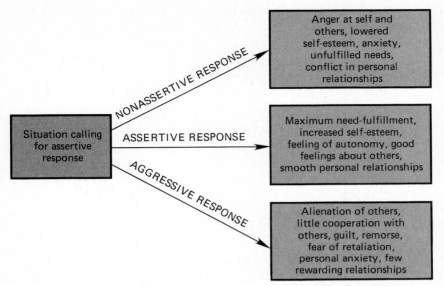

Outcomes of nonassertive, assertive, and aggressive responses to a typical situation in interpersonal relations in which an assertive response would be appropriate.

as rejection. A careless comment, a broken friendship, a bad grade, or a critical remark from someone can produce an hysterical reaction. There may be panic, crying, impulsive acts, or even suicidal thoughts or actions. This type of reaction usually indicates an underlying lack of self-esteem. Great significance is placed on the opinions of others because the individual holds such a low opinion of him- or herself. People experiencing problems related to low self-esteem should seek out things they can do well and concentrate on such ego-boosting activities.

Fear of Leaving School

During their last year of college, many students face decisions regarding career, further education, relocation, marriage, and general life style. If college has served as a prolongation of a comfortable state of dependency, the prospect of graduation can arouse intense inner conflicts. The need to make important decisions plus the threatened loss of security may lead to reactions ranging from minor anxiety to true psychotic episodes. These factors also give rise to "professional students." They constantly change majors or prolong graduate study almost indefinitely in order to avoid leaving the shel-

tered environment of the college. Students experiencing this problem should explore (perhaps with a therapist) the emotional functions college has served and their attitudes toward the future. Most colleges and universities make available career placement services which can be helpful to students seeking employment. This guidance can help overcome the fear of leaving.

DEVELOPING ASSERTIVENESS

An important aspect of our model of emotional health (*see* Chapter 1) is the ability to fulfill our human needs in effective and appropriate ways. Central to this ability is the assumption of personal responsibility for one's own physical and emotional well-being. Assuming responsibility then leads us directly to the necessity for an adequate degree of assertiveness. Rather than passively sitting back and wishing that other people would recognize and fulfill our needs, we actively work toward the fulfillment of our own needs.

Almost every one of us has at times let someone "walk all over us" and later on felt extremely resentful toward *ourselves* for letting it happen. This is a good indication that we should have been

You Can Become an Assertive Person

Assertive behavior can be developed through attending classes or it can be self-taught A good brief book on the subject is *Your Perfect Right* by Robert Alberti and Michael Emmons (Impact 1978). Some of the steps in developing assertive behavior are

1. Learn to recognize the three types of behavior.
2. Observe and analyze your own behavior for each type.
3. Identify situations in which you might act more assertively.
4. Think about ways to act more assertively, and observe how more assertive people act in similar situations.
5. Try out assertive behavior in role-playing situations with a friend teacher or therapist.
6. Evaluate the results getting feedback from the other person.
7. Try assertive behavior in real-life situations.
8. Continue to evaluate and develop your assertiveness.

Long-established behavior patterns do not change overnight. But with practice, assertive behavior can become increasingly spontaneous and effective. Eventually, you may be able to respond assertively without hesitation in any situation.

more assertive. Assertiveness training teaches us to distinguish between three types of behavior; nonassertive, assertive, and aggressive. *Nonassertive behavior* is when we deny our own needs and rights. We fail to express our true feelings and desires. We allow others to make choices for us. As a result we fail to achieve our goals or to fulfill our needs. We feel resentful toward others, but even more resentful toward ourselves.

Assertive behavior takes place when we make our needs and desires known to others and stand up for our right to have these needs fulfilled. We make our own choices and achieve our goals without in any way infringing upon the rights of other people to have their needs fulfilled. As a result, we feel good about ourselves and others, and they harbor no resentment or disrespect for us.

Assertive behavior is sometimes confused with aggressive behavior, but there are distinct differences. *Aggressive behavior* occurs when we accomplish our goals or fulfill our needs at the expense of the rights, needs, or feelings of others. Such behavior may hurt or humiliate others, or it may put them on the defensive. This makes it difficult for us to achieve our goals. Even when goals

are achieved through aggressive behavior, we often generate hatred and frustration in others, and vengeance may result.

Because assertive behavior serves to benefit all parties involved, you may wonder why people are so often either nonassertive or aggressive. Part of the answer may lie in the ambivalent nature of our culture. On the one hand, tact, diplomacy, politeness, modesty, and self-denial are usually praised. This is especially true for children, females, students, employees—in other words, for all of us in at least some of the roles we assume. On the other hand, "getting ahead" is also praised. And it is often acceptable by some portions of society to "step on" or "use" others in the process. Also, nonassertive or aggressive behavior may result from feelings of inadequacy or low self-esteem. Finally, many people have simply never learned to distinguish between these three types of behavior.

DEVELOPING COMMUNICATION AND INTIMACY

Effective communication and the development of emotionally intimate relationships are essential

The inability to form intimate relationships often leads to feelings of isolation.

● ASK YOURSELF

1. How do you personally feel when you have acted nonassertively in a situation that called for some assertiveness?
2. Does it appear that you need to develop more assertiveness? If so, how do you plan to accomplish this?

steps in the fulfillment of many of our basic needs. Thus, many of our problems in human relations and personal emotional health can be traced to our frequent failure to communicate our thoughts, needs, and feelings openly and honestly. Lack of communication ranks high among the causes of sexual problems, family conflicts, and emotional problems of all kinds. Noncommunication and the resulting social isolation are implicated in many suicides and other emotional breakdowns.

Closely related to communication failure is the inability to establish or maintain intimate relationships. In essence, we all share the desire to develop various intimate associations. Through these we can help satisfy our needs to belong, to give and receive affection, and to maintain self-esteem.

To many people, the word *intimacy* is misunderstood as being synonymous with sexual intercourse. Sexual intimacy may imply physical closeness. But emotional intimacy is an important part of

every close relationship in which one person feels free to communicate innermost feelings to another person and vice versa. It is quite possible to have emotional intimacy without sexual implications or to have sexual interaction in the absence of true emotional intimacy.

Friendship, for example, provides more profound psychological benefits than mere companionship or shared interests. Intimate friends help us deal with anxiety by lending us their support. They also enable us to keep our problems in perspective through sharing their troubles with us. And sometimes, just by listening to us in a nonjudgmental manner, friends help us reduce emotional stress as we verbalize our internal conflicts or sort out possible solutions.

Relationships that combine sexual as well as emotional intimacy, such as between married or otherwise committed persons, require the continuous stimulation of effective, sensitive communication. Without it, these relationships eventually disintegrate because they fail to fulfill the needs of the individuals.

Now let us look at some major factors in the communication process, and how we can relate to others more effectively. We will also discuss some of the common causes of communication failure.

Communication is not limited to spoken language. Our nonverbal behavior, commonly called *body language,* conveys much of our message, particularly its emotional content. Our true feelings are often more accurately perceived by others through facial expressions, posture, body gestures, clothing, and general appearance. In other words, feelings that we may not desire to, or are unable to, express verbally can be otherwise communicated. The perception of our verbal messages is further influenced by our tone of voice, its volume, and our inflection of words or sentences. Let us take, for example, the phrase "I love you too." Generally, we would say this sounds like a positive expression of caring for someone. However, try saying it in a sarcastic tone, squinting the eyes, pursing the lips, and clenching your fists. Obviously the entire meaning is changed. Most of our nonverbal communication is not that dramatic. Rather, we reveal our emotions in numerous, more subtle ways. For those who are interested in understanding body language, there are many excellent books available on the subject.

Successful communication requires at least two participants: one who honestly expresses a feeling and one who truly wants and tries to understand that feeling as well as hear the words. Many of us tend to view the speaker as taking an active role while the listener is passive. The listener is also an active participant of whom much is required for communication to be successful.

Many people fail as listeners because they are so wrapped up in their own problems that they are unable to direct their attention and concern to the needs of others. The sounds may enter their ears and be transmitted as nerve impulses to their brains, but they do not hear the message. For example, a person may really be very unhappy and lonely. They attempt to cover up these feelings by talking excessively about trivia, by having many companions, or having too many social commitments. Unless listeners are being sincerely attentive, more than likely they will not perceive the true meaning.

Sometimes people do hear, but fail to understand—or worse, misunderstand—what is being said to them. Often people take messages as being critical when in fact they were not intended as such. Little misunderstandings have a way of growing into big hostilities simply because we fail to ask for clarification of some ill-taken or vague statement. How often do people take remarks to be personal criticism when in fact they were merely a case of misplaced frustration? When someone makes a statement causing you to feel defensive or angry, these feelings should not be concealed in hopes they will disappear. Anger does not go away easily, and most likely it will be expressed at another time —probably not an appropriate time.

In some instances, unsuccessful communication results from an inadequate vocabulary or an inability to express oneself in a particular area. In our society, sexual matters are frequently subject to

flawed communication or even noncommunication because of such inadequacies. Many people have difficulty finding the right words to explain their sexual feelings and desires. Sexual intercourse can easily become an unexciting, ritual chore simply because neither partner communicates what he or she would like to do. Fantasies should be shared; often it turns out both individuals were thinking similar things. People who are raised in homes where sex is an embarrassing topic, do not have an easy time overcoming this feeling. However, it *is* possible to establish communication even if both partners are shy.

Probably the greatest barrier to successful communication and the development of intimacy lies in the reluctance of so many people to make known their true feelings. There are many reasons why this is so. Men, for instance, have generally been raised to hide their emotions. Many men have been taught to equate masculinity with a strong stonelike character. It is seen as unmasculine to reveal sadness in tears, to reveal fear by refusal to compete or strive, or to reveal love through tender words or gentle caresses. Whatever emotions we cannot let escape naturally will nevertheless surface, although they may be cleverly disguised or distorted. Frequently, men substitute material conveniences for verbal communication of affection. Expressions of love can also be displaced to objects such as cars or motorcycles or to sports because these are acceptable masculine love outlets.

Women also hide their feelings. This is particularly true when it comes to feelings of hurt and rejection. By not saying "I feel you are rejecting me" and asking "What do you think makes me feel this way?" women may let resentment build up and begin to communicate hostile feelings.

These are just a few examples of how ineffective communication becomes established. One insensitive response is likely to be met with another. The consequences are a tangled mess of negative emotions: hurt, frustration, rejection, anger, resentment, hostility, and so forth. In time, communication is viewed as a waste of effort, and finally there may be no real communication whatever. Many people withdraw ever deeper into themselves, silently directing hate at others or even toward themselves.

There is, however, no reason why communication cannot be successful. First, you must know as much as possible about your own true feelings. Then you must risk exposing your emotions to another person. Without this risktaking, it is impossible to communicate honestly. Intimacy or closeness can develop only in such an atmosphere. When your defenses are lowered, you are vulnerable. It is easy for someone, as part of his or her own ego defense, to aim a hurt directly at one of your vulnerable areas.

Communication is further enhanced by the ability of both persons to remain as nonjudgmental as possible. Such an attitude allows for continued honesty and assures that vulnerability will not be taken advantage of. Both can thus feel free to present themselves as they truly are, not as they would like to be or would like others to see them as being.

Intimacy requires a mutual concern for the welfare and happiness of each individual. It can develop and be maintained only so long as communication remains open and honest. It also requires the

● ASK YOURSELF

1. Who are the people with whom you currently enjoy a relationship of emotional intimacy?
2. What emotional rewards do you receive from each of these relationships?
3. Do you in some way block the development of emotional intimacy? Do you do this intentionally? If so, are you more or less emotionally fulfilled as a result?
4. If you wanted to experience emotionally intimate relationships, how would you go about developing them?

mutual acceptance of both persons for what they are. Then each can lower any barrier-producing ego-defense mechanisms that may be in the way. Many of the problems in our relationships could be avoided if barriers to open communication could be lowered.

IN REVIEW

1. Continued emotional health requires a lifelong process of adjustment to ever-changing life situations. Every age has its challenges:
   a. Early infancy — trust versus mistrust
   b. Later infancy — autonomy versus feelings of inadequacy
   c. Early childhood — initiative versus guilt
   d. Middle childhood — industry versus feelings of inferiority
   e. Adolescence — identity versus role confusion
   f. Early adulthood — intimacy versus isolation
   g. Middle adulthood — productivity versus stagnation (Career burnout is a problem for many at this time.)
   h. Late adulthood — integrity versus despair
2. College students have their own special adjustment problems:
   a. Identity formation versus identity diffusion
   b. Development of independence
   c. Establishment of intimacy
   d. Sexual decisions
   e. Special problems of new students
   f. Special problems of transferring students
   g. Problems of academic performance
   h. Alienation
   i. Rejection reactions
   j. Fear of leaving school

3. An adequate degree of assertiveness is necessary for need fulfillment and thus is an important contributor to emotional health. Assertiveness can be developed by anyone.

4. The ability to form intimate relationships also fulfills important needs. Effective communication is necessary for the development of intimacy. Also, individuals must have an adequate amount of self-esteem in order to be able to lower their defenses and permit intimacy to develop.

CHAPTER 3
EMOTIONAL PROBLEMS

The increasing complexity of life, greater social mobility, and increased social isolation pose grave challenges to the emotional health of many people in our country. Our society faces many problems in its attempt to create a social and physical environment favorable to good emotional health. It must begin to provide more adequate facilities for the prevention and treatment of the many emotional problems that occur in our complex culture.

The prevalence of emotional problems in the United States is high. There are no exact figures available, only estimates. These vary considerably depending on how emotional illness is defined. According to the concepts set forth in this chapter, everyone has experienced some degree of emotional disturbance at one time or another. Of course, there is a very wide range of degrees. What we are concerned with here is the number of people that have emotional problems serious enough to interfere with normal living patterns.

COMMON EMOTIONAL PROBLEMS

Several types of emotional problems are extremely common in everyday life. Anyone might expect to experience one or more of these problems from time to time. If they are mild in degree and short in duration, then they really are no cause for alarm. If, however, they are persistent or severe enough to interfere with one's happiness, productivity, or relationships with others, then they definitely warrant some form of therapy.

● ASK YOURSELF

1. How often do you feel anxious? Often enough to pose a problem?
2. How do you personally know when you are experiencing anxiety?
3. What are some ways you might reduce your anxiety level?

Anxiety

Everyone feels anxiety at times. It is a normal response to threats directed toward one's body, possessions, loved ones, way of life, or value system. It is not the same as fear, however. *Fear* is the response to an immediate, real danger. *Anxiety* is a response to a vague, obscure, or imagined danger. Some anxiety is normal when we are making an extreme effort or when we are having to adapt continuously to rapid or drastic changes in our lives. In fact, moderate anxiety helps stimulate us to useful action. It may play an important role in beneficial change and personality growth. In contrast, excessive anxiety not only makes a person unhappy, but often prevents any useful action. Severe, disorganizing anxiety is usually called *panic*.

The experience of anxiety is subjective—it is described differently by different people. But its essential feature is the unpleasant anticipation of some kind of misfortune, danger, or doom. It is accompanied by tenseness, restlessness, and the feeling that something must be done to avoid disaster.

TABLE 3.1   CHARACTERISTICS AND INCIDENCE OF DEPRESSION

High incidence of depression	Low incidence of depression
Female	Male
Black	White or "other"
Formerly married	Currently married
Female head of household	Male head of household
Did not finish high school	High school graduate
Low income	High income
Unemployed	Employed

Source: "Basic Data on Depressive Symptomatology," U.S. Department of Health and Human Services, Public Health Department, *Vital and Health Statistics,* Series 11, No. 216 (April 1980).

Since the source of the anxiety is often vague, the "something" that must be done is also vague and thus nothing constructive is or can be done. Persistent anxiety leads to a feeling of helplessness and the fear of collapse.

Anxiety has many physical symptoms as well. Some common indicators are excessive sweating; muscle tension, especially in the back of the neck; sighing; and rapid breathing and/or pulse. Others include digestive symptoms such as indigestion, loss of appetite, diarrhea, or "butterflies" in the stomach. Frequent urination and sexual dysfunctions may also result from anxiety.

Severe, disabling anxiety should receive immediate treatment. This usually involves psychotherapy to get at the basic source of the anxiety and perhaps tranquilizers for immediate relief of unpleasant symptoms.

Depression

One recent U.S. government survey reported that 17.3 percent of the U.S. population, aged 25 to 74 years, was experiencing significant degrees of emotional depression at the time of the survey (*see* Table 3.1).

It could be said that this is a depressing era. We must deal with worldwide conflict and hardship, nationa politica and economic problems, and the individual realities of crime, unemployment, and inflation. Perhaps it is not surprising that vast numbers of people are suffering from depression. What may be surprising is that even in the best of

times, many of these people would still be depressed. For depression, like anxiety, often has no real external cause, but arises entirely within the individual. Depression can result from either psychological causes or biochemical causes (such as improper balance of brain chemicals) or a combination of the two.

Also like anxiety, the degree of depression may range from mild to seriously disabling. Mild depression usually results in a loss of pleasurable interest in life. Spontaneity is gone. Everything one does requires greater effort than before, and provides less satisfaction. One does not feel physically ill, but neither does one feel particularly well ("the blahs"). Fatigue is excessive and much time may be spent sleeping. A person with mild depression such as this still goes to work or school, meets most obligations, and appears normal. But there is no joy in life.

More severe depression may involve deep despondency, a feeling of physical illness, or both. The person feels gloomy, helpless, hopeless, and worthless. Insomnia (sleeplessness) is common. Many lonely hours are spent in deep despair in the late evenings or early mornings. Loss of appetite and weight loss are characteristic, as is a loss of sexual interest or ability. The most serious consequence of deep depression is that it frequently leads to suicide.

As in anxiety, treatment usually revolves around psychotherapy and drugs. In deep depression, suicide prevention must always be part of the plan of treatment.

Depression often includes a sense of loss—real or impending.

● ASK YOURSELF

1. Do you often feel depressed or "down"?
2. Are your periods of depression associated with losses or threatened losses? Or do they appear without any obvious cause?
3. Is depression a problem in your family? Might you have inherited a tendency toward depression?

Emptiness

Many people today complain of a rather vague feeling that they often call a sense of emptiness. Not only are they unsure of what they want out of life, but they also do not even have a clear idea of what they feel. They do express painful feelings of powerlessness and lack of direction, which often results in a great difficulty to make decisions. They know what others—*parents, spouses, employers, professors,* and *society*—expect of them, but not what they themselves want. They may satisfy all of these other people but still feel overwhelming emptiness. Some typical cases include people who spend their days in boring, routine, assembly-line work and who find little additional fulfillment in their leisure activities. However, people in all occupations, including many who are quite successful by most standards, are subject to feelings of emptiness. They do their jobs and do them well, but there is no satisfaction. Emptiness is also commonly felt by many women in traditional housewife roles who have no sense of identity other than being "his wife" or "their mother." Unemployed or underemployed people of both sexes commonly experience similar feelings.

The problem of emptiness is usually caused by a failure to achieve self-actualization. There is a need (perhaps through therapy) to gain a fuller understanding and an acceptance of one's intrinsic nature. These people must come to terms with their needs, interests, feelings, and abilities and separate them from the multitude of values and expectations imposed by others.

Loneliness

Another common feeling is loneliness. Though certainly not a new problem, it is increasing in incidence and severity. One of the many causes of loneliness is industrialization; many jobs lack

● **ASK YOURSELF**

1. Do you often feel that your life lacks meaning? If so, can you identify the causes for such feelings? What can you do to improve your situation?
2. Have you ever felt "alone in a crowd," even though you were with people you knew? Do you often feel lonely?
3. Can you think of some ways to combat your feelings of loneliness? Do your ego defenses keep people away from you? Do you belong to any organizations that provide social interaction?

human contact. Other causes are increased reliance on impersonal electronic media for entertainment and increased mobility, with many people moving away from friends and relatives. Still another is the increasing life-span; many older people are limited in their social contacts because of poor health and small incomes. Among the effects of loneliness are emotional problems, alcoholism and other drug dependencies, and many suicides.

The feelings of emptiness and loneliness often go together. Many people today say they are lonely even in the presence of other people, including their own families. The reason for the close association between loneliness and emptiness is that when people do not know what they want or what they feel, they depend on others for their sense of direction. But when this fails to produce a sense of direction, people feel that they have been failed by the others and are "alone in a crowd." This, of course, does not apply to people who are lonely because their life situations deprive them of needed social contact, as is the case with many elderly people.

Another common factor in loneliness is the inability to establish intimacy — to form close relationships with people. Many people, for reasons of ego defense, keep everyone "at arm's length," and form only the most superficial relationships.

Loneliness caused by personality traits can only be resolved by changing those traits. But the loneliness of many people today results from factors beyond their control. Our society could and should provide more opportunities for social contact for the many people to whom circumstances such as

Loneliness is prevalent in today's depersonalized society.

age, poverty, or physical handicaps deny this important part of life.

EMOTIONAL DYSORGANIZATION

The traditional approach to a discussion of emotional disturbance has been to describe various abnormal behavior patterns and give them definite labels and diagnoses. Unfortunately, these labels mainly describe the way in which the person is acting at the time. They have little to do with the basic cause of the problem or the most effective approach to its treatment. They are only descriptions of signs and symptoms.

We will depart from tradition and instead present some of the ideas of Dr. Karl Menninger, one of the founders of the Menninger Clinic in Topeka, Kansas. This clinic is one of the outstanding psychiatric hospitals in the United States, and Dr. Menninger is recognized as one of the country's leading psychiatrists. In his book *The Vital Balance* (1967), Dr. Menninger proposes doing away with much of the labeling of emotional problems. He places them instead into a range of severity that extends from relative emotional normalcy to complete emotional collapse.

Dr. Menninger has coined the word *dysorganization* to describe the difficult or painful experiences a person must undergo in trying to cope with life situations. (The prefix *dys-* means "painful" or "difficult.") Dysorganization does not mean the same as *disorganization*, which is a common term meaning a lack of organization or a state of disarrangement. In dysorganization, the process of organization is difficult and painful, but there is organization.

According to Menninger's concept, the degree of emotional illness is in proportion to the amount of dysorganization present. There are, of course, degrees of dysorganization, and Menninger divides them into five levels or steps of increasing severity. Although these five levels are convenient for purposes of description, they are steps in a continuum. Other levels of emotional dysorganization lie between them.

First Level of Dysorganization

The first level of dysorganization is what we commonly call "nervousness." This is a condition we all feel at times when we have slightly more than the usual amount of difficulty in coping with a situation. We feel more tension and possibly more fear, anxiety, frustration, and anger. Consequently, we must make a greater effort to keep these internal tensions under control.

Various conscious and unconscious mechanisms are activated at the first level. Consciously, a greater effort at self-control is exerted. Unconsciously, there is automatically an increased use of repression or suppression of anxiety-producing thoughts or ideas. Some repression is, of course, always at work. But this increased use of repression is a step toward eventual separation from reality as one progresses into deeper levels of dysorganization.

"Nervous" people often become unusually sensitive to surrounding events. They may hear small

Dr. Karl Menninger

mysterious noises at night or notice the slightest change in the appearance of something. This alertness is part of the stress response that the body makes in preparation for emergency action. And perceptions are sharpened. The senses of sight, hearing, and smell become keener. Since the body is alert, a nervous person may have difficulty sleeping at night or may awaken at the slightest noise.

Other first-level symptoms include touchiness, tearfulness, irritability, nervous laughter, moodiness, or depression. Restless behavior is common, such as walking the floor, biting fingernails, chewing pencils, twirling hair, cracking knuckles, or drumming with the fingers.

At this level, the person is often worried and spends an excessive amount of time thinking about some topic of concern. Or the mind may wander into a daydream so as to avoid reality, thus preventing effective thought or action.

Psychosomatic disorders are common among people at the first level. Itching skin, upset stomachs, headaches, and various vague pains can occur.

The symptoms of the first level are a response to stress. Anyone who is placed under enough stress

Severity of dysorganization ⟶

	LEVEL 1	LEVEL 2	LEVEL 3	LEVEL 4	LEVEL 5	
Normal emotional organization: Coping with ease	Strong tensions Anxieties Nervousness Mild neuroses	Phobias Hysteria Physical disorders Obsessions Neuroses	Open aggressions Homicides Attacks Assaults Social offenses	Severe emotional illness Paranoias Depressions Delirium Melancholia Psychoses "Insanities"	Loss of will to live Suicide "Human vegetable"	Physical death

Diminishing coping ability ⟶

The levels of emotional dysorganization lie along a continuum of progressively severe emotional distress.

will show some of these symptoms. For this reason, Dr. Menninger states that everyone has experienced emotional dysorganization, for everyone reaches the first level at times. But when the unusual stress is gone, most people return to a level of optimum adjustment.

Other people are more limited in their ability to cope with stresses. They chronically function at the first level, even under normal stresses. If they are subjected to more than the usual amount of stress, or if they are no longer able to control behavior with the first-level mechanisms, they must develop other self-maintenance devices. These new devices lead to the more serious second level of dysorganization.

Second Level of Dysorganization

The essence of the second level is increased detachment of the person from his or her environment and reality. Second-level persons are thus less realistic than those at the first-level. The coping process becomes unpleasant and emotionally painful. The individual is unhappy and feels a sense of failure, uselessness, or depression. Other people are seen as either indifferent or as definitely antagonistic, though they may actually be trying to help. Control over words and actions is reduced. One

says things one does not really want to say and does things one would rather not do. Tolerance for frustration is very low.

The second-level coping devices may be used temporarily by a usually normal person placed under great stress; or they may become a permanent part of the personality. People at the second level are still able to function in society, but their methods greatly reduce the pleasure they receive from life and make them difficult to be with. Although Menninger and others (American Psychiatric Association, 1980) suggest that such terms be abandoned as obsolete, the common name for a person who lives at the second level is *neurotic*.

Let us now consider some of the common symptoms of the second level of dysorganization. These are all ways in which the person tries to restrain unacceptable impulses like hostility or guilt. These symptoms also occur in normal people, but not to this degree.

Withdrawal A person may unconsciously withdraw from contact with the world temporarily by fainting, developing amnesia (loss of memory), refusing to see or hear certain things, or developing *phobias*. Phobias are excessive and often crippling fears of objects or situations, such as high places, closed rooms, insects, cats, thunder, or darkness.

Self-punishment Some people have repressed such intense hostilities toward other people, groups, or situations that they take their hostilities out on themselves. This type of self-punishment is usually quiet and subtle. It often takes the form of an accidental injury or a psychosomatic disease.

Compensating acts—rituals Not all rituals indicate an emotional conflict. For example, there is nothing particularly unhealthy about the ritual of getting up at the same time every morning, showering, combing your hair, and dressing for school. These actions are a means to an end. A ritual indicates conflict when it becomes an end in itself, when the mere act of doing it becomes a kind of release. For example, a student might sharpen a pencil many times while taking a test, not because the pencil point was dull, but because sharpening the pencil relieved some of the anxiety about the test.

Compensating acts— obsessional thinking Obsessional thinking takes the place of action. Much time is spent mulling over the same thoughts. As such thinking becomes more intense, effective action and productivity decrease. Although obsessional thinking does indicate an emotional problem, it may be of value if it deflects persons who are thinking about dangerous aggressive actions from actually going through with them. Too often, however, obsessional thinking takes the place of needed beneficial action.

Compensating acts—compulsions These are irresistible desires or drives to perform acts that seem unreasonable and unnecessary to other people. Compulsive cleanliness is a common example. Some people are so concerned with the cleanliness of their homes that the normal use of the house or apartment becomes impossible. Compulsive overeating is another example. Some compulsions can lead to serious trouble. Compulsive stealing (*kleptomania*) and setting fires (*pyromania*) may be motivated by an unconscious desire to be caught and punished.

Personality deformities These are probably the most common second-level symptoms. Personality deformities result when persons adopt an emergency defense mechanism and make it a per-

Compulsive overeating can be a stress-relieving coping device.

manent part of their personality. Many eccentricities, perversities, dependencies, and offensive personality traits are acquired because of a chronic maladjustment to the stresses of normal life.

A person may become *infantile* (helpless) or *narcissistic* (extremely egocentric); develop into a bully, a braggart, a worrier, or a liar; or develop a dependency on alcohol or other drugs. Alcoholics and drug abusers are often emotionally immature and use these substances to escape from reality. If they had not turned to alcohol or other drugs, they would probably have adopted one or more second-level defense mechanisms.

Hypochondria Another common characteristic of the second level is *hypochondria*—an exaggerated anxiety and concern about one's health. The hypochondriac imagines illness or else makes a big fuss over very minor disorders. Psychosomatic disorders and hypochondria help people avoid stressful situations. Illness can be an escape mechanism, a means of avoiding responsibilities. Many people

Acting out aggressive impulses indicates severely diminished coping ability.

go through life with a series of vague illnesses. They are never really sick, but they are never well enough to assume normal responsibilities. If a physician successfully treats one illness in such a person, the person replaces that illness with another.

Third Level of Dysorganization

This level is characterized by the display of aggressive impulses. The person operating on the third level expresses open and direct aggression to people and things. All—or nearly all—of his or her self-control has been lost.

In general, such an individual has the following traits:

1. Unconcealed aggressive impulses
2. No regard for laws and social customs; little, if any, conscience
3. Reduced judgment, consciousness, and perception during an aggressive act
4. Little or no remorse or guilt after an aggressive act
5. Relief of emotional tension after an aggressive act

As with the first two levels of dysorganization, individuals may drop to the third level for a short time while under great stress and then return to a healthier level when the stress disappears. Or, they may remain at the third level and adopt chronic aggression as a way of life.

Even when individuals reach the third level only temporarily and briefly, the results may be grave. They may commit murder or assault, or may become violent toward property and leave a place in shambles. Suicide attempts sometimes occur when their aggression is self-directed. People who often reach the third level definitely need treatment, as they present a threat to society and to themselves as well.

Fourth Level of Dysorganization

The fourth level represents serious emotional disturbance. The conscious thought processes are greatly disrupted. Because persons at this level have lost contact with reality, their interpretation of the outside world is badly distorted. Their emotional reactions and behavior are inappropriate, exaggerated, and unpredictable. Their behavior may be bizarre. They have lost any effective productivity, at home, school, or on the job. Sometimes their behavior seems scarcely human.

In the past this condition was called *lunacy* or *insanity*. A more acceptable word for this fourth level today is *psychosis*. A psychotic person is out of touch with reality.*

Some psychotic behavior patterns have been described as distinct kinds of emotional illness. But Menninger believes that these patterns are merely different symptoms of the same general disturbance. The fourth level of dysorganization often takes one of the following forms:

1. *Extreme depression.* Depressed people may feel overwhelmingly sad, guilty, hopeless, or despondent. They may feel inadequate, imcompetent, unworthy, or just no good. There is often a slowdown of physical body functions.

2. *Extreme anxiety.* These persons show a great overflow of poorly controlled energy; they talk rapidly, constantly, and incoherently. They may indulge in much bizarre activity, such as walking in a circle day after day or repeatedly stacking and restacking the same objects.

3. *Excessive concern with self.* People may become mute and totally withdrawn into themselves. They may have delusions of their own importance and constantly tell people that they are God, Jesus Christ, the president of the United States, or some other important figure.

4. *Persecution complex (paranoia).* These people may imagine that someone or some group is plotting to get them in some way. They may become very suspicious, resentful, and defensive. Someone in this state of mind can be quite dangerous.

Fifth Level of Dysorganization

The fifth level represents the greatest extreme of emotional dysorganization. The will to live is gone. All that is left is a self-destructive determination to end life or to settle for minimal existence.

These extremely disturbed people either commit suicide or become human vegetables, refusing to eat, drink, or make any contact with reality. Force-feeding may be necessary in order to prevent starvation.

*American Psychiatric Association, *Diagnostic and Statistical Manual of Mental Disorders,* 3rd ed. Washington: American Psychiatric Association, 1980.

● ASK YOURSELF

1. Do you agree that everyone is emotionally disturbed at one time or another?
2. How far down in Menninger's levels have you ever gone in your life? How long did you stay at that level? What stress or stresses put you there? Have you done anything to prevent it from happening again?
3. Would you be able to recognize your symptoms better now?

The earnest suicide attempts of fifth-level people are quite different from the half-hearted attempts of those at lower levels. In the exhibition suicide, the person is consciously or unconsciously hoping and planning for a last-minute rescue. This type includes the person who stands on a rooftop or at an open window while a crowd gathers and the fire department is called. It also includes the person who takes a few sleeping pills and immediately calls for help. These display attempts at suicide stem from various motivations. In general, however, they are desperate cries for help. The person who attempts suicide in this manner should be given psychiatric help. It has been found that such persons often make repeated suicide attempts. The next time they might succeed.

Suicide

Suicide is officially the ninth leading cause of death in the United States today. The official annual suicide rate is about 13 per 100,000 population, and about 30,000 deaths are recorded as suicides each year. The actual suicide rate is probably at least double the official rate, since many suicides are unrecognized as such or disguised as accidents or natural deaths to avoid losing insurance money or embarrassing survivors.

No group or class of people is free from suicide. Every person is a potential suicide, and almost everyone at some time gives it some brief consideration. However, the risk of suicide does relate to certain individual characteristics.

One significant pattern in the incidence of suicide is its increase with advancing age. Suicide is rare among those under 14 years of age. The rate rises sharply in adolescence and sharply again

Though not all suicide deaths are recognized as such, suicide is a major cause of death today.

among college students, for whom suicide is second only to accidents as a cause of death. Several factors are commonly associated with college suicides, the most frequent of which is academic failure. Failure brings not only the disappointment and disapproval of parents, but it shatters a person's self-esteem as well. The second leading cause of college suicide is the end of a love affair. When romance ends, there is more than just disappointment. There is the tumultuous feeling of being rejected and abandoned, and a serious loss of self-esteem. College suicides often involve reserved, introverted, or shy students who, lacking social contacts, tend to internalize their problems.

The suicide rate rises again in middle age, when many men realize that they have not attained their career goals and in fact will never attain them. Middle-aged women may turn to suicide in reaction to menopause if they feel that they have lost their "womanhood." The stimulus can also be their chil-

dren's departure from home, after which they may no longer feel useful and needed.

Finally, the suicide rate reaches its peak among the elderly, who today suffer from a host of emotionally crippling diseases and conditions. Our society idolizes youth and young ideas. After a forced retirement at age 65 or 70, many people feel useless, lonely, bored, and frustrated. In addition, they may suffer great financial insecurity, physical pain from chronic ailments, or may have terminal illnesses.

Nationally, women attempt suicide more often than men. Because of the methods used, women are often saved before death occurs. Men are more inclined to use violent and swift means, such as firearms. Women tend to take pills, leaving a considerably longer period of time during which medical intervention can be sought. The success rate of male suicides is about three times that of females.

In recent years the suicide rate for women has risen sharply, while the rate for men has remained fairly static. This increase has been attributed to various causes. Part of the increase is related to increasing opportunities and expectations for women. The opportunity to succeed in a career is also the opportunity to fail. It has been suggested that women are now committing suicide for reasons that were once typical of males, such as despondency over unfulfilled career goals.

Severe depression is twice as common among females than males. The severely depressed woman is the greatest potential suicide. A woman who spends her days at home in the traditional role of "wife and mother" may see herself as a failure. Her suicide attempt is often an extreme and urgent plea for someone to recognize how alienated and isolated she feels and how little of her potential is being used. There may be conflicts between her desires to succeed in the traditional role of homemaker and in a career as well. Marital dissatisfaction is also an important factor, as evidenced by the rapidly climbing divorce rate.

In general, suicide is less frequent among married persons, with the notable exception of those under 24 years of age, in whom the rate is much higher than in single persons. In single people over 24 and in divorced people of any age, the rate is higher than among married people over 24. The rate among young widows (under 35) is quite high.

Racially, suicide is about twice as prevalent

TABLE 3.2 PSYCHOLOGICAL FACTORS IN SUICIDE

1. Severe depression
2. Social isolation
3. Sense of loss
4. Psychosis involving hallucinations or delusions
5. Hallucinogenic drugs such as PCP or LSD
6. Suicide for spite (to hurt survivors)
7. Poor impulse control (minor frustration causes suicide)
8. Identification with other people who have committed suicide
9. Chronic or terminal illness

People who threaten suicide or state that they would be better off dead are very high suicide risks.

among whites as all others. Among young American Indians, however, the suicide rate is at an epidemic level.

There is a direct relationship between suicide and social status. The higher one stands on the social scale, the more susceptible he or she becomes to suicide. The rate is high among physicians, dentists, lawyers, business executives, and similar professionals.

Theories on the causes of suicide Many psychological theories have been proposed to explain suicide, ranging from the Freudian to the behavioral. Most authorities agree that suicide may have a variety of motivations. They also agree that in a particular case there is rarely a single precipitating cause but that several causes are usually acting in an additive way (*see* "Suicide Danger Signs" below).

Serious losses (or threats of loss) play a major role in the psychodynamics of suicide. These include loss of health, loved ones, money, earning power, job, pride, beauty, status, friends, children, and independence. Depression is predictable in any real, threatened, or perceived loss.

Social isolation is a common factor in suicide. The more intimately one is involved with others, the less likely one is to consider suicide. Unfortunately, the severe depression that commonly precedes suicide is likely to lead to social isolation, further increasing the chance of suicide. Suicide may be considered the final outcome of a progressive failure of adaptation, and of isolation and alienation from the network of human relationships that support us all and give meaning to our lives.

Prevention of suicide Most suicidal people do not fully intend to die. Though their wish to die may be extremely strong, they almost always have an underlying wish to live. But they do not wish to live as they are living, for they see this as being the same or even worse than death. Thus, the crucial element that makes suicide prevention possible is *ambivalence*. Even as suicidal impulses become almost overwhelming, the ebbing wish to live is a root of energy that can be tapped for suicide prevention.

The role of the layperson in the suicide crisis is much the same as in any first-aid situation. It is to preserve life until professional help can be obtained; it is not to play "doctor." Many cities have 24-hour telephone services staffed by mental health professionals or volunteers specially trained in crisis intervention. If such a service is available, one should first encourage the suicidal person to

The success of suicide hot lines depends on open and direct communication.

Suicide Danger Signs

It is *not* true that people who threaten to commit suicide never do so. Most people who attempt suicide actually give warnings, vocal or nonvocal, beforehand. Potential suicides can usually be recognized before they act. Knowledge of the danger signs can save a life. Any of the following indicates a definite risk of suicide.*

1. *Previous attempts.* Over half of those who successfully commit suicide have tried it at least once before.
2. *Previous psychosis.* A history of psychiatric disturbance indicates the possibility of a recurrence.
3. *Suicide note.*
4. *Violent method.* The more violent and painful the method chosen in an unsuccessful attempt, the more serious the intent, and the more likelihood that the attempt will be repeated.
5. *Presence of chronic disease.*
6. *Recent surgery or childbirth.* The birth of a baby leads to severe postpartum depression in some women.
7. *Alcoholism or drug dependence.* Drug or alcohol dependence indicates a need to escape reality. The permanent escape of suicide is often appealing to alcoholics and drug abusers. Also, while under the influence of drugs or alcohol, their ability to resist suicide impulses is weakened.
8. *Hypochondriasis.* Constant physical complaints often indicate underlying depression.
9. *Advancing age.*
10. *Homosexuality.* The suicide rate among homosexuals is high.
11. *Social isolation.* Indicates severe depression.
12. *Chronic maladjustment.*

13. *Financial, employment, or social problems.* A person without money, job, or friends may see little to live for.

14. *No obvious secondary gains.* Many suicide threats are really attempts to manipulate others. For example, people often use the suicide threat to keep a dying marriage or love affair going. When there is no such motive and the threat is truly self-directed, the chance of suicide is much greater.

15. *Signs of grave risk:*
    a. *The wish to die.* Repeated statements by a person that he or she would be better off dead indicate very high suicide risk.
    b. *Presence of psychosis.* The psychotic person who is suspicious, fearful, or hears voices should be regarded as potentially suicidal.
    c. *Depression.* Depression is the most common cause of suicide. Any of the following symptoms indicates severe risk.
       (1) Guilt
       (2) Feelings of worthlessness and despondency
       (3) Intense wish for punishment
       (4) Withdrawal and hopelessness
       (5) Extreme agitation and anxiety
       (6) Loss of appetite, sexual desire, sleep, physical energy
       (7) Sudden well-being, indicating a feeling of relief at having made the decision to die

* Adapted from Solomon and Patch, 1975; Adams, 1981.

call for help. But a crisis intervention service is often not available, or the suicidal person will not cooperate in calling such a service. Then the person must be kept from committing suicide until other professional help can be obtained.

The first step in suicide intervention is to establish a relationship with the suicidal person and maintain contact. As an opening line, the National Save-A-Life League suggests asking, "Are you thinking of killing yourself?" Communication in crisis intervention must be open and direct. You must remain calm and convey attitudes of helpfulness, hopefulness, and genuine concern. You must build upon the suicidal person's ambivalence about dying, and convince the person that he or she really does want to live. You must encourage the person to relate everything possible about his or her problems and show the person that, with help, these problems can be solved.

Everyone has some strengths. The severely depressed person often loses sight of his or her own strengths or ability to use them. In talking to a suicidal person, you should keep emphasizing strengths so as to decrease feelings of helplessness and hopelessness.

As soon as the immediate crisis of threatened suicide seems to have abated, the person should be encouraged to seek competent professional help. *Suicidal impulses recur,* and it is dangerous to assume that there will not be further attempts. As many friends or relatives as possible should help the person by means of support. The chance of suicide is reduced through social interaction and increased through alienation.

● **ASK YOURSELF**

1. Almost everyone considers suicide as at least a remote possibility at one time or another. Have you ever given more than just a passing thought to suicide? Were you under any unusual pressures at that time? Had you suffered a loss?

2. Do you still sometimes seriously consider suicide? Are you aware that a lot of people who were once extremely unhappy are now living very satisfactory lives because they sought help with their problems?

3. You may suspect that a close friend of yours is suicidal. How will you handle the situation?

Recognizing the Severely Disturbed

With today's high incidence of emotional problems, there are few of us who do not occasionally encounter a severely disturbed person. Such a person may be a close relative, a neighbor, or a stranger in a public place. In any case, it is important to recognize the individual as disturbed and to know how to handle such a person. A person suffering from severe emotional disturbance will generally exhibit one or more observable symptoms.

Changes in behavior pattern A normally quiet person may suddenly become very belligerent or overtalkative. Or conversely, the happy, outgoing person may become quiet and moody. Any sudden and radical change in a person's normal mood or behavior may indicate emotional conflict. What is important here is a change in the general pattern of behavior extending over a long period of time, not just a passing reaction to some stress or irritant.

Loss of touch with reality Emotionally disturbed people usually (some authorities say always) lose contact with reality to some degree, from slight to total. They may be unable to recall who they are, where they are, why they are there, the day or date, or similar information. Or they may withdraw from reality to the extent that they are completely unaware of their environment.

Amnesia Temporary or permanent memory loss is a common symptom of mental disturbance. One of the mind's defense mechanisms is to repress memories that are too painful for the conscious mind to bear. Loss of memory is also characteristic of the senile elderly person, often as a result of reduced blood supply to the brain due to atherosclerosis.

Delusions A delusion is a distorted belief or idea. Emotionally disturbed persons often harbor beliefs (called *delusions of grandeur*) that they are famous scientists or surgeons, prosperous executives, secret agents, or even God. Much harm may be done to themselves or to others when they act upon such delusions. Even more dangerous are delusions of persecution in which disturbed persons believe that family members, neighbors, members of racial or religious groups, or just a vague "they" are plotting against them. People suffering from such persecution delusions may become dangerous. They may suddenly react to their mistaken beliefs by attacking those who they feel are "after" them.

One-sided conversations In the folklore of emotional disorders one of the sure signs is "talking" to yourself." Actually, we all talk to ourselves from time to time. But a person carrying on an animated conversation with himself or herself in a public place is very likely disturbed.

Hallucinations Severely disturbed persons very commonly experience hallucinations of one or more of the senses. They hear voices; they see, smell, taste, and feel imaginary things. Though such hallucinations may seem absurd to others, they are very real to those suffering from them, even to the point that such hallucinations are acted upon, with serious or even fatal results.

Helping the Severely Disturbed

With increasing incidences of emotional disturbance, drug abuse, stressful situations, and a growing population, the *psychiatric emergency* is becoming more commonplace. This may be defined as a situation in a public or private place in which an individual's behavior becomes dangerous to that person or others and for which prompt and decisive action must be taken. In helping a severely disturbed person, the immediate regard is for the safety of everyone concerned. In addition, the ability of the patient to recover and the speed of recovery often depend on the handling he or she receives during the acute emergency phase of the illness.

Since every case is different, there can be no standard formula for helping a person in a crisis. There are some things, however, that can be done in almost every case.

The first is to *make contact*. Show a personal concern and a willingness to listen. Then try to *reduce anxiety*. This is done by encouraging the person to talk about the situation and to feel free to express emotions of anger, sadness, grief, or remorse. *Focus on the issues*. Find out what is threatening the person and explore what has been done or might be done about the situation. *Encourage positive action*. The person in a crisis needs help and should be encouraged to seek or accept it.

The handling of the psychiatric emergency should be thought of as first aid and the basic rules of first aid applied. The prime rule is to recognize your own limitations as an untrained person. To at-

● ASK YOURSELF

1. Your father was always hard working and easy to get along with. Now he has changed. He does not go to work and seldom leaves the house because he believes there is a plot to kill him. He has a gun and sometimes threatens to shoot family members. What are you going to do?
2. You arrive home one day and your brother is severely depressed, having just lost his girlfriend to another man. He talks about suicide. What do you do?

Psychotherapy can help resolve many emotional problems.

tempt to play psychiatrist is to risk physical injury, lawsuit, and further deterioration of the condition of the disturbed person. Thus, the best course of action is to immediately call for assistance. A physician, even though not a psychiatrist, can often help calm a disturbed individual. Most police officers have both training and experience in handling disturbed persons; they are generally more readily available than physicians and can immediately respond to a call for help.

While waiting for assistance, keep cool and calm and take only such action as is absolutely necessary to prevent someone from being hurt. Such action should involve a minimum of physical force or harsh words. It should include efforts to reassure the disturbed person that you are a friend and are trying to protect and help the person. If physical restraint becomes necessary, use only as much force as is absolutely necessary.

TYPES OF TREATMENT

Fortunately, the methods and facilities for the treatment of emotional problems have been greatly improved during recent years. The person who becomes emotionally disturbed today, regardless of the level of dysorganization reached, stands an excellent chance of recovery. This assumes that he or she receives the benefit of modern methods of treatment and that this treatment is begun promptly. Most patients today begin to show improvement very quickly. Most treatment is now on an outpatient basis (without hospitalization). When hospitalization is necessary, patients usually return home in only a few weeks, to continue treatment on an outpatient basis.

Psychotherapy

Psychotherapy is an extremely general term. It can include over 200 different approaches to counseling that are in current use. Some of these methods are well known and widely practiced; others are less commonly applied. New methods constantly move in and out of fashion. Psychotherapy can be on either an individual or a group basis.

The various forms of counseling can be classified in several ways. One way is to group the therapies into three broad and overlapping categories—directive, permissive, and interactional. In the *directive* therapies the counselor is viewed as an expert who diagnoses and analyzes a problem, decides on solutions, and communicates them to the counselee. Examples of this include psychoanalysis, behavior therapy, and est (Erhard Seminars Training).

Somewhat opposed to this approach is the *permissive* (also called *evocative*) group of therapies. The counselor here, also an expert in dealing with personal problems, does not make diagnoses or dictate solutions. Instead, the counselor acts as a *facilitator* who stimulates people to solve their own problems and who creates an environment where problems can be solved and personal growth can occur. Some examples include client-centered therapy, Gestalt therapy, and the various encounter therapies.

William Glasser

Somewhere between the directive and permissive approaches to counseling are the *interactional* therapies. Here the counselor and the counselee interact together as more or less equals in a team effort. William Glasser's reality therapy is an example of this.

Even a brief description of the therapies in current use would fill an entire book. Nathaniel Lande, in *Mindstyles/Lifestyles* (1976), has done an excellent job of this. We will mention just one or two examples of each of the above groups of therapies.

Psychoanalysis Psychoanalysis is a system of therapy that dates back to Freud and the early years of this century. Great importance is attached to the role of the subconscious mind in causing emotional conflict. Psychoanalysis attempts to explore the unconscious mind through a long series of sessions in which the patient freely relates anything that happens to come to mind. The therapist, called an analyst, interprets what the patient says, trying to help the patient recognize the subconscious feelings that have led to problems.

Behavior therapy Behavior therapy is the application to human beings of techniques derived from experiments with animal behavior. The behavior therapies differ in several important respects from traditional psychotherapy procedures. Perhaps the most important difference is that behavior therapists work directly toward modifying objectionable behavior. They do not attempt to identify the "underlying unconscious disease process" that most psychotherapists (and all psychoanalysts) believe to be the cause of symptoms.

For this reason the behavior therapist does not deal with the unconscious, ego structures, or defense mechanisms, and does not employ insight as the prime means of treatment. Behavior therapy is mainly used for patients who want to modify specific behaviors, rather than for those whose problems are more diverse.

The rationale of behavior therapy is that the undesirable behavior patterns were learned in the first place and, with proper training techniques, can be unlearned. This may involve either weakening the undesirable responses or learning new, more desirable responses.

Behavior therapy has been most successful in treating specific problems such as phobias, sexual disorders, overeating, gambling, smoking, and other compulsive behaviors. Drug dependence and alcoholism have also been treated, though with somewhat less success.

The advantages of behavior therapy, in cases for which it is suited, include: (1) it often works; (2) it may be administered by nonprofessionals such as psychiatric aides, parents, and school teachers, after only a few hours of training; (3) the procedures are clear-cut, unambiguous, and consistent: results may be achieved in a short time, often just a few weeks; and (4) it is rare for a new behavioral problem to appear in the place of the old one.

Client-centered therapy Client-centered therapy was developed by Carl Rogers. The person, not the problem, is the focus of attention. Emphasis is on feelings and a person's inherent tendency toward self-actualization. Attention is directed to the present situation, not to one's past experiences. The therapist does not attempt to advise or direct the client in the growth process, but trusts the client's ability to provide his or her own direction. Client-centered therapy can be applied in a wide

● ASK YOURSELF

1. Have you ever participated in any kind of psychotherapy? Do you feel that it helped you?
2. Do you know of other people who have benefited from such counseling?
3. Do you feel that it is a disgrace to see a therapist?
4. Are you dealing with any problems right now for which some counseling might be helpful?

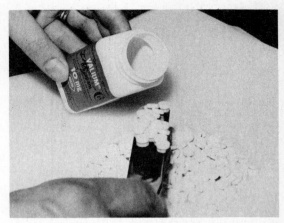

There is growing concern about the overuse of tranquilizers in the United States.

variety of situations where the aims are to increase interpersonal understanding and to enhance personal growth.

Reality therapy An interesting concept of emotional conflict is presented by William Glasser, a psychiatrist, in his book *Reality Therapy* (1965). His belief is that everyone who is emotionally disturbed suffers from the same basic inadequacy—the inability to fulfill his or her basic human needs. The severity of the symptoms reflects the degree to which needs are not fulfilled. Whatever the symptom, it disappears when the person's needs are successfully fulfilled.

Glasser sees another common characteristic among all his patients: They all deny the reality of the world around them. His concept of reality therapy is to lead patients toward reality through helping them find effective ways of fulfilling their needs.

According to Glasser, the two most urgent needs are the need to be loved and to love and the need for self-esteem. We all have these needs, but we vary in our ability to fulfill them. Central to the fulfillment of either need is involvement with other people—at the very least, one person, but hopefully many more. For good emotional health we must, at all times in our lives, have at least one person who cares about us and for whom we care ourselves. Without this essential person, we cannot fill this basic need. One characteristic is essential in the other person—he or she must be able to fulfill his or her own needs and thus be in touch with *reality*. In other words, he or she must be in good emotional health. Of course, it is common for two people to fulfill mutual needs and thus maintain emotional health through involvement with each other. Glasser sees the psychiatrist or psychologist as providing temporary involvement until the patient learns to fulfill needs through involvement with others.

Drug Therapy (Chemotherapy)

The development of new drugs has had an important influence on the treatment of emotional illness in recent years. Drugs have become a primary means of treating the seriously disturbed patient. Their use has reduced the time needed for improvement and recovery, and has lowered the population of many psychiatric hospitals.

The many drugs available fall into several basic groups. Major and minor tranquilizers are used to calm anxiety; sedatives combat overactivity and insomnia as well as anxiety; antidepressants help raise the mood of severely depressed patients; antipsychotics reduce or temporarily remove such symptoms as hallucinations.

Drugs are sometimes used as the primary treatment in cases of mild anxiety or mild depression. But in more serious cases, drugs are used in connection with psychotherapy. Often, drugs help make the patient accessible for psychotherapy. Before these drugs were available, many patients were too agitated or too withdrawn to be reached by psychotherapy.

The old-time psychiatric hospital ward, with some patients wildly agitated and others mute and withdrawn, is fortunately becoming a thing of the past through the use of modern drugs. Such drugs are not curing the patients, but they are making

Too Much of a Good Thing?

There is growing concern about the possible overuse of tranquilizers. One tranquilizer, Valium (diazepam), is the most prescribed medication in the United States. About 45 million prescriptions for Valium are written each year. An estimated 10 to 15 percent of all Americans take Valium at some time during any given year. While the great majority of these people experience no ill effects, there are enough problems to distress many authorities, including the Food and Drug Administration.*

One problem is addication. A "Valium Anonymous" organization has even been formed. Lethal and near-lethal overdoses are another issue. While seldom lethal by itself, Valium in combination with alcohol or other drugs is causing many deaths. Finally, the policy of routinely using tranquilizers for the stresses of ordinary life has been questioned. It would seem much more helpful to apply the stress management techniques discussed in Chapter 1.

*See "Overcoping with Valium," *FDA Consumer* (January 1980).

them calm and rational enough to participate in psychotherapy. After release from the hospital, many patients continue on drugs, and possibly psychotherapy, for varying periods of time.

Today, perhaps 90 percent of all emotional disorders, including major and minor forms, are treated with psychotherapy, drugs, or a combination of the two. Although other forms of treatment have thus become secondary, we will briefly describe two of them.

Shock Therapy

Prior to the advent of tranquilizers and antidepressant drugs, shock therapy was often used in the treatment of mental illness. There are several kinds of shock therapy; a major form, *electroconvulsive therapy (ECT)*, is still being used—but with greater discretion than in years past.

ECT is not indicated in the treatment of all mental disorders and its use should be restricted to carefully selected patients. Because of indiscriminate, and sometimes punitive, use, ECT has been the subject of much controversy; it has therefore become more stringently controlled. If used and administered properly, this procedure is painless and involves little physical risk to the patient. When used to treat severely depressed or manic (agitated) patients, the results may be strikingly favorable. While symptoms such as overexcitement or stupor are significantly reduced, the basic illness does not

disappear. However, with the elimination of these symptoms, patients are more accessible and receptive to psychotherapy.

After receiving ECT, patients usually experience mild confusion or loss of memory. These effects rarely persist for more than a few weeks or perhaps months after the last treatment.

State hospitals for the mentally ill once relied on ECT as an important part of their therapeutic programs. However, with the availability of drugs that produce similar results, they seldom use such treatment anymore. Also, because of the more rigid criteria in selecting patients, the time and paperwork involved make it impractical. Private hospitals still conduct ECT therapy, but it usually requires the consultation and approval of two or more psychiatrists.

Lobotomy

Rarely performed today, lobotomy is the surgical severing of the nerve tracts that connect the frontal lobes of the brain to the thalamus. The frontal lobes are centers for fear and anxiety. Today, the effects of lobotomy can almost always be achieved through drugs.

TYPES OF SPECIALISTS

Many patients today are treated by a specially trained group of people instead of an individual.

This group is sometimes called a *psychiatric team.* In addition to physicians and nurses, it may include psychologists, occupational therapists, social workers, and others who are concerned with specialized problems. Let us consider several members of this team.

Psychiatrist.

A psychiatrist is a physician (M.D.) who has had additional special training in treating mental illness and has been licensed by the American Board of Psychiatry and Neurology. A psychiatrist may use any method of treatment, whether individual, group, drug, or shock therapy.

Psychoanalyst.

A psychoanalyst in a psychiatrist who has had additional special training in psychoanalytic methodology and has fulfilled the requirements for membership in a psychoanalytic association. Most analysts use a variation of Freudian theory.

Clinical psychologist.

The clinical psychologist does not go through medical school, but instead does postgraduate study in psychology and usually holds the degree of doctor of philosophy (Ph.D.) or master of arts (M.A.). Clinical psychologists give psychological tests, make diagnoses, and engage in psychotherapy. They often work in association with a psychiatrist.

Psychiatric social worker.

Psychiatric social workers take two years of postgraduate study and hold a master's degree in social work, with training in psychology. They make contact with relatives, friends, employers, and others connected with the patient. They assist them in making any changes in the environment of the patient that seem necessary for the patient's recovery.

Psychiatric nurse.

The psychiatric nurse is a registered nurse who has had special training and experience with the mentally ill. He or she may supervise a hospital ward and administer treatments under the supervision of a psychiatrist, as well as assisting in therapy.

Mental health aide or assistant.

The aide or assistant, referred to as a *psychiatric technician* in several states, is charged with the actual physical care and custody of the hospitalized patient. The aide or technician is a paramedical staff person who has received on-the-job training. A few colleges now offer training programs for psychiatric technicians. Aides are important because of the great amount of time they spend with the patient and the personal influence they may have over the patient. A good relationship between the aide and the patient can greatly speed the recovery process.

TYPES OF FACILITIES

A wide variety of treatment facilities is available for people experiencing major or minor emotional problems. Many people fail to receive needed treatment only because they or their friends or relatives are not aware of the facilities available to them. Treatment is often available at little or no cost to people who cannot otherwise afford help.

Outpatient Clinics

Many patients with minor to moderately severe emotional illness can be treated today while they are still living at home. Psychotherapy sessions usually play an important part in such treatment. The psychiatrist or psychologist administering this care may be either in private practice, supported entirely by patient fees, or in a community clinic, supported by government or charity. Often these community clinics have a sliding scale of fees, depending on the ability of the patient to pay.

Outpatient clinics are also very important in providing care for the patient who has had short-term hospitalization during the acute stage of an emotional illness and is completing recovery at home. Such follow-up treatment can greatly reduce the chances of needing to return to the hospital.

Psychiatric Sections of General Hospitals

There is a growing trend for general hospitals to have certain sections specifically for the treatment of emotional disorders. These provide the patient with psychiatric treatment in a hospital setting without requiring travel far from home. They also help reduce the social stigma of emotional illness by treating it in the same context as any other disorder.

State Mental Hospitals

In the past, emotional problems often required prolonged hospitalization because no effective

A wide variety of treatment facilities is available for people who experience emotional problems.

methods of treatment were available. Some patients received only custodial care—that is, they were given humane shelter, but no specific treatment. Since hospitalization was so prolonged, the responsibility for its cost was usually assumed by the state government. The main purpose of such hospitals was to confine the ill where they could not bother anyone.

Most states are now applying the newer approaches to the treatment of emotional problems, with results often comparable to those obtained in private hospitals. It has been shown that if a state is willing to spend the money to provide intensive care for newly admitted, acutely ill patients, it will save even more in the long run. Given intensive care, many patients can leave the hospital in a short time. Without such care, the same patient may require hospitalization at state expense for years, and perhaps even for life. The states that recognize this principle have reduced the population of their state mental hospitals during a period in which the population of the country has risen and the incidence of emotional problems has climbed.

Private Mental Hospitals

In recent years, increasing numbers of privately owned hospitals for the exclusive treatment of mental or emotional problems have opened. Some treat all kinds of mental problems, while others are restricted to such groups as adolescents, elderly persons, or alcoholics.

Comprehensive
Community Mental Health Centers

Unfortunately, an enormous gap still exists between the best that could be done and what is actually being done for the emotionally disturbed. Many communities have no facilities for treating emotional problems. Where available, such services are often priced beyond the reach of all but the most fortunate. Treatment for emotional problems is often specifically excluded from payment by health-insurance policies.

In order to stimulate the development of comprehensive community mental health centers, the federal government has made financial aid available to centers that meet certain standards of qualification. The following ten criteria describe the comprehensive community mental health center.

1. Inpatient services
2. Outpatient services
3. Partial hospitalization services such as day care, night care, and weekend care
4. Emergency services available at all times
5. Consultation and education services available to community agencies and professional personnel
6. Diagnostic services
7. Rehabilitative services, including vocational and educational programs
8. Precare and aftercare services in the community, including foster-home placement, home visiting, and halfway houses
9. Training
10. Research and evaluation

To date, the number of such comprehensive centers in the United States remains limited. They are especially scarce outside major metropolitan

● ASK YOURSELF

1. During a routine physical examination, you mention to your doctor that you sometimes feel "nervous." She asks whether you might like a prescription for a common tranquilizer. What are some other ways that you might deal with the nervousness?
2. What types of mental health facilities are available to people in your community? How does a person go about finding help? What costs are involved? Do the facilities seem adequate for the needs of the community?

areas. But these criteria are an excellent yardstick by which a community can evaluate its own mental health facilities. While it is generally impossible for a small community to provide all these services, several neighboring communities can collectively develop an excellent comprehensive mental health center. Adequate community mental health facilities often can be established through the efforts of concerned citizens, working in cooperation with interested professional personnel.

IN REVIEW

1. Maintenance of emotional health is both a personal and a social challenge.
2. Everyone experiences emotional problems at some time. Common problems include:
   a. Anxiety—a response to a vague, obscure, or imagined danger
   b. Derpression—loss of pleasurable interest in life; the cause may be psychological or biochemical
   c. Emptiness—the feeling that one's life lacks meaning.
   d. Loneliness—the feeling of isolation from other people.
3. Emotional problems vary widely in their severity; they can be grouped in levels of emotional dysorganization. Each level represents increased difficulty in coping with life:
   a. First level—slightly increased difficulty in coping, commonly called "nervousness"
   b. Second level—the coping process becomes definitely unpleasant; symptoms include withdrawal, phobias, self-punishment, ritualistic behavior, obsessional thinking, compulsive behavior patterns, personality deformities, and hypochondria
   c. Third level—display of aggressive impulses
   d. Fourth level—loss of contact with reality, often called psychosis
   e. Fifth level—loss of the will to live.

4. Suicide is an important cause of death in the United States; every person is a potential suicide. Various high-risk categories of individuals are recognized.
   a. Depression is the most common emotional factor associated with suicide.
   b. Prevention of suicide depends on taking advantage of a suicidal person's remaining wish to live.
5. Severely disturbed persons may be recognized by changes in behavior patterns, loss of touch with reality, amnesia, delusions, one-sided conversations, and hallucinations.
6. The best way to help disturbed persons is to encourage them to seek professional counseling and assist them in arranging for it.
7. Today's methods of treating emotional problems are highly successful. They include:
   a. Many forms of psychotherapy—for example, psychoanalysis, behavior therapy, client-centered therapy, and reality therapy
   b. Drug therapy (chemotherapy): this form of therapy has revolutionized the management of severe emotional problems
   c. Shock therapy (electroconvulsive therapy): while seldom necessary, this therapy is sometimes highly beneficial
   d. Lobotomy—the rarely performed surgical severing of nerve tracts in brain.
8. Mental health professionals include psychiatrists, psychoanalysts, clinical psychologists, psychiatric social workers, psychiatric nurses, and mental health aides (psychiatric technicians).
9. Emotional problems are treated in facilities such as state mental hospitals, private mental hospitals, psychiatric sections of general hospitals, outpatient clinics, and comprehensive community mental health centers.

AGING AND DEATH

You may wonder why we would devote a chapter of a college health text to the seemingly remote subjects of aging and death. Actually, these topics are important to younger people for several reasons. For one thing, our attitudes and feelings about our own aging and death affect us our entire lives. It is well established that our fulfillment as individuals depends on healthful attitudes toward aging and death. Another reason for studying aging and death is to help us understand and assist our own aging and dying relatives and friends. In addition, it should help us cope with the loss or impending loss of those who are close to us. Finally, through this study we should be better prepared to deal with our own aging and dying. As we study these subjects, it will become apparent that aging and dying are worthy of our attention long before we personally experience these parts of our lives.

A NEW VIEW OF AGING

The active, positive approach to aging has been a welcome trend in recent years. Many colleges and other agencies have established programs to study aging and to develop more effective ways of dealing with it, from both a personal and a social approach. Modern medicine has provided us with many more "good years" of life. Now we are learning how to make the best use of these years.

The biggest problem to be overcome is a residual negative outlook on aging and the elderly. While many other cultures assign a very high status to the elderly, our culture has fostered adoration of the young. We admire the physical perfection of the young, and we hope to solve old problems with new ideas. Most of us have yet to accept the inevitable changes in ourselves as we grow older. We want to stay young forever. We try to conceal from everyone, including ourselves, that we are unable to halt the gradual changes in appearance and physical function. At the same time, we have exaggerated the effects of the aging process. We tend to think of elderly people as being much more limited in their abilities than they actually are. This contributes to our fear of growing old.

Recent studies on aging show that elderly people retain their abilities to a much greater degree than is commonly supposed. Most of them remain near the peak of their intelligence. Many are also capable of much more physical activity than our culture expects of them. And finally, older people are much more sexually capable and more interested in sex than we have assumed.

Because our culture has expected less of older people, they have expected less of themselves. Now that new light is being shed on the subject of aging, much of this may change. We will all benefit from this new, more positive approach to aging.

A Positive Approach to Aging

There is a tremendous need to establish more positive attitudes toward the later stages of life. Our expectations for our old age are likely to become "self-fulfilling prophecies." If we expect to be

Most older people retain a high degree of physical, mental, and sexual ability.

healthy, happy, and active in old age, then we probably will be. If we expect less, we may experience less.

No one age contains all the positive aspects of a fulfilling life. Every age has its own special qualities both positive and negative. For instance, in childhood there is freedom of time but there is little freedom of choice. In youth there is self-determination, physical stamina, and adaptability. On the other hand, lack of experience results in many mistakes, and the pressure of competition is fierce during this period. The middle years often represent the period of greatest productivity. But there may be little leisure time to enjoy the fruits of one's labors. For many people, all of the positive aspects of life first come together in their mature years. This is often the best time in a person's life. There may be freedom of both time and choice. The wealth of experience accumulated over the years helps us enjoy our time and resources to the fullest. Many older people are able, for the first time, to do the things they have always wanted to do.

Growing old should and can be a wonderful period of life, a time of savoring past experiences and looking forward to new ones. Growing old can be a time of continued curiosity and learning, even though physical functions are beginning to fail. It can be a time of sharing one's wisdom with younger people, while having the wisdom to understand youths' limited capacity to accept what they hear.

Children especially love to hear old people talk about the way things were and what it was like in the "old days." Youngsters who have no contact with older people will thus lack a sense of continuity in their lives. Our older people are a source of wisdom that our youth-worshipping culture has left largely untapped. The young must become acquainted with this alienated segment of society and attempt to improve the plight of the aged. At the same time they will be preparing society to receive themselves when they reach their own "golden" years.

The Biology of Aging

What causes aging? Is our maximum potential life span already "programmed" into us at the time of our birth? How important are environmental factors in influencing the aging process? Why do people in some families live much longer, on the average, than people in other families? Will science eventually overcome aging and death?

The biology of aging is still poorly understood. It does seem that both genetic and environmental factors influence the aging process. It also seems certain that no single theory or explanation of aging is adequate. Aging is a complex process with multiple causes, and its rate is similarly influenced by many factors.

The most common biological theory of aging is the *genetic* one. The genes (nucleic acids) of an in-

A life of poverty, hardship, and deprivation contributed to the premature aging evident in this old photo.

● ASK YOURSELF

1. What difference does it make how a young person feels about aging and death? Isn't it better not to think about such things?
2. How does our society's attitude toward aging differ from that of many other societies?
3. How does the concept of a "self-fulfilling prophecy" apply to older people?

dividual are believed to gradually deteriorate, causing the body cells to degenerate and eventually die. Even cultures of human cells growing and dividing in glassware under laboratory conditions gradually lose their vitality and die out after a certain number of cell divisions, or generations.

Another prominent biological explanation of aging is the *immune theory*; this theory attributes aging to the progressive failure of the body's immune system. This process has a twofold effect. First, the body's ability to defend itself against invading germs (pathogens) weakens. But also the immune mechanisms, such as antibodies, begin to attack normal, healthy body cells. This is called the *autoimmune process*.

Another biological explanation is the *"wear-and-tear" theory*. This states that the human body simply wears out with use, just as a machine does. Still another biological concept is that waste materials gradually accumulate, begin to interfere with normal body functioning, and eventually block it entirely.

Environmental factors in aging are important as well. Consider the bodywide premature aging that is evident in a person who has lived a life of poverty, hardship, and deprivation. The biological age of the person's skin and internal organs may appear to be twenty or thirty years older than his or her chronological (actual) age. The person's life span is shortened accordingly. Many other environmental factors have been associated with aging. Stress and diet are prominent among these. Harmful chemicals and radiation are presumed to accelerate aging. Cigarette smoking definitely speeds up the aging process, as shown by the wrinkled skin of heavy smokers.

Longevity

Most of us would like to live a very long life, especially if we could be assured of good health at an advanced age. Hickey (1980) and others have identified some common characteristics of people who have succeeded in doing this. The genetic factor is evident; most very old people had parents who also lived very long lives. Most have had long, happy marriages. Most have been active all their lives—physically, mentally, socially, and sexually—and remain so at an advanced age. Moderation also seems to be important. Few really old people have been heavy eaters or drinkers.

Mental attitude is apparently one key to longevity. People who live very long lives have usually retained a positive and optimistic approach to life. They have been highly adaptable; they have been able to adjust to, and make the most of, the incredible amount of change they have witnessed during their long lives. Finally, they are people who, early in life, developed effective ways of dealing with

Good health in the mature years is the result of a lifetime of healthful living.

stress. They are living proof of the benefits of stress management.

Planning for Your Old Age

Possibly because of our use of denial, many of us fail to make adequate provisions for our old age. We find it very difficult or uncomfortable to think of ourselves as old. We live for the present and have only the vaguest thoughts of how we will spend our mature years. Then suddenly (it seems) we are there, often totally unprepared.

Thus, we must first prepare for old age by changing our attitudes. We must accept the reality that we will grow old. Rather than fight the inevitable aging process, we must plan for it. We should look toward maturity not as something to dread, but as potentially the most enjoyable time of our lives.

Of foremost importance is having the health and stamina to enjoy the full potential of these years. Good health in the older years is the result of a lifetime of healthful living. Those who enjoy the best health in their maturity are those who have had lifelong good health habits. They have been careful about their diets and controlled their weight. They have exercised regularly. They have used alcohol in moderation if at all and avoided tobacco. They have regulated their lives to avoid undue emotional stress, and developed effective means of dealing with life's unavoidable stresses. Much of the physical and mental deterioration we see in some older people could have been prevented or delayed

through healthful living habits during the early and middle adult years.

Making the most of the mature years does require some financial planning. Social Security and other pension plans have eased some of the burden for the elderly. But to live beyond the bare subsistence level requires additional financial preparation. This is especially true if we plan to travel.

Finally, an often neglected part of preparing for maturity is developing enough interests and social contacts to occupy one's expanded spare time and to keep one's mind and body actively functioning. Many people make the mistake of confining their interests to their careers or their children. Then one day they find themselves with neither of these remaining and with few other emotional resources. It is certainly never too late to develop new interests and to cultivate new friendships. But most people adjust to their maturity more easily if they have already established a broad basis of continuing interests and social contacts.

Assisting the Elderly

Even before we reach our own maturity, most of us become concerned with aging as family members grow old. We want to do all that we can to help them enjoy their remaining years.

It should be evident by now that our attitude toward the elderly is important. In all that we do for, and with, the elderly, we must be sensitive to the importance of their maintaining feelings of in-

Meaningful work is important to the well-being of older people.

● ASK YOURSELF

1. How long would you like to live?
2. How long do you think you will live?
3. How are you preparing for your own old age?

exist. Older persons, with their wealth of experience, can make important contributions in various advisory positions. They can serve business and industry, schools, cities, and other government agencies. Numerous opportunities for involvement in volunteer work already exist or could be created. Volunteer workers certainly are needed in hospitals, schools, programs for disadvantaged children, "meals on wheels" programs, and in the homes of ill or disabled persons. Volunteers are important in most hospice programs (*see* page 76).

One thing most older people appreciate is visits by family and friends. Too often, younger people, caught up in busy schedules, do not take time to visit the elderly. But interaction with younger people helps the elderly stay vital and alert. At the same time, the contact with older people helps the young come to terms with the aging process. Some young people find that visits to the elderly cause them to feel uncomfortable. This may be a reflection of their anxiety over their own futures. If they have not come to terms with aging as part of living, seeing an older person can be a vivid reminder of their own mortality and vulnerability.

Most older people like to live in their own homes or apartments as long as they can. They find security in familiar things and places and their sense of independence is enhanced. But eventually the day may come when other arrangements must be made. In doing so, again, it is important to enable the older person to retain as much independence as is possible. Sometimes family members can share their homes with an aging person, often to the benefit of all generations involved. But often space limitations or declining health make this impossible and a retirement home, nursing home, or similar establishment is the only alternative. In helping an older relative select such an institution, remember that the quality of such facilities varies widely. Great care should be taken to find a home that offers more than just custodial care. It should be properly licensed and inspected by the appropriate government agencies. The cheerfulness of the surroundings should

dependence and usefulness. We must remember that most elderly people are quite capable of doing things for themselves and for others. It is important to them to be able to do so. Often our well-meaning efforts to make their lives easier only serve to strip them of self-esteem. Much of what we interpret as mental degeneration in elderly people is merely their reaction to the loss of their sense of independence and usefulness. Thus, in assisting the elderly, there are many cases where "less is better." Obviously, this is not saying that we should do nothing, but it is a plea for sensitivity in what we do.

We can do much for the elderly by providing them with more opportunities to perform significant work and service, not just time-filler activities. Involvement in significant service helps to maintain a person's sense of usefulness and self-esteem. It keeps life "worth living." One possibility is a more gradual, or phased, retirement from employment. Working hours may be gradually decreased, in contrast to the abrupt retirement that is now typical. In addition to the psychological benefits, gradual retirement could ease the financial burden for both the retiree and the retirement pension systems.

Many other possibilities for significant work

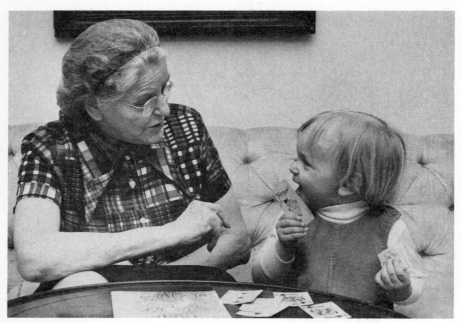

The interaction of younger and older persons benefits both generations.

● **ASK YOURSELF**

1. What is the value of a sense of independence in elderly people?
2. How can we help them retain this feeling?
3. Why do some younger people find that visiting the elderly causes them to feel uncomfortable?

be considered. The number and quality of staff are important, as well as the quality of meals and the extent of recreational, cultural, and social offerings. A physician can often help in this selection of facilities.

In all that we do, it is important to view our elderly citizens as an important resource, rather than as a burden.

COMING TO TERMS WITH DEATH

We all realize that our death is inevitable, yet few of us want to talk about it, face it, or prepare for it. Death has traditionally been a taboo subject for discussion in our culture. This is because most of us have not come to terms with our feelings about death. Even talking about death causes an uncomfortable degree of anxiety for many people.

By studying death we gain an understanding that enables us to deal more comfortably with the death of people who are important to us and eventually with our own dying. At the same time, positive attitudes about death enable us to live our entire lives more fully. In fact, the study of death is one of the best ways to learn what it means to be alive — searching for individuality, meaning, immortality, freedom, and love. The anticipation of our own death puts us in touch with our deepest feelings — our anxieties, hopes, needs, and opportunities as human beings. Much is lost if we wait until death is imminent before we experience this revelation.

Confronting death contributes to our total plan for life. Facing our own death will force us to see our lives in a totality that includes our past, our present, and that indefinite portion of life remaining ahead of us. Then we can organize and understand that totality. Many people think of death as unreal, or as something in the very distant future that they should postpone thinking about. As a result, they are incapable of experiencing their lives as a whole or of forming any total life plan.

Cultures often avoid the reality of death by dealing with it symbolically.

Attitudes Toward Death

Our attitudes toward death are strongly influenced by the culture in which we grow up. In other words, these attitudes are mainly *learned* rather than being inherent. Our culture should teach us how to anticipate death—our own and that of others. It should teach us how to experience the death of others and how to adjust to these losses. However, the fact is that our culture, as a whole, does not deal with death in a healthy way. This is reflected in the difficulty that many of us, as individuals, experience in coming to terms with death.

Cultures often deal with death symbolically—the *grim reaper,* the *gentle comforter,* or even as *sleep.* While they perhaps comfort the survivors, these symbols also keep death a step removed from reality. Such views of death can prevent us from coming to terms with it in a realistic way.

People have always exhibited a fear of death, much of which is attributable to its being an unknown encounter. This fear, however, is a composite of many fears—loss of self, loss of feelings, loss of thoughts, permanent separation, pain and suffering. It also includes emotions such as anger or

● ASK YOURSELF

1. Why do so many people avoid talking about death?
2. How do we acquire our attitudes toward death?
3. What do authorities on death mean when they say that we can live our lives fully only after we have come to terms with our feelings about death?

jealousy toward those who will remain and who may erase even the memory of our existence. Despite our resistance to death, its certainty affects each of our lives. The more fully we can come to face our attitudes toward dying, the less helpless we will feel in crises related to death. It will also be easier to communicate with those who are dying and their survivors. Psychiatrists think that our unresolved feelings toward death affect our daily living more than we realize. These feelings are reflected in the way we drive, eat, work, sleep, and even in our commitment to love.

Dr. Robert Kavanaugh (*Facing Death,* 1974), a leading authority on the subject of death, believes that life can be lived more fully once we have come to terms with our true feelings about death. He suggests we take a reflective journey back through our lives, focusing on those experiences that have formed our death-related attitudes. It is also good to reflect on other people's death fears, by examining the emotions they evoke in ourselves. When there is an occasion to visit a dying person or to be in contact with persons who have recently encountered death, we should not make excuses to stay away or be otherwise uninvolved. Being comfortable around reminders of death is not easy. But if we are ever to reach a healthful attitude toward death, it is necessary to dispel our feelings of helplessness.

Facing Death

There has been a dramatic increase in the number of seriously ill persons who are aware that they are probably not going to live longer than a few months or years. Many diseases still exist for which medicine cannot promise a cure. But life-prolonging treatment is often available and it can give patients time to prepare for death. Also, unlike in the past, there are very few patients who are not made aware

● ASK YOURSELF

1. Suppose that your physician discovered that you suffered from some rare, incurable disease and had perhaps two years to live. Would you want to know that fact? If you did know, how would you plan to spend your remaining time?
2. What if it were only two weeks? How would you spend each day?

of their diagnosis. Although they may not accept it initially, when the mind is ready and able to accept it, they can adjust to the situation.

While the physical needs of the terminally ill are being met, it is obvious that they and their families need special help to cope with their emotions. As patients struggle toward an acceptance of death, they find others are reluctant to listen to fears they need to vent. They find it difficult to obtain "permission" to die. The plight of such persons has had much to do with stimulating interest in the dying process. People like Dr. Robert Kavanaugh and Dr. Elisabeth Kübler-Ross (1970) have devoted much time to helping the dying and studying how we can best handle this part of life. (The study of death is called *thanatology*.) Their efforts have resulted in a wealth of knowledge, giving us new insights into death.

Dying a peaceful death can be helped by the presence of a special confidant who lends support and encouragement. Such a person will help the dying person break off ties to all except that which is part of the self. It is very important for dying people to receive permission to die from those who are closest to them. This permission is tacitly granted when those who are held dear reach the point of acceptance in their own process of adjustment to the impending death. Once permission has been given, they will slowly let go of all persons and worldly possessions held dear until everything except their own person has been relinquished.

Accepting Death

People who die in peace are often observed to have tranquil faces. This is associated with their having successfully gone through certain stages of coping mechanisms that lead to the state of acceptance. (Kübler-Ross, 1970).

The first stage is *denial and isolation*. This oc-

curs following the initial awareness of impending death. It is the stage when the person goes from saying "Not me!" to "Why me?" This is a very trying time for everyone close to the dying person, as the person displaces anger indiscriminately.

After the anger subsides, the person may try to *bargain* or make deals for a longer period of life. Many bargains are made with God; frequently these are promises based on guilt feelings.

The next stage is *depression;* this involves sadness about leaving family behind, losing a job, not seeing children grow up, and perhaps leaving behind many debts or much unfinished business. Another part of the depression involves mourning one's own impending death. This is a vital emotional adjustment in preparing for final separation.

The last stage is *acceptance*. During this stage the person will have already mourned the loss of self, other people, and all earthly possessions. It is not a state of happiness, but one that is nearly devoid of feelings. Hope is usually clung to until the very end and only wavers when signs of death are imminent. It is important to know that these stages are not mutually exclusive of one another. They often coexist and can also be repeated. The amount of time it takes for a person to progress through these stages varies greatly and depends on how much time is actually left. It is possible for an individual to go through all stages and reach the state of acceptance within a matter of days—or it may take many months.

The ease with which the dying patient goes through these stages depends upon his or her attitudes toward both life and death. It also depends upon the kind of support received from those persons who are important to the patient. Let us now examine the reactions of such persons.

Reactions of the Patient's Family

The presence of a slowly dying family member causes many changes in a household. Normal activities are interrupted. Numerous emotional, financial, and interpersonal stresses may result. Depending on how well the family copes with the situation, the end result may range anywhere from strengthened relationships among the survivors to hostilities that cloud family ties for years to come.

Open communication can decrease the problems of both the patient and the family. This includes the sharing of feelings among other family members and with the dying person as well. Our cultural conditioning, unfortunately, leads us to avoid such discussions and to act all the while as if the threat of death were not a reality.

Close relatives of slowly dying persons experience a mixture of emotions. In some respects, the stages of adjustment of family members are similar to those of the person facing death. Here, too, denial is often the first reaction. Family members may deny that there is illness in the family. They may mimimize its severity. Sometimes they shop around among physicians for a more encouraging diagnosis. Their failure to communicate their feelings about the situation is part of this denial.

Anxiety and depression are commonly felt also by family members. Those who are dependent upon the dying person may feel threatened by the loss of security. If the dying person has provided the family's income, there may be anxiety over financial security. If he or she has been involved in raising children, the spouse may feel threatened by the prospect of being a lone parent. Also, the loss of the ill person's companionship may cause anxiety.

As with any anticipated loss, depression is typical. Feelings of irritation or anger are often reported by those close to the dying person. They may become annoyed at the many inconveniences that result from the progressively more severe illness. Normal schedules and activities are interrupted. Seemingly endless arrangements must be made. Financial problems arise. Anger is often directed at physicians and nurses for not doing more for the patient or for failing to keep family members informed of the patient's condition. Anger can be directed even at the patient, whose illness may be interpreted as his or her own "fault" or even as a hostile act. The phrase "die on me" is frequently used in such cases, "How can you die on me when I still have these children to raise?"

While physicians and nurses may be accused of avoiding the dying patient, family members may also avoid visiting him or her. They may feel uncertain about what to say or simply be uncomfortable when they are near the dying person. (For those who have not come to terms with death, this may be an anxiety-provoking reminder of their own vulnerability.)

Finally, guilt plays a prominent role in the feelings of many people who are close to a dying person. There are many potential sources of guilt feel-

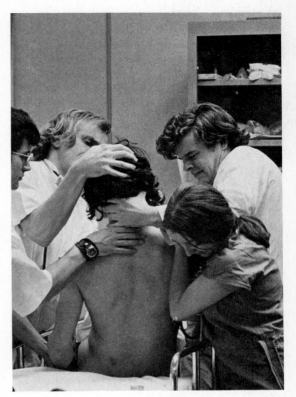

Those who care for the very ill often experience high degrees of stress and are subject to career burnout.

ings. There may be guilt over one's anger at having to care for the dying person. Guilt may be felt for not having treated the dying person better in years past. There may be guilt for not having loved the dying person more or for not having done certain things for him or her. Many people feel guilt for not having urged the patient to seek treatment earlier. Guilt feelings may develop if one is aware of looking forward to the death of the patient with some pleasure. This is common when the dying person has long been a source of emotional conflict. Guilt may even be felt merely because one is not dying and will survive and enjoy life after the dying person is gone. Children commonly feel guilt because they believe they may have caused the patient to die. Though guilt may have absolutely no justification, it can be strong enough to interfere with normal functioning.

● ASK YOURSELF

1. Have you experienced the death of someone close to you? What advice would you have for someone else facing the same situation?
2. If you were the dying person, how would you want your family members to deal with your situation?

The hospice concept includes care at home as well as in a hospital.

Family members who share their emotions will gradually accept the impending death. The most difficult time for the family is often the final phase. Now the patient is slowly detaching himself or herself from the world, including the family. Family members seldom understand that a dying person who has found peace and accepted death must separate himself or herself from everything and everyone around. This includes loved persons. A person can feel ready to die only by "letting go" of these important relationships. Family members often misinterpret this letting go as rejection. Yet it is the final, essential step that enables the patient to die in peace.

Care of the Dying

The past few years have brought considerable improvement in the ways in which health professionals deal with dying and death.

One significant recent change is honest communication between professionals and their patients. The old tendency was to shelter the patient from the knowledge that his or her condition was likely to prove fatal. A few physicians still take this approach. This is unsatisfactory for a number of reasons. If the deception is successful, it denies the patient the opportunity to make necessary plans and adjustments. More often, the patient knows the truth but is prevented, by the attempt at deception, from discussing his or her impending death with family and friends. Communication is blocked even with the health professionals attending the dying person. We now recognize that dying persons need to discuss their feelings with others. Incidentally, the deception also sheltered the unprepared health professionals from the uncomfortable task of participating in such discussions. Today, most people in the healing professions are better prepared to discuss death with their patients. The ideal approach now is to give the patient a realistic appraisal of his or her condition, and at the same time

maintain a hopeful or optimistic outlook. Studies at Stanford University and other locations have shown that such an attitude actually extends the patient's survival time.

Hospice Care

The hospice is a concept whose time has come. *Hospice care* helps dying persons round out their lives meaningfully. The hospice concept includes care at home as well as in a hospital. The concept is not a new one; hospices existed in France, Germany, and other European countries in the 1500s (Hardt, 1979; Garfield, 1979). However, they were forgotten until St. Joseph's was organized in London in 1902; and it was only in the late 1970s that the hospice movement gained momentum in the United States and Canada. There are still relatively few hospices in operation in North America.

The goal of the hospice is to keep the patient free of pain, comfortable, and alert through the final stages of life. Pain can arise from physical, psychological, social, spiritual, or any combination of sources. It can be controlled by carefully analyzing these sources and applying sophisticated combinations of medication and psychological support. Heroic life-sustaining measures such as resuscitators, cardiac massage and the like, are not part of the hospice concept.

Family members are included in the entire process, whether the patient is at home or in the hospice facility. Family services usually include:

1. Training family members to participate in the patient's treatment

2. Encouraging them to do such things as cooking special meals for the patient (familiar or favorite foods stimulate the appetite and are very important psychologically)

3. Unlimited visiting hours so that the entire family, including children, has access to the dying person

4. Providing family meeting rooms and living facilities when death is imminent

5. Providing social and educational programs for the family and patient, and continuing them after death has occurred

Prolonging Life

As death draws near and it becomes obvious that there is no hope for recovery, questions may arise as to how long life should be prolonged through the use of machines or other extraordinary measures. There are no easy, standard answers. Each case has to be considered independently, according to the circumstances. The dying person's needs and desires have to be respected as much as possible. However, under the stress of illness and pain, and while perhaps disoriented by pain-killers or other medications, sometimes he or she is not able to make wise decisions. Ideally, understandings can be reached between patient, physician, and close relatives in advance of the patient's final deterioration. Such agreements, made while the patient is still coherent, ease the burden for all concerned as death approaches.

Defining Death

Before the advent of modern medicine, defining death seemed easy: When someone's pulse and breathing stopped, he or she was dead. Now many people are successfully revived following such a "death." And in many patients, it is now possible to stimulate breathing and heartbeat with machines long after there is any conscious awareness or any hope of recovery.

We now understand that there really is no one moment of death. Death is a process, not a point in time. A person dies cell by cell and organ by organ. It is quite possible for one or more of a person's organs, such as the brain, to die while the remainder of the body remains alive. This, of course, is the basis for transplanting organs from persons who have died (for example, in accidents). Death is now

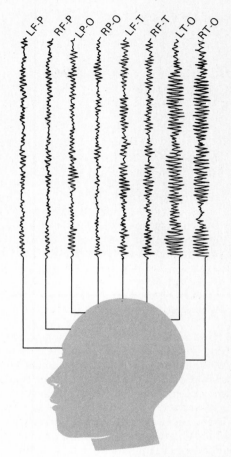

A normal electroencephalogram (EEG), a recording of brain waves.

defined by the absence of brain activity. This is determined by taking an electroencephalogram (an EEG, or recording of the electrical activity of the brain). A person's body can be artificially sustained on a long-term basis, even in the absence of brain activity. But most people now agree that when the brain is dead, the *person* is dead, regardless of whether the remainder of the body continues to function. Physicians still debate how long they should wait after the brain waves disappear (which produces a "flat" EEG) before they abandon hope and deactivate the life-support systems. The current edition of *The Merck Manual* (Merck & Co.) states that after twelve hours the condition should be considered irreversible.

After death has occurred, survivors often experience a deep sense of loss.

What's It Like to Die?

As we have discussed, there is really no clear-cut moment of death. For a particular dying person, the time or even the date of death can vary considerably, depending upon the definition of death being applied. In any case, though, medical technology now allows many persons to survive situations in which they have apparently died or been extremely close to death. Many interesting studies have been published in which the experiences of these people are described.*

There are, of course, many interpretations of the experiences surrounding death. The same experience, for example, might be explained in spiritual, psychological or physiological terms, depending on the interpreter's frame of reference. But regardless of the explanation, the conclusion reached by most authorities on death is that it is not an unpleasant experience.

Moody has interviewed many people following their revival from apparent death. About a dozen distinct elements in the death experience are reported with some frequency. The average subject reports about eight of these elements. The order in which we list them is typical of how they occur, but the exact sequence is somewhat variable.

1. *Hearing the news.* Many people remember hearing physicians or others pronouncing them dead. The exact words used may be recalled and their accuracy confirmed by those who spoke them.
2. *Feelings of peace and quiet.* This sensation is also referred to as one of extreme comfort and relief.
3. *Hearing an unusual sound.* The sound is often described as a very loud ringing, buzzing, or clicking, or as beautiful or strange music.
4. *Passing through a dark tunnel.* The passage is also described as being through a well, a cave, a vacuum or void, or through a valley.

5. *Leaving the body*. Many people report memories of standing or floating a few feet away from their bodies and looking at the body and the other people present.

6. *Meeting others*. After leaving their bodies many people remember being aware of the presence of other spiritual beings, either previously known deceased persons or strangers.

7. *Seeing a bright light*. The light is recalled as being very bright but not hurting the eyes. People with a religious background often describe this as a "being of light" who radiates love and warmth and communicates knowledge and understanding.

8. *The life review*. This is probably the most widely known experience of dying persons. It is expressed by the classic statement, "My whole life flashed before my eyes." Even though it is instantaneous, each image is extremely sharp and vivid and is clearly remembered following resuscitation or revival.

9. *Coming back to life*. There is usually little memory of this process, other than just that of waking up or regaining consciousness. However, the moods and feelings associated with the "death" experience linger on for some time.

10. *Telling others*. The people who have been through this experience have no doubt about its reality and importance to them. They may be eager to talk about it if they find a receptive listener. On the other hand, if people seem skeptical, they may be reluctant to share these experiences for fear of being seen as peculiar or emotionally unstable.

11. *Effects on lives*. As might be expected, there is usually an intense feeling of being lucky to be alive. There is often a sense that life has been given more meaning by the experience.

12. *New views of death*. Almost without exception, there is a loss of the fear of death. Death is no longer seen as an unpleasant event. While most people who have experienced death want to continue to live as long as possible, they are able to do so with a new freedom from the fear of death. Certainly, their experiences can serve as a valuable lesson for us all.

* A good introduction to the subject is *Life After Life* by Raymond A. Moody (Mockingbird Books, 1975). For several contrasting viewpoints, see "After Resuscitation, Is Afterlife Recalled?" Medical World News, *19*, No. 4 (February 20, 1978), 37.

Mourning: The Stages of Grief

After death has occurred, the survivors experience a deep sense of loss, and a feeling of sadness penetrates their very being. Dr. Kavanaugh has identified distinct phases in the grieving process. They include *shock, disorganization, volatile emotions, guilt, loneliness, relief,* and *reestablishment*. For those who have been expecting death to occur, the period of grief is not as long. However, even when all mourning is apparently finished, grief frequently reappears. Grief can be so intense that the body can become seriously ill, or mental stress may culminate in psychiatric disorders. The bereaved usually receive some comfort and help from others, but most support ends soon after the funeral. The need for caring friends who will listen and understand

● **ASK YOURSELF**

1. Is there a hospice in your community? If so, what services does it offer? Do you feel that a hospice is a good idea?
2. Assuming that you are in good health right now, what instructions would you give your family or physician concerning the treatment you would want in case you suddenly became seriously ill or were severely injured in an accident?

may go on for a matter of months. Mourning is not complete until the bereaved have adjusted to living without the one they have lost.

Children are often left out of the mourning experience. However, such attempts to protect children from the grief and finality of death can lead to con-fusion and the establishment of strange perceptions about death. Children should be made aware of the reality of death. Deaths other than those of family members can provide the occasion to approach the subject of death—for example, the death of a loved pet or a national figure, children may bring up the subject themselves.

Children should be included in discussions of an impending death in the family and in the events that follow death. The truth should always be told, within the limits of the child's comprehension. It is never good to tell a child that a dead person is only going away temporarily "on a long trip." Given the opportunity, children adapt to crises and unpleas-ant realities as well as adults do. Often a child will ask for further explanations or clarifications of the concept and actual event of death. This is common, and usually all that is needed is a reaffirmation that his or her ideas are correct.

IN REVIEW

1. Our entire lives are affected by our attitudes and feelings about aging and death; our fulfillment as individuals depends on having healthful attitudes toward aging and death.
2. People now take a more active and positive approach to aging; they realize that older people are much more capable than they formerly assumed.
3. We should emphasize the positive, not the negative aspects of aging.
4. The biology of aging is still poorly understood. Genetic deterioration, weakening of immunity, wear and tear, accumulation of wastes, and environmental factors all possibly contribute to the aging process.
5. Good health in one's old age is often the result of a lifetime of healthful living habits.
6. In assisting the elderly, we must be sensitive to their need to feel independent and useful.
7. Unresolved feelings and fears about death influence our daily lives more than we realize.
8. People who know that they are in the process of dying pass through some iden-tifiable emotional stages:
   a. Denial and isolation
   b. Bargaining
   c. Depression
   d. Acceptance
9. The presence of a slowly dying family member can cause numerous emotional, financial, and interpersonal stresses in a household. Many of these problems can be prevented through greater understanding and improved communication.

10. The hospice concept includes home and hospital care for dying persons and assistance to their families.
11. Death may not occur in a single moment, but may be a prolonged process. People are usually considered to be dead when brain activity ceases.
12. Those who have been revived from apparent death report that dying is not an unpleasant experience.
13. After a death, most survivors experience several stages of grief: shock, disorganization, volatile emotions, guilt, loneliness, relief, and reestablishment. Children should not be sheltered from the mourning experience.

TWO
DRUGS, ALCOHOL, AND TOBACCO

Reaching our life potential depends on making the most of environmental, psychological, and sociological factors. Arising both from within ourselves and from without, such factors help determine our happiness and productivity. We can modify the effects of such factors upon us with the use of drugs.

Some drugs are used to treat disease, arising either from pathogenic or emotional causes. Designed for use by medical professionals, these drugs are to be dispersed on a limited basis. Other drugs are used socially. Designed for use by nonprofessionals, they are dispensed by profiteers, both legally and illicitly. The effects of some of these social drugs are merely mood-modifying, while others affect basic behavioral patterns.

Drugs used properly may help to restore us to wholeness, improving our sense of well-being and productivity. Drug misuse (abuse) carries as great a potential for destroying our sense of well-being as proper drug use has for restoring it. This section informs us on the difference.

CHAPTER 5

DRUGS

What is your definition of a drug? Many individuals consider drugs any substance used to treat a disease. Others feel a substance is not a drug unless it is prescribed by a doctor. However, a more complete definition is: any substance, other than food, *that alters the body or its functions.*

Some drugs are used only for their medical value; others have medical value but are also used for recreation. Drugs used for recreation can sometimes alter behavior far beyond the range of normality. Many social problems with drugs are caused by *drug-altered behavior.* The problem of drug use becomes complicated when the drug in question has more than one use. For example, barbiturates and narcotics are beneficial when used as sleeping pills and pain-killers, but can be very destructive when used for recreation as "downers" for their depressant characteristics.

MEDICAL USE OF DRUGS

Present knowledge of the chemical structure of drugs often enables scientists to predict what a drug will do and what the results of its actions will be. Any drug that is powerful enough to be effective has the potential to produce some adverse reactions that are not always predictable. These actions can be classified as either *side effects* or *untoward effects.* A side effect of a drug is any action or effect other than the one for which the drug is administered. Side effects are not necessarily harmful. Morphine is usually given to relieve pain, not for its

WARNINGS: May cause drowsiness. May cause excitability especially in children. Do not exceed recommended dosage because at higher doses nervousness, dizziness, or sleeplessness may occur. If symptoms do not improve within 7 days or are accompanied by high fever, consult a physician before continuing use. This product contains aspirin and should not be taken by individuals who are allergic or sensitive to aspirin. Do not take this product if you have asthma, stomach distress, bleeding problems, glaucoma, difficulty in urination due to enlargement of the prostate gland, high blood pressure, heart disease, diabetes, thyroid disease, or are presently taking a prescription medicine, either an antihypertensive or antidepressant drug containing a monoamine oxidase inhibitor, or especially for diseases of the heart, blood vessels, diabetes or gout, or if symptoms persist; or give this product to children under 13 years except under the advice and supervision of a physician. Stop taking this product if your ears ring or new symptoms appear. Use during pregnancy ONLY under your doctor's direction. Keep this and all drugs out of the reach of children. In case of accidental overdose, seek professional assistance or contact a poison control center immediately.

Americans are symptom oriented. TV and magazine advertising has made many people think that any symptom needs treatment.

ability to constrict the pupil of the eye. This constriction is a side effect. An untoward effect is any reaction regarded as harmful. The untoward effects of morphine—nausea, vomiting, constipation, and addiction—are obviously undesirable and harmful.

● ASK YOURSELF

1. Define the word "drug." Ask five people to define "drug" and compare their definitions with the book's definition.
2. What is a *side effect*? Have you ever experienced any side effects from a drug?
3. What is an *untoward effect*? Have you ever experienced any untoward effects from a drug? If your answer is Yes, were you taking this drug for medical or recreational purposes? Explain how these untoward effects took place.
4. Has advertising influenced you in the drugs you use? What influence does advertising have on drug use?

● ASK YOURSELF

1. What is a drug's *generic name*? Who assigns a drug its *generic name*?
2. What is the difference between a drug's *generic name* and its *trade name*? How do you and your family order drugs — by their *generic names* or their *trade names*?
3. What are *common* or *street names*? Give a drug's *generic name* and then list as many of its common names as possible.

Drug-induced behavior is often extreme and bizarre.

It is important to remember that most Americans are symptom-oriented. This may be a direct consequence of the standard methods of advertising medicines. Television, radio, and magazine advertisements are designed to focus our attention on symptoms, not the underlying causes of illness or pain. Thus, many people seek immediate relief of symptoms and want their physician's assistance in doing so. But if a physician prescribes a drug that eliminates symptoms, most patients do not feel the same urgency to continue with diagnosis and treatment. These are the phases of medical attention that are really most important for an individual's long-range health.

How Drugs Are Named

A drug or other chemical substance may have many names. This causes confusion unless you know why the different names are used. The *official name* of a drug is the name listed in an official government publication such as *The Pharmacopeia of the United States*. The *chemical name* (mainly used by chemists) is a precise description of the chemical makeup of the drug. For example, the chemical name of the antibiotic *tetracycline* is *4-dimethylamino-1,4,4a,5,5a,6,11,12a, -octahydro-3,6,10,12,12a, pentahydroxy-6-methyl-1, 11-dioxo-2-naphthacenecarboxamide*.

The American Medical Association Council on Drugs and the World Health Organization jointly assign a *generic name* to every drug. The name *tetracycline* is a generic name. This name is never capitalized and it is the same in all countries. Many drugs are also sold under *brand names* or *trademark names*. These names are always capitalized and they usually appear with the sign ® to the upper right of the name. The sign ® indicates that the name is registered and its use is restricted to a specific drug company that is the legal owner of that name. For example, tetracycline is known by the brand names Achromycin, Panmycin, Polycycline, Tetracyn, and Tetracyn V. Sometimes a single drug may be sold under ten or twenty different brand names.

Common, street, or *slang names,* terms, or expressions are the language of illegal drug use. These vary from area to area and are constantly changing. Teen-agers and young adults often use a drug's street name.

Drug Labels

The labels on prescription and nonprescription drugs should be followed to protect against misuse and harmful effects. According to law, all nonprescription drugs must list these six points on the label:

1. Name of the product
2. Name and address of the manufacturer, packer, or distributor
3. Net contents of the package
4. Quantity and generic names of active ingredients
5. Names of any habit-forming (addicting) drugs contained in the product
6. Cautions and warnings needed for the protection of the user (This information may also be on an insert in the package: *Read all inserts*)
   These cautions and warnings include:
   a. Avoidance of use by children
   b. Avoidance of chronic use
   c. Avoidance of use during pregnancy (or breast feeding)
   d. Avoidance of use in the presence of specific diseases or conditions
   e. How to use medication safely
   f. When not to use medication
   g. When to stop taking the drug
   h. When to see a physician

Self-medication

Today drugs can be bought in supermarkets and restaurants and from vending machines. Sales promotion by radio and television encourages self-medication for real or fancied illnesses. Although self-medication may harm an individual, the hazards can be minimized by educating everyone about drugs.

The overuse of any drug can cause detrimental effects. Continued self-medication for a recurring ailment may mask a serious condition, endanger your life, and create a need for prolonged and expensive medical treatment. Often an individual does not realize the potential dangers of taking prescription and nonprescription (over-the-counter) drugs together.

The medical use of nonprescription drugs A nonprescription drug, like a prescription drug, must be proven safe and relatively effective before it can be sold. Nonprescription drugs pose no threat to the average person when *used as directed*. The key to using nonprescription drugs is *reading, understanding, and following the instructions on* the label. The illustration shows a typical nonprescription label. Label warnings help protect you against misuse and harmful effects; follow them closely.

How to Use the Physicians' Desk Reference (PDR)

Two *PDR* books are available for sale or are in most libraries. One *PDR* lists prescription drugs, and the other *PDR* lists nonprescription, or over-the-counter, drugs. These *PDRs* are the best references on drugs, their uses, effects, and side effects. Each book is divided into the following sections:

1. *Manufacturers' Index* (white section). This section contains names and addresses of manufacturers, and where you may address inquiries.
2. *Product Name Index* (blue section). This section lists drugs according to product categories—for example, oral contraceptives, cough medicines, sedatives, vitamins and antibiotics.
3. *Product Name Index* (pink section). Products are listed in alphabetical order by *brand names*. Page numbers of other sections have been included to help you locate more information on the drug.
4. *Active Ingredients Index* (yellow section). Drugs are listed in alphabetical order under their *generic name* (of the principle ingredients contained in the drug).
5. *Product Information Section*. This is an alphabetical listing of manufacturers and the drugs they manufacture. This section contains the product information *given out by the manufacturer* such as actions, uses, administrations, dosages, precautions, forms supplied (pill, capsule, and so on), and any warnings, hazards, contraindications (reasons for not using the drug), and side effects.

If you want to find . . .	And you already know . . .	Here's where to look . . .
The brand name of a product	The manufacturer's name	White section: Manufacturers' Index
	Its generic name	Yellow section: Active Ingredients Index*
The manufacturer's name	The product's brand name	Pink section: Product Name Index*
	The product's generic name	Yellow section: Active Ingredients Index*

Essential product information such as:
  active ingredients
  indications
  actions
  warnings
  drug interaction
  precautions
  symptoms & treatment
    of oral overdosage
  dosage & administration
  how supplied
	The product's brand name	Pink section:
		Product Name Index*

	The product's generic name	Yellow section: Active Ingredients Index*
A product with a particular chemical action	The chemical action	Yellow section: Active Ingredients Index*
A product with a particular active ingredient	The active ingredient	Yellow section: Active Ingredients Index*
A similar acting product	The product classification	Blue section: Product Category Index*
The generic name of a brand-name product	The product's brand name	Pink section: Product Name Index, Generic name will be found under "Active Ingredients" in Product Information Section.

* In the pink, blue, and yellow sections, the page numbers following the product name refers to the pages in the Product Identification Section where the product is pictured and the Product Information Section where the drug is described.

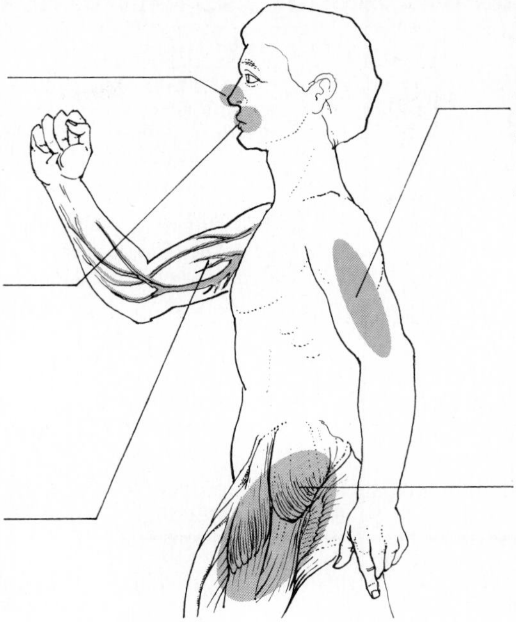

INHALATION

For treating specific lung disorders and administering anesthetics

Misues of cocaine and volatile
solvents

Called "snorting" or "sniffing"
in drug slang

ORAL DOSAGE

Method most convenient for a
medical patient

Digestive system modifies actions
of drugs too powerful or fast-
acting if injected

INTRAVENOUS INJECTION

For injecting drugs directly into
bloodstream

Permits extremely fast action,
often used in emergencies

Permits large dosages, for both
medical and drug abusers

Called "mainlining" by addicts;
permits the most extreme effects
from certain drugs

SUBCUTANEOUS INJECTION

For injecting drugs just beneath
the skin

For drugs rendered ineffective by
digestive juices acts more quickly
than oral route but more slowly
than intramuscular injection

Called "skin popping" in drug
slang

INTRAMUSCULAR INJECTION

For injecting drugs into the muscle layers

For drugs that cannot be injected
directly into the bloodstream

Because muscles are richly supplied with blood vessels produces
an effect intermediate between
subcutaneous and intravenous
injection

Called "muscling" in drug slang

Methods for introducing drugs into the body.

When two or more drugs are taken together or within a few hours of each other, they may interact and alter the expected action of any one of them. These interactions occur between drugs, prescription and nonprescription, and at times between drugs and foods. Because of the possibility of drug interactions, your physician should always know the names of *all* drugs, both prescription and nonprescription, that you are using at any one time.

Here are three basic principles for using nonprescription drugs:

1. Most over-the-counter drugs do not cure any disease or speed healing. Most only temporarily relieve *symptoms*. If a symptom persists or returns often, you should see a physician to find the *cause* of the symptom.

2. The best over-the-counter drugs are the ones designed to relieve a specific symptom. Take the recommended dosage and avoid nonprescription drugs that are combinations of ingredients. These are the most expensive and usually do not contain an adequate amount of any specific drug that you need.

3. Avoid using all drugs if possible. No drug is totally free of harmful side effects. The ideal way to avoid (or treat) illness is by adopting healthful holistic living habits: good diet, exercise, rest, and stress management.

Drug Administration and Actions

As the next illustration shows, drugs can be introduced into the body in various ways. Some drugs (penicillin, for example) can be taken orally in the form of capsules or can be injected. Other drugs, such as insulin, which is used in the treatment of diabetes, cannot be exposed to the digestive system's chemical action. They must be injected into the body.

After being taken into the body, drugs are distributed by the bloodstream to the many organs, tissues, and cells. The action or effect of a drug may occur on the surface of cells, within cells, or in the body fluids surrounding the cells. In most cases, the action occurs within individual cells and has either a direct or indirect effect on the central nervous system.

Many drugs contain molecular parts similar to

TABLE 5.1 DEFINITIONS OF ADDICTION AND HABITUATION

Addiction	Habituation
Drug **addiction** is a state of periodic or chronic intoxication resulting from the repeated consumption of a drug (natural or synthetic) and producing in the individual:	Drug **habituation** is a condition resulting from the repeated consumption of a drug that produces in the individual:
1. An overpowering desire or need (compulsion) to continue taking the drug and to obtain it by any means	1. A desire (but not a compulsion) to continue taking the drug for a sense of improved well-being or other effect
2. A tendency to increase the dose (tolerance)	2. Little or no tendency to increase the dose (little or no tolerance)
3. Both psychic (psychological) and physical dependence on the effects of the drug and hence presence of *abstinence syndrome* (withdrawal illness)	3. Some degree of psychic dependence on the effect of the drug, but absence of physical dependence and hence of abstinence syndrome
4. A detrimental effect on the individual and on society	4. Detrimental effects, if any, primarily on the individual

those found in the cell's normal "diet." This permits the drugs to participate in a few stages of the cell's normal chemical processes. Ultimately, of course, the differences between the drug and the normal chemical will be detected by the cell's systems. But by this time the drug's work has been done. The cellular processes are no longer normal; the cells, the organs, and the interrelated body systems have been altered—the drug has taken effect.

A drug affects an individual in three ways:

1. The *therapeutic,* or disease-treating, effects alleviate a disease or its symptoms. These are the effects and actions sought when treating a disease or condition.
2. The *side effects* or *actions* and *untoward effects* or actions are any effects other than the one for which the drug is administered. These are often the effects an individual seeks when using a drug recreationally.
3. The drug may have a *symbolic effect* or *meaning.* Often a drug's effectiveness is enhanced when doctors, nurses, pharmacists, or advertising indicate that it is beneficial. Also, the symbolic meanings of a drug can produce irrational effects, "fantasies," and false physical and emotional effects. These are based upon inaccurate information from friends, hearsay, or "street knowledge" ("I know

what that drug can do because I have used it").

THE DRUG-ABUSE PROBLEM

Most drugs have legitimate and useful places in the treatment of disease and alleviation of pain and discomfort. This is the "proper" use of drugs. Theoretically, any drug can be abused—that is, used for purposes other than those intended by a physician. Drug abuse is defined as the self-administration of drugs in excessive or otherwise inappropriate dosages (amounts), which causes damage to an individual and/or society. Despite the usefulness of this definition, the term is still a difficult one to define completely and adequately. The spectrum of responsibility—from individual to social—frequently clouds important scientific and medical debates about drugs. Attitudes toward drug abuse have also taken on political significance, again adding to the scientific confusion.

Habituating or Addicting?

In the past, certain commonly abused drugs were said to be *habituating* or *addicting.* The explanations of these terms most often quoted today are those of the Expert Committee on Addiction-Producing Drugs of the World Health Organization. These definitions are given in Table 5.1. But

We live in a drug-oriented society. Almost everyone uses some drugs for a variety of reasons.

the terms *addict, user,* and *habitual user* can be ambiguous. In common usage, these words are defined in terms of the drug involved. For example, an *addict* has been described as someone who is dependent on physically addicting drugs. A *user* has been described as one who has an habituation to a non-addicting drug.

Dependency A physician must know if a person is addicted to (physically dependent on) a specific drug in order to direct proper medical treatment. Whether a drug is addicting or habituating makes no difference to the law or to the legal control of a drug. But society does not accept addiction, while it may accept the periodic, *recreational,* or "weekend," use of a drug. The behavioral problems of addiction are treatable emotional problems of an individual. The social and legal problems in the use of a certain drug are not determined by whether or not it has physically addicting properties.

The terms habituating and addicting are now being replaced by *psychological dependence, drug dependence, substance dependence,* or *dependency.* The repeated use of any substance (food, tobacco, aspirin, alcohol, narcotics, and so on) may cause some individuals to develop a *dependence* on it.

Large amounts of food are consumed by individuals dependent on food as an emotional stabilizer. Large amounts of mild pain-relievers (analgesics) with aspirin or aspirinlike ingredients are consumed every day. Tranquilizers, sedatives, and hypnotics are the most widely prescribed and consumed drugs in the United States and the world. Many people believe they cannot get up, do their work, keep their nerves quiet, or go to sleep unless they have a drug to help them. There is a sound medical basis for the temporary use of these drugs by some, and

Pregnant women may unwittingly risk the health of their unborn babies. Cigarettes, caffeine, and alcohol can harm the fetus. Even cold remedies and pain-killers taken by many women during pregnancy can pose a risk. This informative pamphlet can be obtained from: Marketing Services, Dept. 126F, Addiction Research Foundation, 33 Russell Street, Toronto, Canada, M5S 2S1.

● ASK YOURSELF

1. Explain the various methods used to introduce drugs into the body. Have you taken any drugs lately? By what method?
2. How do drugs affect the body? Have any drugs you have used affected you in these ways?
3. Define habituation, addiction, and dependency. What types of dependency are there? Have you ever been dependent upon anything?
4. What is meant by the "recreational" use of a drug? How often do you use drugs for recreational purposes? Under what circumstances?

● ASK YOURSELF

1. How do drugs create problems for society? Describe some drug-induced behavior you have witnessed. How was this behavior dangerous to the individual or society?
2. What is a "stimulant" drug? What is a "depressant" drug? Have you ever used a depressant or a stimulant drug? Describe your physical and psychological reactions to it. Was your behavior dangerous to you or to society while you were under its influence? If so, in what way?
3. What is meant by the term "tolerance"? Have you ever shown tolerance to a drug? Describe your reactions.

---

the permanent use by a few. But in general this is a form of "socially acceptable" drug dependence.

Problem Drugs

Problem drugs are those that cause death, marked personality changes, or abnormal social behavior. Most drugs can cause death, and some, such as tobacco, cause a great many deaths. Drugs that cause personality changes and marked abnormal social behavior are known by a number of names. They are often described as *mood-modifying, psychoactive* (mood-, perception-, or consciousness-altering), *psychotropic* ("mind-changing") or *psychotoxic* ("mind-poisoning"). The most accurate of these terms is psychoactive. Psychoactive drugs are used for recreational purposes because they can cause *euphoria* (an extreme or exaggerated sense of well-being), *hallucinations* (perceptions that have no relationship to objective reality), or recognizable changes in personality or behavior. The recreational use of psychoactive drugs becomes a social problem when it causes unpredictable or dangerous behavior or personality changes.

Depressants and stimulants In general, psychoactive substances act in or on cells. They increase or decrease cellular activity of nerve centers and their conducting pathways (nerves and nerve tracts). *Depressant* substances temporarily depress cellular and, consequently, body functions. Such drug-induced depression of the central nervous system is frequently characterized by a lack of interest in surroundings, inability to focus attention on a subject, and a lack of motivation to move or

talk. The pulse and respiration become slower than usual, and as the depression deepens, the ability to use senses, such as touch, vision, hearing, smell, and taste, diminishes progressively. Psychological and motor activities decrease. Reflexes become sluggish and finally disappear. Depressant drugs are often quite accurately called *downers*. They literally slow down the cellular activity of an individual's nervous system. If a strong depressant is used, or if abusively large doses are consumed, depression progresses to drowsiness, stupor, unconsciousness, sleep, coma, and death.

A central nervous system *stimulant* is a drug that temporarily increases cellular processes. This in turn increases body or nerve activity. Stimulant drugs quickly produce a dramatic effect, but their medical usefulness is limited because of the complexity of their actions and the nature of their side effects. Such side effects may include hallucinations, euphoria, anxiety, extreme nervousness, and tremors.

Tolerance The repeated use of most psychoactive drugs, such as the narcotics, can produce biochemical and physiological changes. These changes cause the user to keep increasing the dosage to maintain the same mood-modifying effect desired. At this point, a drug user is said to have developed a tolerance for the drug. *Tolerance* is an acquired reaction to a drug that necessitates an increase in dosage to maintain a given action or effect. As tolerance increases and more of the drug is used, body cells are gradually exposed to greater and greater quantities of it.

For a period of time, the body will adjust to these slowly increasing dosages of a drug. However, it is always possible that a user will take a larger dose than the body can tolerate (an overdose), leading to an extreme reaction or death.

If you are to deal with psychoactive drugs, or understand individuals who use them, it is wise to learn something about the mind-altering, or psychotropic, effects of some of the more commonly abused psychoactive drugs.

PSYCHOACTIVE DRUGS

Today much research on drugs is concerned with how drugs alter states of mind. Researchers are attempting to discover how certain chemcial substances modify people's moods, personality, and behavior.

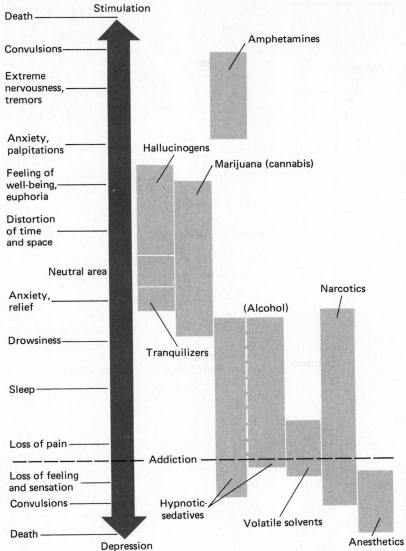

The effects of increased doses of drugs. As doses of drugs increase, the effects progress along a continuum until a lethal dose, producing death, is taken. Also, as shown by the second color, the effects of drug groups overlap when equivilant doses are given.

The abnormal social behavior of some drug abusers, moreover, causes other problems. One of these involves the issue of personal responsibility during times of drug-induced behavioral changes, which may be called "times out." During such an interval, a person might act in ways that depart from normal rationality and morality.

The depression and stimulation of the central nervous system are independent actions. The illustrations showing the continuum of drug effects and responses was suggested and formulated by Dr. Robert W. Earle of the University of California at Irvine.

The continuum of drug effects reaches to over-

● ASK YOURSELF

1. Define the term "psychoactive." How does your definition differ from the book's definition?
2. Explain the Continuum of Drug Actions diagram. Have you ever experienced any of the feelings shown? If so, when and how were they produced? Can other substances produce such feelings in an individual?
3. After reading the following section, place the commonly abused psychoactive drugs on the Drug Actions diagram. Explain why you placed specific drugs at specific points.

stimulation and death at one extreme and to depression and death at the other. The neutral area of this continuum is the range of stimulation and depression an individual encounters normally. Often drugs in different groups have similar actions. For example, narcotics are used to relieve pain, but they may also, as a side effect, induce sleepiness. Barbiturates are used for their ability to induce sleep, but they do not relieve pain. Thus, the sleep-producing effects of these two depressants, narcotics and barbiturates, overlap on the continuum chart. In fact, many of the drugs that affect the central nervous system have similar actions.

As we progress along the chart from the neutral area outward, we can locate the specific points where the effects of drugs overlap. These points show where the continuum moves from the major effective area of one group of drugs into the area of another, more powerful group. The weaker drugs are nearer the center, while the most powerful drugs are at the two extremes.

If dosages are increased, any of the drug groups listed can produce the complete range of effects of stimulation or depression. Such overstimulation or extreme depression is the effect the drug abuser is seeking. Consequently, dosages taken by drug abusers far exceed those normally used in medical practice. The complete range of effects produced by increased dosages is represented in the next illustration.

The drug continua in the chart point out the fact that the extreme effects of many different drugs are actually very similar. For example, any of the stimulants will produce hallucinations if the dosage is large enough. This is why many individuals, while

usually preferring one drug over another, will abuse any available psychoactive drug. As the specific actions of a drug become more familiar and less spectacular, the individual may experiment with new ways to use the drug. A user often combines drugs of the same type or of different types (poly-drug use) to produce a more intense effect. Examples of common combinations are alcohol and marijuana, alcohol and methadone, and methadone and heroin. Or a user progresses from taking the drug orally, to injecting it under the skin, to injecting it directly into a vein. Some drug abusers are not satisfied with experiences from one drug at one consistent dosage and progress toward the extremes of the continuum.

Anesthetics

Anesthesia means the loss of feelings and sensations. Although the term strictly means the loss of *all* sensations, it usually refers to the loss of the sensation of pain. Anesthetics are the most potent nervous system depressants and thus permit the performance of surgery or other painful procedures. They are so dangerous that physicians must be specially trained in order to administer them safely. Such specialists are known as *anesthesiologists*.

Anesthetics can depress all body cells. While being anesthetized, most people pass through a "stage of excitement." Individuals using anesthetics for recreation are usually looking only for these feelings. After the excitement stage, however, they may easily pass into the next stage, "medullary paralysis," in which respiration ceases and death occurs. The recreational use of anesthetics is obviously very dangerous.

PCP (Angel Dust) The most widely abused anesthetic is *phencyclidine*. This drug is known by many street names such as PCP, "angel dust," "peace pills," and many others. PCP is among the most widely used drugs on the street today. It is often dissolved in a liquid, then sprayed on marijuana, parsley, oregano, tobacco, or other plant leaves and smoked. It has also been drunk, taken in pill form, or injected. These are the most dangerous methods of taking the drug, and the danger of an overdose is extremely high.

Phencyclidine was developed as a potent animal tranquilizer and anesthetic. It has never been used legally on humans because the range between an effective dosage (the amount needed to relieve pain) and a possibly lethal overdose is too narrow for safety.

Persons taking PCP will usually hallucinate and experience feelings of depersonalization (that is, "spaced out"). They become extremely aggressive, lose their muscular coordination, and may be unable to speak. A person who takes a heavy dosage, or two doses close together, will experience these symptoms, eventually goes into a coma, and often dies. In Los Angeles county in 1980, 345 known deaths occurred from the use of PCP.

Narcotics

Narcotics are used in medicine primarily for their analgesic effect — that is, their ability to relieve pain. Administered in controlled dosages, they cause insensitivity to pain without producing loss of consciousness or even excessive drowsiness. Physicians generally use anesthetics or members of the hypnotic-sedative group when they want to anesthetize patients for operations. They never use narcotics alone for this purpose, because narcotics used in dosages large enough to produce sleep or stupor could depress the respiratory center sufficiently to cause death.

An individual who abuses narcotics may develop a physical dependence on the drug. Tolerance to narcotics develops very quickly. When a person whose body has developed a tolerance is cut off from a supply of drugs, he or she develops a condition called *withdrawal illness* (in this case, *narcotic-solvent-type abstinence syndrome*). The following are the major narcotics that are abused in North America.

Morphine The major derivative of chemically refined opium is morphine. Morphine is the best analgesic available. It may be pure white, light brown, or off-white in color and comes in many forms: cubes, capsules, tablets, powder, or liquid solution.

Very little morphine is sold on the streets. Rather, morphine is used when heroin is not available and is usually stolen from physicians' offices or pharmacies. Morphine is also often obtained by forging prescription forms that have been stolen from physicians.

Heroin Heroin is produced from morphine. During this process the total amount of the drug is reduced in volume, which facilitates smuggling and makes the heroin more potent per ounce than the mor-

Ghetto poster: "It's just so good, so out of sight. There's nothing in the world like it." It is impossible to lift oneself out of the ghetto while fighting drug addiction.

phine that went into the process. In studies carried out in England by Dr. J. H. Willis, nondrug users and drug users were unable to tell whether they were being given morphine or heroin. Heroin enters the central nervous system more easily than morphine. Consequently, heroin seems to be a more potent drug. Heroin may vary in color according to where it is produced. Mexican heroin is grayish-brown to brown and is usually sold in a powdered form. White heroin is produced in an area of Southeast Asia known as the Golden Triangle, which includes parts of Burma, Thailand, Laos, and China. Most heroin is diluted with milk sugar (lactose), mannite (a mild laxative), procaine (a local anesthetic used by dentists), or quinine. Very often dangerous chemicals and drugs such as strychnine, LSD, phencyclidine (PCP), or amphetamines are used to adulterate heroin. Street heroin often contains no more than 1 to 4 percent pure heroin. Purer heroin is sold to middle- and upper-class users at much higher prices. Such heroin is too strong to inject, and users "snort" it (inhale it through the nose) or smoke it in a water pipe.

● ASK YOURSELF
_____

1. What is an anesthetic? How are anesthetics used in medicine? Have you ever had to have anesthesia for medical purposes? Can you describe your emotions and feelings while it was taking effect?

2. Why is PCP classified as an anesthetic? Have you ever been exposed to PCP? Explain the reactions of the individual under the influence of PCP.

3. Explain why anesthetics are so dangerous to use for recreational purposes.

TABLE 5.2 CONTROLLED SUBSTANCES AND EFFECTS

Drugs	Schedule*	Often prescribed brand names	Medical uses	Dependence potential§	
				Physical	Psycho-logical
Narcotics					
Opium	II	Dover's Powder, Paregoric	Analgesic, antidiarrheal	H	H
Morphine	II	Morphine	Analgesic	H	H
Codeine	II III V	Codeine	Analgesic, antitussive	M	M
Heroin	I	None	None	H	H
Meperidine (Pethidine)	II	Demerol, Pethadol	Analgesic	H	H
Methadone	II	Dolophine, Methadone, Methadose	Analgesic, heroin substitute	H	H
Other narcotics	I II III V	Dilaudid, Leritine, Numorphan, Percodan	Analgesic, antidiarrheal, antitussive	H	H
Depressants					
Chloral hydrate	IV	Noctec, Somnos	Hypnotic	M	M
Barbiturates	II III IV	Amytal, Butisol, Nembutal, Phenobarbital, Seconal, Tuinal	Anesthetic, anticonvulsant, sedation, sleep	H	H
Glutethimide	III	Doriden	Sedation, sleep	H	H
Methaqualone	II	Optimil, Parest, Quāālude, Somnafac, Sopor	Sedation, sleep	H	H
Meprobamate	IV	Equanil, Meprospan, Miltown, Kesso-Bamate, SK-Bamate	Antianxiety, muscle relaxant, sedation	M	M
Other depressants	III IV	Dormate, Noludar, Placidyl, Valmid	Antianxiety sedation, sleep	P	P
Stimulants					
Cocaine†	II	Cocaine	Local anesthetic	P	H
Amphetamines	II III	Benzedrine, Biphetamine, Desoxyn, Dexedrine	Hyperkinesis, narco-lepsy, weight control	P	H
Phenmetrazine	II	Preludin	Weight control	P	H
Methylphenidate	II	Ritalin	Hyperkinesis	P	H
Other stimulants	III IV	Bacarate, Cylert, Didrex, Ionamin, Plegine, Pondimin, Pre-Sate, Sanorex, Voranil	Weight control	P	P
Hallucinogens					
LSD	I	None	None	N	?
Mescaline	I	None	None	N	?
Psilocybin-psilocyn	I	None	None	N	?
MDA	I	None	None	N	?
PCP‡	III	Sernylan	Veterinary anesthetic	N	?
Other hallucinogens	I	None	None	N	?
Cannabis					
Marijuana Hashish Hashish oil	I	None	None	?	M

* Scheduling classifications vary for individual drugs since controlled substances are often marketed in combination with other medicinal ingredients.
† Designated a narcotic under the Controlled Substances Act.
‡ Designated a depressant under the Controlled Substances Act.

§ H = High, M = Moderate, N = None.
P = Possible. ? = Unknown.

    Some individuals are able to use heroin off and on for years without becoming addicted. Regular users, however, become addicted faster than users of any other narcotic. The body's tolerance to heroin builds up very rapidly. Thus, an addict requires increasingly larger doses to get the desired effect and to prevent withdrawal illness.

Synthetic narcotics The synthetic narcotics differ

Tolerance	Duration of effects (in hours)	Usual methods of administration	Possible effects	Effects of overdose	Withdrawal syndrome
Yes	3 to 6	Oral, smoked			
Yes	3 to 6	Injected, smoked	Euphoria, drowsiness, respiratory depression, constricted pupils, nausea	Slow and shallow breathing, clammy skin, convulsions, coma, possible death	Watery eyes, runny nose, yawning, loss of appetite, irritability, tremors, panic, chills and sweating, cramps, nauseau
Yes	3 to 6	Oral, injected			
Yes	3 to 6	Injected, sniffed			
Yes	3 to 6	Oral, injected			
Yes	12 to 24	Oral, injected			
Yes	3 to 6	Oral, injected			
Probable	5 to 8	Oral	Slurred speech, disorientation, drunken behavior without odor of alcohol	Shallow respiration, cold and clammy skin, dilated pupils, weak and rapid pulse, coma, possible death	Anxiety, insomnia, tremors, delirium, convulsions, possible death
Yes	1 to 16	Oral, injected			
Yes	4 to 8	Oral			
Yes	4 to 8	Oral			
Yes	4 to 8	Oral			
Yes	4 to 8	Oral			
Yes	2	Injected, sniffed	Increased alertness, excitation, euphoria, dilated pupils, increased pulse rate and blood pressure, insomnia, loss of appetite	Agitation, increase in body temperature, hallucinations, convulsions, possible death	Apathy, long periods of sleep, irritability, depression, disorientation
Yes	2 to 4	Oral, injected			
Yes	2 to 4	Oral			
Yes	2 to 4	Oral			
Yes	2 to 4	Oral			
Yes	Variable	Oral	Illusions and hallucinations (with exception of MDA); poor perception of time and distance	Longer, more intense "trip" episodes, psychosis, possible death	Withdrawal syndrome not reported
Yes	Variable	Oral, injected			
Yes	Variable	Oral			
Yes	Variable	Oral, injected, sniffed			
Yes	Variable	Oral, injected, smoked			
Yes	Variable	Oral, injected, sniffed			
Yes	2 to 4	Oral, smoked	Euphoria, relaxed inhibitions, increased appetite, disoriented behavior	Fatigue, paranoia, possible psychosis	Insomnia, hyperactivity, and decreased appetite reported in a limited number of individuals

Source: Drug Enforcement, 2 (Spring 1975), 20–21.

from the opium derivatives and their compounds in that they are laboratory products made not from opium, but from coal tar or petroleum products. Some of the more common synthetic compounds are marketed under the names Darvon, Percodan, Demerol, Percobarb, methadone, and Nalline. Their chemical properties resemble those of various opium derivatives; their narcotic effect (and addictive potential) varies. But all narcotics, including synthetic ones, are addictive.

Darvon (Propoxyphene) Dangers

Between 1,000 and 2,000 deaths a year have been traced to the use of Darvon and other propoxyphene-containing compounds. Some are intentional overdose deaths, while others are caused accidentally by taking propoxyphene with other drugs or alcohol. Darvon is one of the most widely prescribed pain-killers in the United States today. In 1977, 33.5 million prescriptions were written for Darvon or other drugs containing propoxyphene. Propoxyphene is also found in some over-the-counter drugs.

These drugs are extremely dangerous when overused or taken along with tranquilizers, sedatives, or alcohol. Many overdose deaths occur when people take Darvon for pain and then drink alcohol. Or, they may take Darvon to alleviate the pain of a hangover. Because of the many overdoses, the U.S. Food and Drug Administration cautions all individuals to follow these three recommendations:

1. Never use Darvon and other propoxyphene-containing products unless there is no alternative, and then only with care.
2. Pharmacists should warn customers of the dangers of using Darvon along with tranquilizers, sedatives, or alcohol.
3. Patients should not ask for Darvon.

Source: U.S. Food and Drug Administration, "Darvon Study Continued," *FDA Consumer, 13,* No. 5 (July-August 1979), 3.

Volatile Chemicals

The practice of inhaling the vapors of *volatile chemicals* (chemicals that evaporate readily at room temperature) is also a major social concern. The solvents in plastic or model-airplane cement are volatile chemicals and are often inhaled for their mood-modifying effects. These effects are primarily feelings of well-being, cheerfulness, euphoria, and excitement—feelings that closely simulate the early stages of alcohol excitement. As a person inhales more, he or she begins to appear "drunk," becomes disoriented, and speaks in a slurred manner. Such behavior may continue for 30 to 45 minutes, followed by drowsiness, stupor, or unconsciousness. Unconsciousness may last for as long as an hour. If the person has inhaled too much glue vapor, or if exposure to the vapors has been prolonged, the person may die.

Several toxic solvents are used in the manufacture of airplane cements. Common to many brands are isoamyl acetate and ethyl acetate. Other toxic solvents used in many products include benzene, toluene, and carbon tetrachloride. High concentrations of these solvents may be found in cleaning fluids, paints, and paint thinners. Also, the hydrocarbons in gasoline (such as butane, hexane, and

pentane) may cause solvent intoxication when inhaled. Prolonged inhalation of the fumes of any of these fluids may cause death. Labels on many types of solvents and gasoline include the warning "Use only in a well-ventilated, open area."

Tolerance to solvents develops rapidly. The user must soon inhale the vapors from the contents of several tubes of cement to experience the effects desired.

Nitrites *Amyl nitrite* is a volatile liquid, sold under prescription, in thin glass ampules covered with a nylon net so that they can be broken ("popped") and inhaled (they are known as "poppers"). Amyl nitrite dilates blood vessels and relieves the chest pains (angina pectoris) associated with some heart conditions. It also dilates the blood vessels in the brain and throughout the body. Some individuals crush and inhale poppers in an attempt to prolong the intensity of the sexual orgasm. The blood flooding the brain is said to give this sensation. This may also cause a throbbing headache, facial flushing, nausea, or vomiting; it is almost guaranteed to reduce the beauty of the orgasm.

A drug very similar to amyl nitrite, *butyl nitrite,* is being sold over-the-counter as liquid "room odorizers" under trade names such as Locker Room and Rush. Butyl nitrite is inhaled and produces effects and side effects similar to those of amyl nitrite. Both drugs can be dangerous, and even deadly, if the person using them has a heart condition or is asthmatic.

Hypnotic-Sedative Drugs

Each drug in this group depresses the central nervous system and results in a condition resembling sleep. The difference between a hypnotic and a sedative is the degree of depression. A hypnotic drug, given in a moderate or even a small dose, will produce sleep soon after it is given. Such reduced dosages of sedative drugs, even when administered several times a day, will calm a person without producing sleep. With increasing dosages, all of the drugs in this group produce a continuum of effects that range from tranquilization to sedation (the allaying of excitement), to the loss of psychomotor efficiency, to sleep, and then to coma and death.

Barbiturates The barbiturates were first used in the United States as hypnotic-sedatives in the form of barbitol, developed in 1903. Since that time, many new barbiturate compounds have been produced and put on the market. Physicians prescribe them mainly to help patients sleep. On the illegal market, barbiturates are known as "sleeping pills," "goof-balls," "reds" (because of the usual capsule color), "downers," or "stumblers."

Barbiturates produce a surprisingly variable effect in the brain and nervous system. After taking these drugs, many users undergo a variable period of hyperactivity and excitement. Then, as the drug depresses the central nervous system, they become relaxed, euphoric, and sleepy. But a user may take a dose of barbiturates at bedtime and discover that it has no sedative or hypnotic effect at all. The initial period of hyperactivity and excitement may last throughout the night, and no sedative or hypnotic effect ever takes place. For some users, certain barbiturates produce a "truth serum" effect in which long-forgotten events are remembered. With abusive dosages, drastic and sudden mood changes may occur. Users are often described as friendly one minute and mean the next, much as with another sedative drug—alcohol (see Chapter 6).

Barbiturates are addictive drugs when abused. Tolerance develops quickly, and physical dependence eventually develops when large dosages are taken over a prolonged time. Barbiturates are usually taken orally ("dropped"); however, users can dissolve the compound and inject it with a hypodermic needle, but it is very destructive to body tissues when injected. Sometimes the capsules are dropped in combination with a stimulant such as Benzedrine, Dexedrine, or Methedrine. This combination overcomes the depressing effects of the barbiturates and extends the excitement and euphoria. The use of a stimulant drug to counteract the depressant effect of a barbiturate is extremely dangerous. The cardiocirculatory system often cannot withstand such drastic "ups" and "downs," and a *stroke* or *heart attack* may ensue.

Combining barbiturates with alcohol is a very risky practice. Because the barbiturates interfere with the body's normal disposal of alcohol through the liver, the two drugs taken together have a total depressant effect far greater than the sum of their individual effects. Often, an overdose of either drug is taken unknowingly; the person is "too drunk" or "too doped up" to realize what he or she is doing. Because of this confused state, it is difficult to tell whether a fatal drug overdose was suicide or accident. The use of alcohol and barbiturates in combination, even in small amounts, is extremely dangerous and often results in death.

Since both are hypnotic-sedative drugs, the effects of barbiturates and alcohol are very similar. A small amount of barbiturates makes the user feel relaxed, sociable, and good-humored, but less alert than usual. After taking more of the drug, the user becomes sluggish, gloomy, and quarrelsome. The tongue becomes "thick," and the user gradually falls into a deep sleep. If a large amount of the drug has been taken, especially in combination with alcohol, the deep sleep may progress into a coma. At this point, only prompt medical attention can save the person's life. Such attention has saved the lives of people who showed no sign of life after lapsing into a barbiturate-induced coma.

The chronic user of barbiturates eventually finds that the dosage must be increased in order for the drug to be effective and to prevent withdrawal. Without a regular, daily dose, an addict will experience the *alcohol-barbiturate abstinence syndrome*. This includes hallucinations, mild-to-severe *delirium tremens*, and convulsive seizures that resemble *grand mal* epileptic convulsions. Often these are severe enough to cause death. Alcohol-barbiturate withdrawal is far more serious than nar-

● ASK YOURSELF

1. What does the term "volatile" mean? Have you ever had any experiences with volatile solvents? If so, explain the sensations they produced in you.
2. Explain the differences between a hypnotic and a sedative.
3. How dangerous is the withdrawal from barbiturates?

cotic-solvent withdrawal. A physician treating someone in barbiturate withdrawal must know the name of the drug (or combination of drugs) the individual was using.

Sedativelike drugs These substances are highly psychoactive and produce an extremely wide range of effects, many of which are similar to the barbiturates. Often, they depress the central nervous system as much as an anesthetic would. Because of their marked, varying effects, it is difficult to place them into standard depressant categories. Thus, they are not found in the continuum of drug actions. The most widely abused of this group is methaqualone.

Methaqualone A nonbarbiturate hypnotic-sedative, methaqualone was first introduced in 1950 and widely used, mostly in Europe, as a sleeping pill and sedative. It is marketed under brand names such as Quaalude, Sopor, Parest, Optimil, and Somnafac. The commonest street names, such as "Q's," "ludes," "luding," "Sopers," and "soapers," are derived from the brand names.

Methaqualone gained popularity because it was believed to be safe and nonaddictive. This is not true. Individuals do become addicted, tolerance is produced; and when withdrawn they suffer from insomnia, abdominal cramps, headaches, and nightmares. An overdose can result in delirium, restlessness, and muscle spasms leading to convulsions and death.

Methaqualone has mistakenly been called an *aphrodisiac,* or "love" drug. Users feel relaxed, friendly, and uninhibited (which effects are mistaken for improved sexual pleasure and sexual performance). Research has found, however, that it actually lowers the ability to perform sexually.

Tranquilizers

Tranquilizing drugs prevent or relieve uncomfortable emotional feelings. They relieve tension and apprehension and promote a state of calm and relaxation. They have a dramatic effect in calming violent, overactive, psychotic individuals.

In the early 1950s, the term *minor tranquilizers* was introduced to distinguish those that reduce anxiety, tension, and agitation from the *major tranquilizers,* which are used to control psychotics.

Major tranquilizers These drugs do not cure mental illnesses, but because they can modify moods and alleviate many symptoms, they are extremely important in making mental patients easier to manage and control. Major tranquilizers are in reality "antipsychotic" drugs, reversing the processes of mental illness in specific groups of individuals. They are not abused to any extent because of their mild actions in all but this one group.

Minor tranquilizers These drugs are the most widely used prescription drugs in the United States. At present, there are more minor tranquilizers sold than sleeping pills, amphetamines, and narcotics put together.

Meprobamate, synthesized in 1950, was the first minor tranquilizer. Today, under its own name, and the trade names Miltown, Equanil, Kesso-Bamate, Meprospan, and SK-Bamate, over 250 tons are sold each year. Other frequently used minor tranquilizers are Librium or Librax (chlordiazepoxide), and Valium (diazepam). Medically, these drugs are used to relieve tension, anxiety, behavioral excitement, and insomnia; they have been effective for acute periods of depression (as after the death of someone close, divorce, and so on).

After repeated use, tolerance develops and dosages must be increased to obtain desired results. But no tolerance to a potentially lethal dosage develops. Users must therefore be cautious because only a limited quantity of these drugs can be taken safely. Abuse by the general public is becoming so widespread that the American Medical Association has warned doctors about overprescribing and allowing prolonged unsupervised use.

The Cannabis Drug Family

Tetrahydrocannabinol (THC) is the psychoactive substance obtained from the common hemp plant (*Cannabis sativa*), grown throughout the world.

● ASK YOURSELF

1. What are the differences between major and minor tranquilizers? Have you ever used a tranquilizer? Was it a major or minor tranquilizer?
2. What form of cannabis is marijuana? What is the most potent form of cannabis?
3. What is the current status of marijuana use in your group of friends? How do you feel about marijuana use?
4. Explain the relative effects of marijuana as shown by the Continuum of Drug Effects diagram. Have you observed these effects in people who use marijuana?

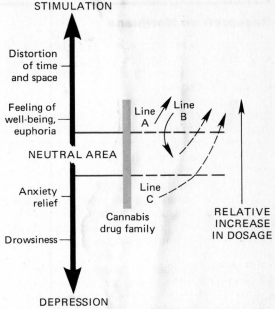

The effects of cannabis consumption lie on a continuum. Various reactions to cannabis are shown by Lines A, B, and C. The effects of cannabis depend on its strength, the amount consumed, the emotional state or "set" of the user, and the setting in which it is taken. Line A represents the experience of taking either a large dose of cannabis or a potent form of it. The solid section of Line B shows the effect of a moderate, "social" dose on a person already stimulated by drugs or the surroundings. The broken section of Line B shows the effect of consuming even more cannabis. Line C shows the effect of any amount of cannabis on a person who is relaxed at the start.

The plant is grown extensively in Jamaica, Mexico, Colombia, Africa, India, and the Middle East. Many botanists consider all hemp plants to belong to a single species with many varieties. The leaves and flowering tops of the plant contain an amber-colored resin, a mixture of many chemicals, one of which is THC. The potency of the intoxicating drugs produced from the cannabis plant varies widely, depending on which plant variety, plant part (stems, roots, and seeds do not contain tetrahydrocannabinol), method of preparation, and storage are used.

The effects of this drug family, more than any other, cannot be described accurately without specifying dosage levels. *Marijuana,* a popular name for the plant itself, is the drug prepared by drying the leaves and flowering tops of the plant to make a tobaccolike material. The most potent preparation of cannabis is *charas* (prepared mainly in India) or *hashish*. Charas is the pure tetrahydrocannabinol resin obtained from the dried flowers of *Cannabis indica* (a specific variety of cannabis). Hashish, as the term is used correctly, is a powdered and sifted form of charas. It is a chalky brown or black substance.

The different forms of cannabinol can be used in many ways. While marijuana and hashish are usually smoked, they can also be baked into foods or added to drinks. Taken in large, strong doses, the cannabis drug family bears many similarities to the hallucinogenic drugs, such as LSD. This is why, for the last few years, it has been classified as a *hallucinogenic drug* and will continue to be classified as such by many experts. The effects from a low, "social" dose of marijuana are quite different. They tend to approximate mild intoxication, with some reactions similar to those produced by alcohol. Because of these findings and recent research on and reports of the effects of cannabinol drugs, the authors feel it should be given a classification distinct from all other drug families. The cannabis drug family has been placed across the "neutral area" of the drug continuum because of its wide range of actions and effects. These depend on: the form of the drug used, the amount used at one time, and many "variables" such as emotional state, "set," "setting," personality, and social factors that affect the individual's response to THC. Depending on these factors, cannabis intoxication resembles that produced by either hallucinogens (stimulants) or sedatives such as alcohol (depressants).

Research on Marijuana

Much research on the effects of marijuana has been done recently. Robert Petersen, assistant director of research at the National Institute of Drug Abuse (NIDA) reported to a session of the National Drug Abuse Conference on some implications of this research.

As Dr. Petersen explained, "one of the unfortunate aspects of the marijuana controversy, to a certain extent, is that the value of marijuana as a therapeutic drug, in my judgment at least, is overblown." Marijuana had been used for centuries for a variety of conditions. Its use declined at the end of the nineteenth century for reasons that are still relevant today. Marijuana in any form is fairly unstable; it breaks down very rapidly in the body and is insoluble in water. Thus, it produces unpredictable and highly variable results.

Several hundred cancer patients have taken marijuana (or its derivatives) to help ease the nausea that often accompanies chemotherapy. In almost every case it was used in conjunction with, and not as a substitute for, conventional nausea therapy. Alone, marijuana's effectiveness to ease nausea is as low as 30 percent effective.

Marijuana's uses in reduction of nausea in cancer therapy is irrelevant to the desirability of its use as a recreational drug. The public should be cautioned against simplistic arguments for removing marijuana as a scheduled drug because of its medical benefits.

Source: Harvey McConnell, "Pot's Medical Value 'Overblown,' " *Journal of the Addiction Research Foundation,* 8, No. 10 (Toronto) (October 1, 1979), p. 3.

The most immediate effect of marijuana, besides its alteration of sense perception, is on thought processes. It is not uncommon for marijuana smokers to begin a sentence and fail to end it because they have forgotten what they were talking about. Additionally, research has shown that long-term male marijuana users suffer from a reduction in the hormone testosterone. This reduces their ability to produce effective sperm and may cause some of these individuals to be sterile. Women marijuana smokers are very likely to have abnormal menstrual cycles, which can interfere with their ability to conceive. Whether these affects are permanent or transitory is not yet known. One permanent detrimental effect that has begun to appear in many long-term marijuana smokers is impaired lung function. Younger and younger individuals are suffering from emphysema—and permanent lung damage—as a result of marijuana use. Controlled research studies concerning marijuana are currently in progress. As time passes, additional information will provide a more complete picture of the implications of cannabis use at various dosages and patterns of use.

Hallucinogens

Hallucinogens are drugs that create vivid distortions of the senses without greatly disturbing the individual's consciousness. Such distortions (hallucinations) may cause persons to see, hear, or smell things that are not really there. Persons who abuse the hallucinogens may, like some of the mentally ill, experience hallucinations when they do not want to or when they are no longer under the influence of a drug. This is why some hallucinogens have been termed *psychotomimetic* (psychosis-mimicking) or psychotogenic (psychosis-producing) drugs. Such drugs, some authorities feel, are capable of temporarily turning a normal person into a psychotic. Users have often been hospitalized to prevent them from doing harm to themselves or others during what seems to be a temporary psychosis.

Mescaline Mescaline is found in the small buttonlike cactus plant called *peyote*. Peyote plants are usually dried and then chewed; sometimes they are boiled in water to make a broth. Mescaline can produce intense hallucinations and euphoria that

Brain Damage from Cocaine

In 1978, Dr. Nils Noya, director of the Bolivian National Institute for Investigation of Drug Dependence, reported that he had treated 500 individuals with severe brain damage from using cocaine. He estimates that there are at least 3,000 more not yet detected.

Cocaine *traffickers* (sellers or those who possess drugs for sale) in Bolivia and Peru have access to almost pure cocaine paste. These individuals inhale this paste after it dries or mix it with tobacco and smoke it. Dr. Noya reports that within a short period of time (at most six months) after they start smoking or inhaling pure cocaine at least four times a week, users lose control of their intellectual powers. Their deductive reasoning powers disappear. Even a simple mathematical problem confounds them. They suffer from depression (cocaine paranoid psychoses) and *formication* (the sensation of small insects crawling on the skin); very severe scars are caused by the scratching. They become very dependent, suspicious, and aggressively defensive.

Dr. Noya adds: "Our theory is that cocaine attacks directly the cells in the brain, especially the alpha cells. We think the intracellular glucose is extracted by the cocaine," killing the cells. "We have to face the fact that cocaine is a hard drug. It can collapse you as a human being. From what we have seen, a person who becomes a cocaine addict cannot live normally again."

Source: "Pure Cocaine 'Proves Devastating'; Nils Noya," *Journal of the Addiction Research Foundation* (Toronto) 7, No. 1 (January 1, 1978).

last between eight hours and two days. These hallucinations may consist of fantastic geometric patterns; users may experience a distortion in their sense of time and space, and feelings of depersonalization.

Pure mescaline, which is extracted from peyote, is available on the illegal market in capsule form. The caps, of course, are much easier to take, although some users experience slight nausea at the beginning of their "trip." Mescaline is not physically addictive.

LSD LSD, commonly referred to by users as "acid," is a tasteless, colorless, and odorless drug derived from lysergic acid diethylamide. The primary danger in taking LSD is that it may cause temporary psychosis and a wide range of behavioral disturbances. Some users experience panic or depression, while others feel euphoria and a sense of great mental clarity or comprehension. Visual hallucinations are commonly experienced.

LSD dilates the pupils of the eyes, raises the blood pressure, stimulates the brain's sensory centers, and blocks off its inhibiting mechanisms. It intensifies hearing, increases the ability to differen-tiate among textures, and may produce a tingling sensation and numbness of the hands and feet. Subjects often report crossovers of sensation; for example, they may seem to hear colors or smell the scent of music (an experience known as *synesthesia*).

Cocaine

Cocaine is extracted from the flowers of the coca plant (*Erythroxylon coca*). This plant grows to heights of 12 or 15 feet. It is cultivated in the Andean highlands of Bolivia, Chile, Colombia, Ecuador, and Peru. In these areas its leaves have been chewed, or drunk as a tea, for relief of fatigue since prehistoric times.

Illicit cocaine is extracted from coca leaves in the Andean highlands and converted into coca paste. This paste is then refined into processed cocaine in Bolivia and Peru. Local users of pure cocaine suffered brain damage so severe that they are unable to care for themselves for the rest of their lives.

Cocaine is the most powerful natural stimulant known. Its general effect on the body is to stimulate, alter mood and behavior, and induce excite-

Death from Cocaine Use

Cocaine can cause death by three routes:

1. *Circulatory collapse.* This is an allergic reaction to cocaine; it is a type of shock in which fluids leave the circulatory system, causing it to collapse. This reaction can occur in any person at any time while using cocaine.
2. *Overdose.* Death is caused by respiratory failure (the person just stops breathing). The person should be given mouth-to-mouth breathing with external heart massage or cardiopulmonary resuscitation [CPR] until he or she can be connected to a respirator.
3. *Inability to control body temperature.* Cocaine can stimulate the temperature-control center in the brain and cause heavy sweating. When, due to the constriction of blood vessels in the skin, the individual can no longer sweat, his or her body temperature can rise dangerously and cause brain damage or death. Any cocaine user who begins to sweat should be taken to a hospital emergency room at once.

ment. Cocaine users usually sniff it into their nostrils. A few take it by hypodermic injection. Sniffing is the more popular method, but it is highly destructive to the tissues lining the nose and respiratory tract because the drug is absorbed slowly through the membranes of the nose. When sniffed, its effects last longer and are less violent than when injected.

Some users mix cocaine with a volatile solvent solution and burn out the impurities in a process called "free basing." This almost pure cocaine is then smoked in a water pipe. The free-base mixture is highly explosive, and a number of persons have been burned while preparing it.

● ASK YOURSELF

1. Where is cocaine located on the Continuum of Drug Effects diagram? Why is it placed at this point?
2. What is the extent of cocaine use in your group of friends, college, neighborhood, or city? What do you think of this use?
3. How is cocaine usually taken? Can death occur from the use of cocaine? Have you known of someone who died from cocaine use? How did death occur?

Amphetamines

Included in the stimulant group is a large number of drugs that mimic the actions of *adrenalin.* In general, the physical reactions of amphetamines include increased heart rate, a constriction of certain blood vessels, increased breathing rate, increased perspiration, and a cottonlike dryness of the mouth. At the same time, amphetamines act on the brain by increasing bodily activity and elevating mood. Feelings and behavior aroused by amphetamines include increased confidence, euphoria, fearlessness, talkativeness, impulsive behavior, loss of appetite, and a decrease of fatigue.

Amphetamines have been used for years for appetite control because they suppress hunger. There are two dangers in prolonged use of amphetamines for weight control. First, tolerance to the appetite-suppressant characteristics of the drug develops very quickly. Even moderate dosages lose their ability to control appetite within four to six weeks. Second, overeating is a behavioral problem, just like drug abuse. Overeating is primarily controlled by psychological factors, not by the physiology of the body. Consequently, amphetamines reinforce the dependent behavior of the individual, simply transferring it from food to the drug.

The amphetamines used for weight reduction, Dexedrine and Benzedrine, are the most widely abused. On the illegal market they are known as

TABLE 5.3 CATEGORIES OF DRUG ABUSE

Experimenters	Occasional users	Regular users	Compulsive abusers
Usually use drugs not more than three times.	Use drugs for social, personal, and emotional gratification.	Use drugs one or more times a week.	Have abnormal personalities or severe emotional problems.
Half never use drugs again.	Are very socially conscious.	Belong to drug subcultures.	Use drugs to cope with stress.
May have hidden emotional problems.	Use any current "in" substance.	Defend their personal right to use drugs.	When using drugs, manifest dramatic personalities, mood, and behavioral changes.
May be convicted for drug possession for one-time use.	Are "social drinkers."	May be occasional heavy users.	Represent a serious drug problem.
Have no drug problem.	Have no serious drug problems.	Can easily become compulsive users.	

"bennies," "dexies," "pep pills," or "whites" (because Benzedrine is often sold as a white tablet), or as "uppers" or "leapers" because of the mood elevation they produce. Several drug companies, without showing substantial evidence, claim that their particular compound suppresses the appetite without stimulating the central nervous system. No amphetamine or amphetaminelike compound has only one of these two effects on the body. The usual circumstance is that while users lose weight, they also lose sleep.

Many do not consider amphetamines to be addicting because they do not produce a classical withdrawal syndrome. This distinction results from a disagreement over the definition of addiction, physiological dependency, and withdrawal. During amphetamine withdrawal, the individual exhibits extreme apathy, decreased physical activity, and sleep disturbances that can last for weeks or months. Suicides have occurred during amphetamine withdrawal, and users should therefore be under medical supervision during this time.

CLASSIFYING THE DRUG-ABUSE PROBLEM

The term *mood-modifying* describes the behavioral effects of psychoactive drugs and is not a chemical or medical classification. Not everyone who takes psychoactive drugs will follow the same predictable pattern of behavior. But all people who try psychoactive drugs belong to one or another of the categories outlined in Table 5.3. Some may only experiment because of social influences (peer pressures), curiosity about the reported pleasant effects, or because of short-term emotional problems.

These individuals should be termed *drug experimenters*. Unfortunately, they can be arrested for experimenting and may carry a criminal record for the rest of their lives.

Individuals who experiment and continue to use psychoactive substances find they desire the mood-modifying effects, or the accompanying social environment drugs give them. This is a form of psychological dependence. These people use such substances infrequently, yet consistently, which is usually more than once a month but less than several times a week. (Individuals who use alcohol in this manner are called "social drinkers.") Many use marijuana in relatively low dosages and in a similar pattern of use; this is called the *social*, or *recreational use* of marijuana. These individuals should be termed *occasional*, or *intermittent, users*.

People who use psychoactive drugs daily or several times a week, extending over a long period of time, are *regular* users. For example, the cigarette smoker who smokes between eight and twenty cig-

● ASK YOURSELF

1. Explain why amphetamines are placed at their specific point on the Continuum of Drug Effects diagram.
2. What are the most commonly abused amphetamines? Have you or any members of your family ever used amphetamines, either medically or for recreational purposes? How do individuals behave when using these drugs? What physical effects do these drugs cause? What effects does withdrawal produce?
3. How dangerous is amphetamine abuse.

● ASK YOURSELF

1. What are the "mood-modifying" effects of drugs? What purpose do they serve? Have you ever used a drug for these effects? Do you use drugs regularly for these effects?
2. Explain the differences between experimenters, occasional users, regular users, and compulsive users. Have you ever known a compulsive user? If so, explain her or his reactions and behavior while using drugs.
3. If you fit into any group other than experimenters, explain why you use the drugs you use.

arettes a day is a regular user. A two-pack-a-day smoker is a heavy user. Someone who takes tranquilizers daily to reduce the effects of the environment is a regular user. Persons who use amphetamines daily because they like the extra energy produced are regular users. People who have a beer or a mixed drink regularly to "relax" or "slow down" are also regular users.

Often regular users lose their ability to *control* their use of drugs. Many become *poly-drug users* and begin to experiment with a wide variety of drugs. Or, they use the drug currently in fashion. For them drug use becomes not just an intermittent episode in the pursuit of a happy life but an essential part of a "turned-on" ideology, life style, and subculture. Often these users are the greatest defenders of the personal right to use drugs. They seldom have had bad experiences with drugs and have closed minds concerning facts about drugs because they have evaluated the situation solely on the basis of their own experimentation. The person who says, "I know because I have been there," is in a dangerous phase of regular drug abuse.

The most dangerous aspect of regular drug use is *personality deficiency*—that is, individuals are unable to control their drug use or deal constructively with life. The self-destructive dependency of a cigarette smoker is a good example. Increased dependency on psychoactive drugs eventually leads to compulsive drug abuse, very heavy drug use, addiction, or alcoholism—all of which are different names for a single basic problem.

Compulsive abuse (addiction, alcoholism) of mood-modifying substances is always associated with an abnormal personality. Whether this behav-

ioral distortion is the cause or the effect of the drug abuse is not known. The abnormal social behavior and modification of moods in such individuals is more of a problem to society than is the drug abuse itself. Responsibility for drug abuse lies solely with the individual.

In all cases of compulsive drug abuse, the person who misuses drugs chooses to do so. Once begun, such abuse leads to psychological dependency as a response to the individual's personality needs. On the other hand, the person who does not find a pleasurable experience in drug use will not continue to use them. Also, if someone's only reason for experimenting was strong social pressure, he or she will probably not continue to use drugs. These individuals are not seeking what drugs have to offer and therefore will reject them.

DRUG-ABUSE BEHAVIOR

The use of legal or illegal psychoactive drugs for recreational purposes follows the same rules and principles as any other type of human behavior. Behavior patterns persist when they either increase pleasure or reduce discomfort. People do not choose just any drug—they want substances that give them a pleasurable experience.

Drug use is an important *coping mechanism,* and it becomes a problem when it is the only coping mechanism the person uses. In humans, adaptation to the environment is much more complicated than in simpler animals. We have the ability to reason and make decisions based upon learned behavior and our physiological and emotional needs. Why do we do the things we do? Why do we feel the way we do? The answers to these questions come largely from our understanding of how basic human needs are fulfilled and how we *cope* with the frustrations of unfilled needs.

Basic Human Needs and Drug Abuse

The most basic human needs are the *physiological needs* (food, water, sleep, and sexual satisfaction). Those in whom these needs are unfulfilled will sometimes turn to drugs as a substitute. They may consume only the amount of food and water needed to stay alive. They often go for days without eating. Drug abusers often suffer from insomnia and wander about at odd hours. Sexual drive decreases

Drug-abuse behavior. Individuals do not choose just any drug — they want a drug that gives them a pleasurable experience.

in regular and compulsive drug abusers and alcoholics.

Curiosity, or the *need to know and understand,* is often the major reason for first drug experiments (especially among young people). Often, their only knowledge of drugs has been obtained from the street and is highly inaccurate.

The *need for comradeship and belonging* is very important to all of us. People feel a strong need for friends, companionship, and acceptance by a group. Group acceptance based on drug taking may cause an individual to progress from experimentation to being an occasional, regular, or compulsive user. Drug use is a means of establishing and reinforcing *self-esteem.* Being accepted as an "adult" confers status. Many young people also use drugs to annoy and upset their parents; others do so as a token rejection of accepted social and moral standards.

Personality and *emotional maturity* can be divided into three areas: (1) personal pleasure and gratification, (2) mature, appropriate behavior, (3) self-image ideals and personal judgment of right and wrong. Maturity acts as a regulator between the need for pleasure and satisfaction and the demands of conscience (personal evaluation of acceptable behavior). An individual matures as he or she learns to respond to emotional, social, and environmental pressures while still attaining satisfaction and pleasure.

Aggressive or dominating behavior is seen by many authorities as a childish means of obtaining pleasure. Young children enjoy being aggressive. In maturity, aggression can be expressed in both harmful and beneficial ways. Drug abuse is often a self-inflicted form of aggression.

Mature people learn to balance their social and personal responsibilities with desires for pleasure. Many regular and all compulsive drug abusers cannot control their behavior adequately and operate on an infantile personality level, a significant aspect of mental illness.

Coping

Drugs, especially depressant drugs such as alcohol and heroin, are used as a *coping mechanism* by individuals who are frustrated and unfulfilled in their basic emotional needs. Drug use is directly related to the "rage of impotence" of those who are powerless to change the social and economic forces that have shaped their lives and removed all opportunities for building self-confidence and self-esteem. The ability to tolerate emotional tension and stress varies considerably among individuals. When frustrations exceed a person's ability to cope, he or she either must have relief or suffer extreme consequences — emotional illness.

Emotionally ill people experience an abnormal amount of tension and show it through nervous, exaggerated, or irrational behavior. The occasional drug taker who becomes a regular drug user is establishing a pathological pattern of drug use. These individuals, when not under the influence of drugs, become increasingly alert (hyperalert). Sounds are exaggerated, lights seem more intense, and perception is keener. As the illness progresses, they look for possible dangers and become restless (they often walk or drive aimlessly). There is increased touchiness, tearfulness, irritability, nervous laughter, moodiness, or depression.

During prolonged periods between drug taking (when "straight" or "on the wagon"), these individuals may appear to be normal and may function ac-

● ASK YOURSELF

1. Which human needs are involved in drug abuse? Interview someone who has used drugs, and compare their needs with those associated with drug abuse.
2. What is a "personality deformity" or "personality disorder"?

ceptably within society. But, under stress, they readily turn to drugs. Changes in personality are often apparent. This is why drug abuse is termed a *personality deformity* or *personality disorder*.

Before an individual ever comes into contact with drugs, he or she has an attitude either for or against drug use. This attitude generally corresponds to that of family, friends, social group, or neighborhood. Where the use of drugs is discouraged, both the availability and the abuse are low. Equally crucial to the person's continued use of drugs after experimentation is the reaction of the social group to the initial drug experience—approval or disapproval, reward or punishment, praise or ridicule. However, even a mature person needs strong social pressures and guidance to reverse the course of drug abuse after becoming dependent on drugs.

Selection of a specific drug (heroin, marijuana, alcohol) is generally dictated by factors such as availability, practices of friends, and social environment. Individuals tend to have drug experiences that fit their own basic emotional needs. Drug users who never progress beyond occasional social use may suffer physical discomfort (hangover, withdrawal, shakes), but they are able to control their emotional reaction to drugs. Psychological processes can override physical drug effects. Emotional factors become more important in drug abuse and social and physical factors less important as use becomes heavier and more compulsive. For example, depressant drugs remove an individual from stresses and anxieties. Heroin does this very quickly; alcohol, slowly, but for a longer period of time. Hallucinogenic drugs and cocaine can compensate for a lack of "peak" or "high" experiences. Amphetamines help increase physical energy.

There is no single reason for abusing drugs, no single pattern of abuse, and no inevitable outcome. Many individuals are able to use psychoactive

TABLE 5.4   SYMPTOMS OF DRUG ABUSE

The following behavior changes and patterns are common in drug abusers:

1. Drastic mood swings with no apparent cause. Individuals will swing from a normal to a euphoric state and back again and not be able to say why.
2. Preoccupation with drugs, drug language, or drug information
3. Erratic school or job attendance
4. Withdrawal from school, extracurricular activities, or hobbies
5. Erratic school performance
6. No immediate goals

drugs without harm to themselves or society. Others become chronic abusers or frequent relapsers. People desperately trying to stop abusing drugs must progressively fulfill their emotional and social needs in order to resume control of their lives.

DRUG-ABUSE TREATMENT

All successful drug-abuse treatment programs are ultimately based on individual motivation. These programs place final responsibility and hope for "cure" on the will of the individual. Changing the external environment will not be sufficient to reverse the tragic patterns of abuse that have become established in recent decades. Treatment programs must provide incentives that fulfill basic human needs and supply alternative methods of coping.

The Goals of Treatment

The usual point at which the drug abuser and society first really meet is when the user runs up against the law. Enforcement of the drug-use laws temporarily takes users out of their environment and, for the period of the jail sentence, forces them to do without drugs. Without further help, this method will not discourage individuals from using drugs. A jail sentence does not increase an individual's self-control.

A successful treatment program must change two factors and develop a third. Treatment must change the physical environment, or the *setting*. It

Drug-abuse treatment. Group treatment in a therapeutic community has forced many individuals to confront their drug-abuse problem and change their behavior.

must also change the emotional coping mechanisms, or *set,* of the individual. A program must also develop an individual's *self-control.*

Treatment Laws and Facilities

Two kinds of agencies are available for the treatment of drug abuse: private and public. The private programs mainly emphasize treatment through family members or peers. Individuals lacking the self-control to stay within the framework of a private program are placed within the legal controls of public programs. However, too often we commit people to public treatment programs out of anger and vindictiveness rather than humane concern. This is certainly shown by the penalties for drug possession. The predominant method of dealing with drug abuse has been to place abusers in jail. Following their release, they generally resume abusing drugs. We still do not have effective methods of treatment for this kind of drug abuser.

Under a federal statute, eligible individuals charged with a drug crime may have the charge against them "held in abeyance." Court proceedings stop at this point and do not proceed until later. To obtain this, offenders must submit to a medical examination to determine the extent of their addiction and whether they can be rehabilitated. Offenders have five days in which to make this decision. If they are found to be suitable candidates for treat-

● ASK YOURSELF

1. Discuss the reasons for public laws controlling drug-abuse treatment.
2. Explain the differences between a civil commitment procedure and a criminal commitment procedure.
3. Call your local district attorney's office or your local police station and ask them to explain the implications of having a felony drug conviction on your record.

ment, they are placed in a hospital through a civil commitment procedure.

The civil commitment procedure ensures control over drug abusers during treatment. Drug offenders then go to a hospital or drug treatment institution. Later they spend time in a halfway house, and still later, at home under the close supervision of a probation or parole officer. If the treatment is successful, they are brought into court, and the judge can dismiss the charges against them at this point. No criminal record is established. If the treatment is unsuccessful, the trial begins where it left off.

A judge can also require drug offenders to be treated through the criminal commitment procedure, which may or may not result in a criminal record. If offenders have been convicted of nonviolent

Society's predominant method of dealing with drug abuse is to jail the abusers. Following release, they generally resume their drug habits.

crimes, a judge can order them to be examined for drug use. If examination reveals them to be drug addicts, or in danger of becoming addicted, the judge can place them in a treatment program. The outcome of the procedure, at this point, differs from state to state. Such offenders usually end up with a criminal record for the crime they committed but not for the use of drugs.

Modes of Drug-Abuse Treatment

The jail experience can change the behavior of drug experimenters and occasional users. Given ade-

quate supervision and psychological guidance, they will give up the use of illegal drugs, if only to avoid jail. But, the treatment of regular and compulsive drug users is a completely different problem. Medical authorities find that compulsive drug users are emotionally disturbed and often physically ill from toxic drug effects. They may need emergency treatment and then long-term psychological help.

Drug-abuse therapy must be aimed at the social and psychological problems of abusers and their families. Many drug abusers (including alcoholics) would prefer not to stop using drugs, but to return to controlled, moderate drug use or drinking. But most authorities are convinced that at the present time, the return to controlled drug use is often an unrealistic or most often an impossible goal. Very few compulsive drug users, including alcoholics, have been able to return to controlled drug use. Such individuals must also be wary of their use of prescription and certain over-the-counter drugs for medicinal purposes. Many of these drugs have properties, similar to the drugs they were abusing, that could cause the drug abuse problem to recur.

No single method of treatment has been developed that significantly reduces the complex problem of drug abuse. All treatment programs consider compulsive drug abuse, addiction, and alcoholism an individual problem. Thus, all treatment programs should be available to all individuals. If offenders do not progress with one type of treatment, they should try other types until they find an appropriate one. This is called a *multimodality treatment program,* and many professionals feel it is essential to the control of drug abuse.

A great deal more research is necessary in order for our society to attain any effective cure and rehabilitation. The damage done by drug abuse is so deep and widespread that the only practical long-range solution is prevention of drug abuse.

Drug-use prevention The prevention of drug use is dependent upon some basic changes in American society.

Time Adults and society must spend more time with children and adolescents. There must be more social, recreational, and educational programs for youths. These must involve the adults of a community. Involvement by a wide range of adults gives children more *role models* to follow and reduces their fears of adulthood. Young people would thus have a more realistic view of *basic human needs* and ways, other than drug use, to fulfill them.

Drug use in the home We live in a "pill society." Pill use in the home must be more controlled. Over-the-counter drugs are so ineffective that millions of people take them without effect. Only drugs proven to be effective should be available to consumers. And young people must learn that the use of any drug is dangerous.

Drug advertising should be less dramatic and more realistic. Advertising causes young people to view all drugs as being the same as ineffective nonprescription drugs. People do not respect drugs. People should learn to view drugs as a last resort in treating disease. Drugs should be used only when the natural defenses of the mind or body are over-whelmed.

Drug-use prevention is dependent upon basic changes in society which can affect young people before they are exposed to drugs. These changes could be the beginnings of the prevention of drug abuse.

THE RETURN TO SOCIETY

Drug treatment programs can only succeed when people recognize that the compulsive user or addict has no ability to combat the ordinary stresses in life. Such a user has relied for months, perhaps years, on the external solace provided by drugs. A cure of this dependency must help redirect attitudes toward personal weaknesses. Thus, a gradual return to society is the ideal goal of successful drug-abuse treatment programs. After an individual has been treated for drug abuse and returns to society, he or she faces many personal problems, These can be social, legal, economic, and medical. Often, with drug problems, the offender and the victim are the same person. Also, the social pressures that encouraged the drug abuse in the first place are still present in this environment when the patient returns to it.

If facilities can provide for a gradual reentry into society, there is a much better chance of adjustment. Short visits home should be made first, Then a halfway house, work camp, parish house, or a day-night hospital may help after the person leaves the therapeutic community. Any of these

The ideal method of returning someone to society after treatment is a gradual reentry through controlled environments such as this halfway house.

settings is potentially useful in providing the abuser with social, therapeutic, educational, and vocational services. These give controlled contacts with the community.

But all too often, treated drug abusers leave the hospital (or more often, the jail) and are literally "dumped back on the street." Consequently, it is a very short time before they are again abusing drugs.

CONTROL AND ENFORCEMENT

Our society is seeking sound solutions to the basic causes of drug abuse. It presently relies on a certain body of laws to protect itself. Laws can be preventive measures, but the current controls are not meeting present needs. Some type of control—legal, social, or individual—is needed for the treatment of drug abuse.

Up until the middle 1960s the major laws controlled the illegal possession, manufacture, and sale

TABLE 5.5   SCHEDULES AND PENALTIES FOR VIOLATION OF THE COMPREHENSIVE DRUG ABUSE PREVENTION AND CONTROL ACT OF 1970*

Drug schedule	Potential for abuse	Dispensing controls	Example of substances in each schedule	Maximum penalties for illegal trafficking	Maximum penalties for personal possession
I	High	Research use only	Nonmedical opium derivatives, cannabis, hallucinogens	Narcotics: First offense—4 to 15 years; $25,000 fine. Second offense and subsequent offenses—6 to 30 years; $50,000 fine Nonnarcotics: First offense—2 to 5 years; $15,000 fine. Second and subsequent offenses—4 to 10 years; $30,000 fine	First offense—up to 1 year (probation possible); $5,000 fine. Second offense—up to 2 years—; $10,000 fine.
II	High	Written prescription. No refills	Medically used narcotics and injected amphetamines		
III	Moderate to low	Written or oral prescription. With M.D.'s authorization, refills up to 5 times in 6 months	Mild narcotics, noninjected amphetamines, methadone, barbiturates, and minor tranquilizers	First offense—2 to 5 years; $15,000 fine. Second and subsequent offenses—4 to 10 years; $20,000 fine	
IV	Low	Written or oral prescription. With M.D.'s authorization, refills up to 5 times in 6 months	Mild sedatives, hypnotics, narcotics, and some stimulants	First offense—1 to 3 years; $10,000 fine. Second and subsequent offenses—2 to 6 years; $20,000 fine	
V	Low	Over-the-counter. Prescription by oral order by M.D.	Restricted over-the-counter drugs. Low-percentage mixtures of narcotics, sedatives, and amphetamines	First offense—up to 1 year; $5,000 fine. Second and subsequent offenses—up to 2 years; $10,000 fine	

*Schedule and penalties may be changed by the U.S. attorney general at any time.

of drugs rather than their abuse. Treatment programs were left out completely. During the middle 1960s laws were established that governed research into the effects and actions of drugs. This permitted greater flexibility in the treatment and control of drug abuse, and the rehabilitation of drug abusers. Such trends separated treatment laws from those governing the possession and sales of drugs.

Federal Drug-Control Laws

In 1970 a schedule of federal drug penalties was established. It is known as the Comprehensive Drug Abuse Prevention and Control Act. Since 1970 there have been periodic minor changes in the law. An example is the Heroin Trafficking Act that was signed into law in 1973. It increased the penalties for traffickers in heroin and made it much more difficult for them to obtain bail for such offenses. In addition, it established five classes, or "schedules," of drugs whose manufacture, distribution, possession for use, and sale are controlled by the federal government. Table 5.5 presents an outline of the drug schedules of the Comprehensive Drug Abuse Prevention and Control Act.

Drugs contained on this table are termed *scheduled drugs*.

Schedule I

Schedule I drugs are those that have a "high potential for abuse because of their mood-modifying effects," and are not currently used in medicine in the United States. Drugs within this group are further divided into "narcotics," and "nonnarcotic" drugs. These drugs may be used in a government-approved research project but may not be possessed for any other purpose.

Schedule II

These drugs (narcotic and nonnarcotic) have the same potential for abuse as Schedule I drugs, but are currently being used in medical practice. As you can see, the fines, controls, and penalties for both Schedule I and II drugs are the same.

Schedule III

Substances placed in Schedule III are considered to have a potential for abuse that is less than Schedule I and II drugs. These drugs are

used in medical practice and are considered to produce moderate or low physical dependence. But, Schedule III substances are considered to have a high psychological dependence potential. The penalties for the illegal trafficking, distribution, and sale of Schedule III substances are less severe than for higher schedules.

Schedules IV and V

These substances have the lowest potential for abuse, the lowest penalties, and the least controls. The major difference between Schedules IV and V drugs is that Schedule IV drugs require a prescription from a physician. Most Schedule V drugs are sold over-the-counter.

As shown by Table 5.5, there are major differences in the prescription requirements of the different schedules. Also, the penalties for trafficking are different for different classes. The possession for one's own use of any controlled substance in any schedule is always a misdemeanor. The first offense is punishable by one year in jail, and up to a $5,000 fine. For a first offense, an individual user 21 years of age or younger, who is convicted of possession, may be placed on probation. If he or she successfully completes the probation, the official arrest, trial, and conviction can be erased from the record.

The director of the Drug Enforcement Administration (DEA), under the direction of the attorney general, decides which schedule a new drug belongs in on the basis of its "potential for abuse." This is also the procedure for moving drugs from one schedule to another.

State Drug-Control Laws

Since 1970 most states have followed the example of the federal government and established five schedules of controlled substances. This brings their laws into conformity with the federal Comprehensive Drug Abuse Prevention and Control Act. The penalties vary from state to state, but most are either more severe or the same as the federal penalties.

The major departure from the federal law has been in the states' control over marijuana. A number of states, led by Oregon in 1973, have "decriminalized" the individual possession of small amounts of marijuana for personal use. Oregon's law basically states that persons found in possession of up to one ounce of marijuana can be charged with a *violation,* which is similar to a parking ticket.

They face a fine of not more than $100. Transportation and possession of more than an ounce and sale or cultivation of marijuana remain a felony with a maximum penalty of up to ten years in prison.

IN REVIEW

1. A drug is any substance, other than food, that alters the body or its functions. Drugs are used for their medical value and also for recreational purposes. Drugs used for recreational purposes alter behavior.
2. Drugs are administered to cure a disease or alleviate a specific symptom such as pain or discomfort. Adverse reactions to a drug are known as side effects, or untoward effects.
   a. Drugs go under many names:
      (1) Official names
      (2) Chemical names
      (3) Generic names
      (4) Brand or trademark names
      (5) Common, street, or slang names
   b. Although self-medication may be harmful, the hazards can be minimized by education. The overuse of any drug can cause detrimental effects. Nonprescription drugs pose no threat when used as directed.
   c. Drugs can be introduced into the body in various ways. Individuals experience three kinds of effects:
      (1) Therapeutic effects
      (2) Side effects or untoward effects
      (3) Symbolic effects
3. Any drug can be abused — that is, it can be used for purposes other than those for which it was intended.
   a. Certain drugs are habituating or addicting. An addict is physically dependent upon an addicting drug. A drug user has an habituation to a nonaddicting drug. Both have a drug *dependency.*
   b. Drugs that cause problems for society can result in death, marked personality changes, or abnormal behavior. Drugs causing marked personality changes or abnormal behavior are most often referred to as psychoactive drugs. These drugs are classified as either depressants or stimulants. Their repeated use usually builds tolerance.
4. The depressing or stimulating effects of drugs on the central nervous system are independent actions. A continuum of drug actions extends from stimulation and death at one extreme to depression and death at the other. Mood-modifying drugs can be placed at specific points on the continuum.
   a. Anesthetics are the strongest depressants on the continuum. Anesthetics produce a loss of all sensations. They are so dangerous that physicians take special training to qualify to administer them safely. PCP (angel dust) is the most widely abused anesthetic.
   b. Narcotics are used in medicine to relieve pain. The major narcotics abused in North America are: morphine, heroin, and the synthetic narcotics.

    c. The solvents in plastic or model-airplane cement are volatile chemicals and are often inhaled for their mood-modifying effects. Amyl and butyl nitrite are inhaled in order to prolong the intensity of sexual orgasm.

    d. Hypnotic-sedative drugs depress the central nervous system into a condition resembling sleep. The major groups of drugs in this classification are the barbiturates and alcohol.

    e. Tranquilizing drugs prevent or relieve uncomfortable emotional feelings. Major tranquilizers make mental patients easier to manage and control. Minor tranquilizers are used to relieve tension, anxiety, behavioral excitement, and insomnia during periods of depression.

    f. The cannabis drugs are obtained from the common hemp plant (*Cannabis sativa*), grown throughout the world. The effects of this drug family, more than any other, cannot be described accurately without specifying dosage levels. Marijuana is a popular name for the plant itself. The most potent preparation is charas or hashish.

    g. Hallucinogenic drugs create vivid distortions of the senses without greatly disturbing the individual's consciousness. The most common hallucinogenic drugs are mescaline and LSD.

    h. Cocaine is extracted from the flowers of the coca plant (*Erythroxylon coca*). Cocaine is the most powerful natural stimulant known.

    i. Amphetamines are a large group of stimulants that mimic the action of adrenalin. Amphetamines have been used for weight control. They produce drastic mood elevations.

5. People who use drugs for recreational purposes can be classified into four categories.

    a. Experimenters: those who often do not use a drug more than three times; experimentation can lead to a conviction for drug possession.

    b. Occasional users: those who use drugs during social situations; they have no serious problem with drugs.

    c. Regular users: those who use drugs at least once a week; they are on the dividing line between those who have a drug problem and those who have none.

    d. Compulsive abusers: those who use drugs as the only means of coping with social stresses; they have the most serious drug problems.

6. People use psychoactive drugs either to increase their pleasure or reduce their discomfort.

    a. Drugs are used to fulfill basic human needs.

    b. Drugs, especially depressant drugs, are abused as a coping mechanism for individuals who are frustrated and unfulfilled in their basic emotional needs.

7. All successful drug-abuse treatment programs are ultimately based on individual motivation.

    a. For a treatment program to be successful, it must change two factors and develop a third.

       (1) It must change the physical environment—the setting.

       (2) It must change emotional coping mechanisms—the set.

       (3) It must develop an individual's self-control.

    b. There are two kinds of treatment agencies—private and public.

    c. Drug-abuse therapy must be aimed at the social and psychological problems of the individual abusing drugs and his or her family.

8. Any treatment of drug dependency must redirect attitudes toward personal weak-

nesses. A gradual return to society is the ideal goal of successful drug-abuse treatment programs.

9. Society presently relies on a body of laws to protect itself from drugs and drug abusers. Some laws control treatment while others govern the possession and sale of drugs.

   a. The federal Comprehensive Drug Abuse Prevention and Control Act regulates the possession and sale of drugs.

   b. The states also have laws controlling the possession and sale of drugs. Except for the control of marijuana, these laws follow federal laws. Most states have "decriminalized" the individual possession of small amounts of marijuana for personal use.

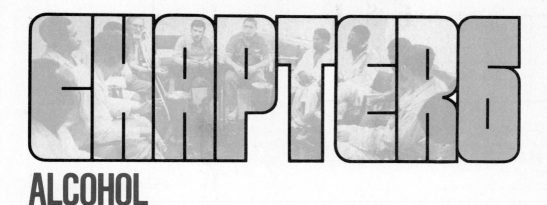

CHAPTER 6

ALCOHOL

Currently there is great public interest in the effects of mood-modifying drugs. Some of these drugs are illegal to possess; others are legal and easily purchased. However, the most prevalently used and potentially dangerous mood-modifying drug is *alcohol*. This dangerous drug can be purchased as a beverage in any country in the Western world.

Alcohol is used in many ways. Most individuals find it a pleasant and generally enjoyable part of dinner parties, social gatherings, celebrations, and so on. Unlike the use of other mood-modifying drugs such as hallucinogens or opiates, the use of alcohol is not *necessarily* questioned or condemned. Alcohol becomes a problem to society and

Alcoholic beverages are often drunk at social occasions. Most Americans enjoy an alcoholic beverage, such as a glass of wine, with dinner, at social gatherings, or as part of a celebration.

● ASK YOURSELF

1. How do most of the people you know feel about drinking? How do you feel about it? What is your family's attitude toward it?
2. What is an "empty" calorie?

individuals only under specific conditions. These include driving under its influence, appearing in public when intoxicated, and drinking excessively or compulsively.

The effects of alcohol are part of a complex web. Some alcoholism is caused by problems of body chemistry and may be inherited. Some is caused by a person's lack of self-control or evasion of responsibility. Some cases of alcoholism are complicated by severe psychological problems. Just as in the abuse of any other mood-modifying drug, alcohol abuse is a *behavioral* problem with *physiological implications*.

ALCOHOLIC BEVERAGES

Alcohol is a sedative (*see* Chapter 5), the most widely used and abused one in the United States.

Because of legal availability, the average person has a much greater chance of becoming physically dependent on alcohol than on any other psychoactive drug.

The alcoholic content of distilled beverages is expressed as "proof," a figure that is exactly double the alcoholic percentage. Thus, 86 proof whiskey is 43 percent alcohol by volume. The alcoholic content of wine, ale, and beer is usually expressed directly as a percentage (*see* Table 6.1).

In addition to alcohol and water, alcoholic beverages contain mainly flavorings and coloring agents. They have almost no food value except calories. As shown in Table 6.2, there are no vitamins, minerals, fats, proteins, or usable carbohydrates in most alcoholic beverages. The exceptions are beer and wine. However, in calorie-reduced "light" beers all carbohydrates have been turned into alcohol.

Calories, however, are abundant in all alcoholic beverages. Most of the caloric content of alcoholic beverages is derived from the alcohol itself. These are termed "empty calories" because there are no other nutrients along with the calories. Another problem is that alcohol must be changed into fat and then into body sugar before it can be used. This

TABLE 6.1 SOURCE AND ALCOHOLIC CONTENT OF ALCOHOLIC BEVERAGES

Beverage	Source* of alcohol	Distilled	Percent alcohol by volume	Content by proof
Beer	Malted barley	No	4–6	Not expressed in proof
Ale	Malted barley	No	6–8	
Wine	Grape juice	No	12–21	
Whiskey	Malted grains	Yes	40–50	80–100
Brandy	Grape juice	Yes	40–50	80–100
Rum	Molasses	Yes	40–50	80–100
Vodka	Various sources	Yes	40–50	80–100
Gin	Various sources	Yes	40–50	80–100

*That is, *ethyl alcohol*, the only form of alcohol that can be consumed safely. Ethyl alcohol is produced from various starches and sugars during fermentation.

Equal amounts of alcohol. These three drinks, 12 ounces of beer, 4 ounces of dry wine, and 1 to 1½ ounces of distilled liquor, contain roughly equal amounts of alcohol.

process provides no nutrition and can cause severe liver damage. The calories in alcohol, however, are filling; and heavy drinkers often stop eating nutritious food.

Alcohol is a mood-modifying drug and can temporarily produce a state of euphoria and an appar-

ent stimulation. This, undoubtedly, is the basis of its attraction. The stimulant effect of alcohol is an illusory one, however. Actually, alcohol is a *depressant*. It slows down the functions of the brain and central nervous system. The first part of the brain to "go" is the center that controls judgment

TABLE 6.2 NUTRITIONAL VALUES OF ALCOHOLIC BEVERAGES

Food nutrient	Type of beverage and quantity		
	Beer (12 ounces)	Gin, Rum, Vodka, Whiskey (86 proof) (1½ ounces)	Wine (dessert) (3½ ounces)
Calories (total)	150.0	105.0	140.0
Calories from alcohol	90.0	105.0	108.0
Protein (grams)	1.0	0.0	trace
Fat (grams)	0.0	0.0	0.0
Carbohydrate (grams)	14.0	trace	8.0
Thiamine (milligrams)	0.01	0.0	0.01
Nicotinic acid (milligrams)	2.2	0.0	0.2
Riboflavin (milligrams)	0.11	0.0	0.02
Ascorbic acid (milligrams)	0.0	0.0	0.0
Folic acid (milligrams)	0.0	0.0	0.0

Source: U.S. Department of Agriculture, "Nutritive Value of Foods," *Home & Garden Bulletin,* 72 (1977), 29.

● ASK YOURSELF

1. Is alcohol a stimulant or depressant?
2. How does alcohol produce the apparent stimulation in drinkers?
3. What controls the amount of alcohol absorbed into the bloodstream?

and inhibitions. Thus, paradoxically, alcohol stimulates drinkers for a brief period by *depressing* the inhibiting mechanisms in their personalities. They become talkative and happy, and assume they are being witty and charming. Frequently, the stimulation will cause people to say and do things they would, if sober, prefer were left unsaid and undone.

The best quantitative measure of what is happening as a normal, healthy person drinks is the concentration of alcohol in the bloodstream. After a drink, alcohol shows up in the bloodstream very quickly. At first, in small amounts, alcohol is absorbed into the blood through the lining of the stomach, but this process slows and then stops just as quickly. The presence of food in the stomach impedes absorption. This is the reason many people prefer a light snack at cocktail parties—it helps modify the effects of the alcohol. Unlike the stomach, the intestinal lining absorbs alcohol rapidly, re-

Alcohol is a drug. It is actually a sedative drug—the most widely abused sedative in the U.S.

How are light beers made? The calories in beer come from alcohol, carbohydrates, or unfermentable sugars that are left in the beer after the brewing process is over. Some calories are lost when the alcohol content is lowered. Light beers contain between 3 and 3.5 percent alcohol, compared with 3.5 to 4 percent in regular beer. But, alcohol supplies the taste in beer. So reducing the alcohol content causes a "watered down" taste. Rather than lose taste, brewers add enzymes that break the unfermentable sugars into alcohol while the beer ages. This reduces the total caloric content while maintaining the taste.

gardless of the amount of alcohol or the presence of food. Consequently, moderate and high blood-alcohol concentrations depend on the emptying time of the stomach. Anything that decreases the emptying time of the stomach reduces the blood-alcohol concentration by spreading its absorption over a long period of time. Carbon dioxide (CO_2) has the opposite effect; it speeds up the passage of alcohol into the intestines. A carbonated mixer with a whiskey drink is more potent than a whiskey-and-water highball. Dissolved CO_2 is what gives champagne its extra kick.

Table 6.3 shows the relationship between body size, number of drinks, and resultant blood-alcohol concentration. Body size is a factor because the larger the bloodstream into which the alcohol passes, the more dilute it will be. Blood-alcohol concentration levels are the basis of drunk driving determinations in almost all states. The appearance and demeanor of a suspected drunk driver has little bearing on whether he or she can be found guilty of driving while intoxicated.

Although alcohol is not officially listed as the

A fatal traffic accident. Alcohol is believed to be a contributing factor in 25 to 50 percent of all fatal traffic accidents.

cause in many fatal traffic accidents, it is believed that many that are due to "high speed" or "failure to negotiate a curve" are actually caused by excessive drinking. Research has shown that alcohol starts to be a factor in accidents at blood levels as low as 0.03 percent.

Alcohol and Your Body

Because all the voluntary muscles are under the control of the brain and nervous system, muscle control is impaired at all blood-alcohol levels. This results in a loss of coordination and a lengthened reaction time; these changes are especially detrimen-

TABLE 6.3 BLOOD-ALCOHOL LEVELS (PERCENT ALCOHOL IN BLOOD)

Body weight	Drinks*											
	1	2	3	4	5	6	7	8	9	10	11	12
100 lb.	0.038	0.075	0.113	0.150	0.188	0.225	0.263	0.300	0.338	0.375	0.413	0.450
120 lb	0.031	0.063	0.094	0.125	0.156	0.188	0.219	0.250	0.281	0.313	0.344	0.375
140 lb	0.027	0.054	0.080	0.107	0.134	0.161	0.188	0.214	0.241	0.268	0.295	0.321
160 lb	0.023	0.047	0.070	0.094	0.117	0.141	0.164	0.188	0.211	0.234	0.258	0.281
180 lb	0.021	0.042	0.063	0.083	0.104	0.125	0.146	0.167	0.188	0.208	0.229	0.250
200 lb	0.019	0.038	0.056	0.075	0.094	0.113	0.131	0.150	0.169	0.188	0.206	0.225
220 lb	0.017	0.034	0.051	0.068	0.085	0.102	0.119	0.136	0.153	0.170	0.188	0.205
240 lb	0.016	0.031	0.047	0.063	0.078	0.094	0.109	0.125	0.141	0.156	0.172	0.188

Under 0.05	0.05 to 0.10	0.10 to 0.15	Over 0.15
Driving is not seriously impaired†	Driving becomes increasingly dangerous	Driving is dangerous	Driving is *very* dangerous
	0.08 legally drunk in Utah	Legally drunk in many states	Legally drunk in any state

*One drink equals 1 oz of 100-proof liquor or 12 oz of beer.
† There is substantial evidence from recent studies that drivers below the age of twenty-five may experience serious impairment of their driving skills even if their blood-alcohol level is below 0.05 percent. Also, studies show that some persons with a blood-alcohol level below 0.05 percent are involved in accidents. The table presents "average" figures based on the "average" person under "average" conditions. Individual differences—both physiological and psychological—must be considered.
Source: New Jersey Department of Law and Public Safety, Division of Motor Vehicles, Trenton, N.J.

Staying sober. During a four-hour party, drinking at the rate of one drink (or one can of beer) per hour will probably not cause excessive intoxication.

● ASK YOURSELF

1. What is the relationship between body size and blood-alcohol level?
2. What is the major way alcohol impairs driving ability? Can you tell when someone has had too much alcohol and would be a poor risk driving an automobile? Explain some of the things you would look for in this poor-risk driver?
3. Have you known someone who was killed or hurt in a traffic accident? Was alcohol involved?
4. Can a person die from an overdose of alcohol? How does this occur?

tal to automobile drivers. Some people feel that their driving ability is improved by small amounts of alcohol, but the truth is that alcohol only makes these people *think* they are driving better. Drinking drivers seldom realize to what extent their driving ability has deteriorated, because the same effects on the brain that cause them to be dangerous drivers also make them unaware of how poor their driving has become.

At low blood-alcohol levels, the main effect on driving is a reduction in the level of judgment and care used. Most people can still drive straight enough, but they may take chances they might otherwise not risk. With higher blood-alcohol levels, there are the additional factors of poor vision and slowed muscular reactions. Further alcohol intake begins to depress the lower parts of the brain. At extremely high blood-alcohol levels, the primitive reflex centers that control breathing and other body functions may be depressed to the point that the person dies. Such high blood-alcohol levels are seldom reached through normal drinking, however, because a person usually vomits or becomes unconscious first. Nevertheless, a fatal dose of alcohol could be consumed if a person very rapidly drank a large quantity of distilled liquor.

Sight is the first sense affected by alcohol. Small amounts of alcohol increase a person's sensitivity to light. He or she is less able to distinguish between two different intensities of light. Focusing the eyes also becomes a problem when people drink. Increasing amounts of alcohol cause greater losses of vision. Hearing is affected also. Have you ever noticed how a party becomes louder and louder as people drink more?

Alcohol interferes with both the storage and retrieval of information. A person under the influence of alcohol has a decreased ability to learn and to recall past events and information. Problem-solving ability is also greatly diminished. Even simple puzzles and arithmetic problems may be difficult or impossible for the intoxicated person to solve.

Probably the most serious organic damage caused by alcohol abuse is to the liver. The liver is the primary chemical organ of the body. Almost 90 percent of the ethyl alcohol taken into the body is metabolized by the liver and converted into a chemical known as *acetaldehyde*. This chemical is toxic to many organs of the body. The acetaldehyde is then converted into carbon dioxide and water, which can then be exhaled and excreted. The liver is the basic mediator of the chemical processes that remove the alcohol from the body. The ability of the liver to handle this process determines the speed at which a person will "sober up." Generally, the liver can process alcohol at the rate of one drink per hour.

Liver ailments are especially common among alcoholics. Some are brought on by deposits of fat in the liver, which kill liver cells and lead to *alcoholic hepatitis*. The lesion produced by alcoholic hepatitis is fatal to about one in ten patients; however, it may heal and become *cirrhosis* (occurring in one of two cases), a hardening of the liver. Cirrhosis is six times as common among alcoholics as among the general population. Cirrhosis and alcoholic hepatitis may be only contributing factors in alcoholism death. Recent research has shown that when large amounts of alcohol are consumed over a long period of time, the liver cells expand, squeezing the blood vessels of the liver. This blocks the liver's blood flow, causing high blood pressure in the liver (*portal hypertension*), which may lead to rupture and internal bleeding. Also, the hypertension can interfere with the filtering function of the liver. This can cause toxins to build up, leading to hepatic coma and death. While cirrhosis is not reversible, the swelling of cells and portal hypertension are. When an alcoholic stops drinking, the cells and blood vessels return to normal.

Besides the burden it places on the liver, alcohol taxes other body systems as well. It may cause serious damage when the person is taking certain prescription drugs. Of the 100 most frequently prescribed drugs, more than half contain at least one ingredient known to interact adversely with alcohol. Adverse interactions between commonly used drugs and alcohol produce 47,000 emergency-room admissions and 2,500 deaths a year. Because of the great range of dosages and interactions, we shall list only the most common (*see* Table 6.4).

● ASK YOURSELF

1. How is the body affected by alcohol? What are the visible effects of alcohol intoxication?
2. Which organ of the body is most seriously damaged by alcohol? Look up the implications of this damage in a medical pathology textbook.
3. Can some of the damage to the liver be reversed by stopping drinking?
4. What is acetaldehyde? How does it affect the body?

FETAL ALCOHOL SYNDROME (FAS)

Alcohol passes freely across the placental barrier between a developing child and the mother. Concentrations of alcohol in the fetus are at least as high as in the mother. Consequently, the consumption of alcoholic beverages by women during pregnancy can cause significant growth problems in a developing child.

In 1973, two physicians discovered that they could diagnose a mother's alcoholism from the

Don't drink to the health of your baby

Drinking alcoholic beverages while pregnant can be harmful to the baby. Why take chances? For information or for help, write: National Clearinghouse for Alcohol Information, Rockville, Md. 20857

Remember, drinking any amount of alcohol can damage a developing baby!

TABLE 6.4   ALCOHOL-DRUG INTERACTIONS

Drug	Effect When Combined with Alcohol
Analgesics (aspirin & aspirin-containing drugs)	Aspirin can cause stomach irritation and bleeding. Alcohol also irritates the the stomach and intestines. Together they can cause severe gastritis and bleeding. Combined with aspirin, alcohol delays blood clotting and can cause possible hemorrhage.
Anesthetics	Both cross-tolerance and synergistic effects are produced. People with alcohol in their systems need greater amounts of an anesthetic to produce sleep. Anesthetics last longer in the system, deeper and longer sleep is produced, and far less anesthetic is needed to produce death.
Antialcohol drugs (Antabuse (disulfiram)	Consuming alcohol can produce: high blood pressure, flushing of face, increased heart beat, headache, dizziness, rapid breathing, nausea, vomiting, weakness, and fainting. Depending on the amounts of drug and alcohol consumed, effects can be fatal.
Antidiabetic drugs (oral *tolbutamide* or *chloropropamide*).	Individuals may experience same symptoms as those taking antialcohol drugs, but milder, if they consume alcohol while taking their medication.
Orinase, Diabinase, Dymelor, Tolinase	Alcohol causes unpredictable fluctuations in blood-sugar levels. Seriously low blood-sugar levels may develop. (*See also* Antiblood-clotting drug below.)
High Blood Pressure drugs	Alcohol increases the blood pressure-lowering effects, leading to faintness and loss of consciousness.
Antiblood-clotting drugs Warfarin (Athrombin-K, Coumadin, & Panwarfin)	Increases anticoagulant effects and thus the danger of hemorrhage. Chronic alcohol abuse, however, can decrease anticoagulant effects. Such changing effects are dangerous.
Anticonvulsants Dilantin (phenytoin)	Alcohol speeds up removal of the drug from the body, making normal doses inadequate.
Antidepressants (tricyclic compounds)	Alcohol increases susceptibility to convulsions. Certain alcoholic beverages (Chianti wine & beer) that contain tyramine (an amino acid) can cause lethal heart attacks when taken with these drugs.
Resipramine Amitriptyline	Alcohol reduces effects. Alcohol increases effects. Alcohol and the two drugs above can lower blood pressure seriously.
Stimulants	Alcohol and stimulants have variable effects. May increase or decrease effects of each. Such inconsistent actions can be dangerous.
Antihistamines	Alcohol increases drowsiness to such an extent that it is dangerous to perform any hazardous task such as driving or operating any type of machinery.
Antimicrobial & Anti-infective drugs	Alcohol may cause nausea, vomiting, headache, and possible convulsions.

Barbiturates	With alcohol, lethal dosages of barbiturates are nearly 50 percent lower than when drugs are used alone. Symptoms of severe interaction are: vomiting, severe motor impairment, unconsciousness, coma, & death.
Narcotics Darvon (propoxphene)	Alcohol greatly increases chance of lethal overdose when taken within 4 hours of Darvon.

Source: Information from *Physician's Desk Reference*, 34th ed., (New York: Medical Economics Co., 1980) and L. S. Goldman and A. Gilman, eds., *The Pharmacological Basis of Therapeutics*, 5th ed. (New York: Macmillan, 1975).

facial characteristics, growth deficiencies, and psychomotor disturbances of their children (Jones & Smith, 1973). They termed the condition *Fetal Alcohol Syndrome* (FAS). Others since then have been able to diagnose FAS in adults as well. Three major signs are considered in diagnosing this condition: (1) Mental retardation and/or central nervous system problems; (2) growth deficiencies; and (3) specific facial characteristics.

Mental retardation, learning problems, hyperactivity, and perception and attention problems are common in FAS children. The average reported IQ is around 68 (mildly retarded), but the range of IQ scores is wide (50 to over 100). Some children show minimal brain damage and are hyperactive; others have short attention spans and do not perceive objects, but are otherwise normal in intelligence.

FAS children are small, grow slowly, and show the facial characteristics shown in the accompanying illustration. FAS has been observed in children of all races.

The mechanisms that produce these effects remain unclear. The question of whether FAS is the direct effect of alcohol or the effect of a breakdown product, such as acetaldehyde, is not known. The more alcohol consumed the greater the damage to the developing fetus. A safe level of alcohol use during pregnancy has not been established.

Fetal alcohol effects (FAE) A range of physical and mental problems in newborn children may result from social drinking during pregnancy. Recent studies have revealed that even moderate alcohol consumption during pregnancy can have less dramatic, but nevertheless deleterious, effects on the development of a child. Fetal alcohol effects may include one or more of the characteristics of FAS. Some researchers believe hyperactivity and learning disabilities not apparent until later in childhood may result from the mother's social alcohol consumption during pregnancy.

Public programs In 1977 the Bureau of Alcohol, Tobacco, and Firearms requested Congress to require that a label warning of the dangers of FAS be put on bottles of alcoholic beverages. In January, 1979 (after a year of hearings), it was decided that such a label would be too complex and should not be required. But the alcoholic beverage industry was to fund an educational campaign warning mothers of the dangers that drinking poses to their

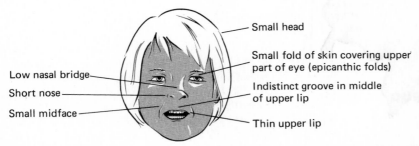

Low nasal bridge
Short nose
Small midface

Small head
Small fold of skin covering upper part of eye (epicanthic folds)
Indistinct groove in middle of upper lip
Thin upper lip

Characteristic facial features of Fetal Alcohol Syndrome children. (Adapted from Streissguth et al., 1980, fig. 1, p. 355. By permission.)

TABLE 6.5   SIGNS COMMON IN FETAL ALCOHOL SYNDROME CHILDREN

Area	Frequent signs	Occasional signs
Eyes	Drooping upper eyelid; crossed eyes (strabismus); upper eyelid covering (epicanthic fold)	Nearsightedness; abnormally small eyes (microphthalamia); abnormal thinness of opening of eye near the nose (blepharophimosis)
Ears	Rotated backward	Poorly formed outer ear (concha)
Mouth	Prominent ridges in upper palate	Cleft lip or cleft palate; small teeth with poorly formed enamel
Heart	Murmurs, especially in early childhood	A combination of congenital heart defects (tetralogy of Fallot); defects involving larger blood vessels
Kidneys		Small, rotated kidneys; water retention
Genitals	Abnormally small labia in females	
Skin	Benign tumors of blood vessels at birth (strawberry birthmark)	Abnormal amounts of hair in infancy
Skeleton	Caved-in chest	Limited joint movement, especially fingers and elbows; small nails; Klippel-Feil anomaly (sign of brain damage); scoliosis (abnormal lateral displacement of the spine)
Muscles		Hernias of diaphragm, umbilicus, or groin

Source: Information in "Fetal Alcohol Effects Linked To Moderate Drinking Levels," *Information and Feature Service,* NIAAA, IFS, *72* (June 9, 1980), p. 4.

● **ASK YOURSELF**

1. How easily does alcohol penetrate the fetal bloodstream?
2. How much alcohol can a woman *safely* consume during pregnancy?
3. What is the best method for combating Fetal Alcohol Syndrome?

unborn children. Also, the National Foundation-March of Dimes is very active in educational programs warning of FAS. They have introduced the information pamphlet, "When You Drink, Your Unborn Baby Does, Too!"

ALCOHOLISM AS DRUG ABUSE

Alcoholism is America's number one "hidden" health problem. Behind the public attitudes and condemnation, alcoholism does extensive damage to individuals, families, and to society as a whole. The alcoholic has difficulty holding a job, continu-

ing an education, and maintaining a stable family life. An important factor in bringing a formerly hidden disease out into the open is the potential for cure. To many people, the available treatment methods seem nebulous and difficult to understand. There is no one method, no one drug, that can provide a cure. Since it appears that there is little to gain from open discussion of the problem, many people will choose to avoid dealing with the disease of alcoholism.

The Making of an Alcoholic

Many psychologists believe that serving a child a colorful nonalcoholic drink made of ginger ale and maraschino cherry juice (a "Shirley Temple"), which simulates the appearance of an adult alcoholic cocktail, is the first step in the making of an alcoholic. Serving children such a drink encourages them to adopt an adult habit in order to gain adult approval. Many alcoholics report learning, at an early age, that alcohol is a symbol of fun. Children should be taught that alcohol is not a harmless social beverage, but a dangerous drug. Also, in

When you drink, your UNBORN BABY does,too!

...To protect the unborn and the newborn

A pamphlet on Fetal Alcohol Syndrome from the National Foundation–March of Dimes. You may obtain this and other pamphlets from your local chapter of the National Foundation–March of Dimes.

If you drink a lot of beer, you drink a lot.

"If you drink a lot of beer, you drink a lot." Many alcoholics drink only beer!

The "Shirley Temple" drink for children. Many psychologists feel this is the first step toward drinking alcoholic beverages later in life.

recent years, parents have seemed relatively unconcerned about their children's drinking habits. They are panic-stricken when their child uses marijuana, but merely send a drunk youngster to bed quietly. This inconsistent behavior is caused by the abundance of scare information concerning marijuana and the legal and social disapproval of marijuana. Actually, many teen-agers use both marijuana and alcohol.

Young people are often multiple or poly-drug users. Many experts feel this situation occurs because of the social acceptance of some psychoactive drugs other than alcohol, such as marijuana. Also, the widespread misuse of methadone treatment programs has changed the pattern of abuse of alcohol and other drugs. Some narcotics addicts now turn to alcohol as an alternative, legally available drug. Sedatives, stimulants, and minor tranquilizers are being mixed with alcohol. Such poly-drug use has greatly increased overdose deaths and complicated withdrawal syndromes that often are not recognized until it is too late to save the patient.

Most teen-agers who begin to drink do so at

Community Task Force Opposes Sale of Alcohol to Minors

A thirty-member committee of parents, teachers, community leaders, and police has mounted an alcohol abuse prevention effort in New York's Community School District 22. This task force was formed by individuals concerned about the easy access young people have to alcoholic beverages. The school district has 24,500 students, and more than 500 merchants are licensed to sell alcoholic beverages in the area.

The task force established three immediate goals:

1. Heighten parental awareness of the extent of teen-age drinking
2. Increase community understanding of the nature of teen-age drinking
3. Prevent the sale of alcoholic beverages to minors in the community

To implement the first two goals, a New York television station produced a half-hour documentary on alcohol and the adolescent. To implement the third goal, the task force mounted a prevention campaign aimed at area retailers. A poster was designed that discourages the sale of alcohol to minors. Junior high school students silk-screened copies for area merchants. Task force members visited merchants about whom complaints were received. If cooperation was not forthcoming, the police were contacted. The task force members contacted the state liquor authority and the courts to use their fullest powers to enforce applicable liquor laws.

U.S. Department of Health and Human Services, "Community Task Force Opposes Sale of Alcohol to Minors," *NIAAA Information and Feature Service*, IFS No. *81* (March 3, 1981), 3.

home under parental supervision and approval. Much of the drinking occurs on holidays or as part of family and social occasions. Whether and how much a person drinks are strongly associated with sex, age, ethnic background, religious affiliation, education, socioeconomic status, occupation, area of residence, and degree of urbanization.

What is Alcoholism?

There are no clear-cut widely accepted definitions of the words *alcoholic* or *alcoholism*. Some people would suggest so simple a definition as "an alcoholic is someone who drinks too much." But how much is too much? With alcohol, as with any other mood-modifying drug, there are levels of use. The following is such a classification.

1. *The moderate drinker, or occasional user.* This is someone who enjoys the social, personal, and emotional gratification of alcohol. Moderate drinkers enjoy the company of others and the "social" use of the drug.
2. *The heavy drinker, or regular user.* The regular user may drink daily for at least a year, or

have six or more drinks on one occasion at least once a week or twice a month. However, regulars users show no personal or physical problems because of their alcohol consumption.

3. *The problem drinker.* This person is a heavy drinker who has alcohol-related problems, but they are not severe enough to classify him or her as an alcoholic.
4. *The alcoholic. This is a person whose drinking interferes with a major aspect of his or her life on a continuing basis.** An alcoholic has alcohol-related problems in at least three of the following four areas:
   a. Marital problems, or social disapproval of drinking by spouse, friends, or parents
   b. Job trouble, traffic arrests, or other police trouble
   c. Frequent "blackouts," tremors, or other physical symptoms of severe alcohol toxicity and physical dependence

* This is the definition of alcoholism of the American Medical Association and the World Health Organization.

d. Inability to control amount of alcohol consumed at one time; morning drinking

Causes of alcoholism Despite years of research efforts costing millions of dollars, the causes of alcoholism are still not definitely known. Many theories have been presented, some of which are backed by extensive scientific evidence, while some are pure speculation. It has not been clearly determined whether alcoholism is caused by physical factors, psychological factors, or a combination of the two. Each theory has strong supporters. There is certainly reason to believe that personality problems are a facet of alcoholism. Yet there is also evidence that some people simply respond differently to alcohol. This variability may be related to some metabolic problem such as the lack of an enzyme that prohibits the normal processing and removal of alcohol from the liver. Possibly, some factor in brain action is also involved in alcoholism. Many authorities today agree that alcoholism should be thought of as a disease, regardless of its cause. The alcoholic should be treated as an ill person, rather than condemned as a sinner or a good-for-nothing. Public acceptance of other drug abusers as "ill" individuals is taking a longer period of time.

Certain personality traits are commonly found among alcoholics. The alcoholic typically has a low sense of personal worth, and therefore feels insecure and isolated from other people. These feelings cause emotional pain, and drink helps wipe out this pain. These feelings are also found among drug abusers and overeaters.

In the past, alcohol was automatically held responsible for family poverty, divorce, child neglect, juvenile delinquency, and most other family problems. Today alcoholism is often recognized as one of several complex emotional problems. However, a vicious circle often develops in which personal and family problems lead to excess drinking, which leads to further family problems, which leads to the eventual destruction of the family unit.

Stages of alcoholism Society has a number of difficulties in dealing with alcoholism. The patterns are not consistent, and the problems produced are unpredictable.

No one ever decides to become an alcoholic. Almost every new drinker assumes that he or she will always be able to handle liquor. This assumption is

Jesse Jackson, founder of Operation Bootstrap. Jackson advocates that individuals help themselves to improve their lifestyles.

almost always right. But there is no way to predict which drinker will develop the disease of alcoholism. The great majority of those becoming alcoholics do not even realize what is happening to them until it is too late to stop.

● ASK YOURSELF

1. Do you, or does any member of your family, drink alcoholic beverages? If so, how would you classify their drinking patterns?
2. Do you know an alcoholic? Look at the definition and criteria for an alcoholic and explain why you think this person is an alcoholic.
3. Could you add more alcohol-related problems to the ones listed above? What makes them problems?

Fortunately, there are signs of developing alcoholism. Recognition of the beginning problem requires both knowledge of the warning signals and the honesty to admit the seriousness of what is happening. There is often only one remedy—*to stop drinking.* It requires a good deal of strength and determination to follow this one cure.

As they become dependent on alcohol, many people learn to appreciate and rely on the feelings of relief from tensions and escape from reality that alcohol can provide. The first step toward alcoholism occurs when a person starts to drink specifically for these effects. About one-fifth of all drinkers can be classified as occasional *"escape drinkers."* These people are not yet alcoholics, but they should be aware of the possible development of the condition. In those who are progressing toward alcoholism, escape drinking becomes more and more frequent. It may quickly develop into a pattern of heavy drinking every night or every weekend.

Another pattern of developing alcoholism is the occasional *"binge,"* or *"periodic,"* drinker. The binge drinker may go for weeks or months without drinking any alcohol. Then they go on a drinking spree that can last for days or even weeks. At these times they become either *regular* or *compulsive* (alcoholic) drinkers. The difference lies in the number of social, economic, and legal problems their drinking causes them or their families. The periodic drinker may be just as much an alcoholic as the regular compulsive drinker. He or she is even more likely to lose a job because of the habit of staying drunk for days at a time.

An *alcoholic blackout* is a period of temporary amnesia. It should not be confused with passing out, which involves unconsciousness. Anyone who drinks too much will pass out. He or she will then be unconsciousness or asleep. Passing out is not a sign of alcoholism; it merely indicates the drinker's own poor judgment or lack of experience with alcohol—or, of course, a desire to reach a state of temporary oblivion.

A blackout is something else entirely. It may occur after the drinker has taken just a few drinks. The drinker remains conscious and appears fully aware of what is going on. He or she may appear normal to others, and may seem fully capable of walking, talking, driving, dancing, and drinking as usual.

But after finishing drinking, the drinker who has had a blackout will have no memory of what took place while drinking. He or she will remember neither the major events nor the minor details. The memory will have "blacked out" everything that happened after the first few drinks. A blackout usually lasts for several hours. During a binge, however, it may last for several days.

Anyone who has had such a blackout either is an alcoholic or is very nearly so. Blackouts usually occur after several months or years of drinking, but some alcoholics report experiencing blackouts from the very beginning of their drinking.

The most important symptom of alcoholism is *loss of control.* This means that the alcoholic cannot stop at a reasonable number of drinks, but must continue until drunk or sick. Depending on the drinking pattern of the individual, such drinking will continue for hours, days, or even weeks. Loss of control does not mean that the alcoholic cannot choose whether or not to drink on a certain day. But if the alcoholic does take a single drink, he or she cannot really determine when to stop.

Alcoholism is a progressive disease. Every case of alcoholism develops at its own pace. Some alcoholics reach an advanced state in just a few months. Others take many years to reach a pattern of problem drinking.

True alcohol addiction The basis of the physical addiction in alcoholism seems to be an altered metabolism that is alcohol-induced and that produces chemicals in the body similar to those produced in opiate addicts. Alcoholism and drug addiction are similar processes. The major differences are the length of time and the dosage required for development of physical dependence. In 1970, the Expert Committee on Alcohol and Alcoholism of the World Health Organization stated that "recent evidence makes it appear that there is more resemblance between the responses of the withdrawal from alcohol and from opiates than was previously realized. . . . When serious symptoms follow the withdrawal of alcohol, they persist almost as long as do those following the withdrawal of opiates." Consequently, the World Health Organization now defines physical dependence as either *narcotic-solvent abstinence syndrome addiction* or *alcohol-barbiturate abstinence syndrome addiction.*

When a physically dependent alcoholic is suddenly withdrawn from alcohol, extreme hypersensitivity to all external stimuli usually appears within a week after the alcohol blood levels return to normal. Such hypersensitivity in its most extreme form (*delirium tremens,* or convulsions) is *alcohol-barbiturate abstinence syndrome* and requires emergency medical treatment. This is caused by a return of function to previously anesthetized neurons, aggravated by prolonged magnesium and potassium deficiency. Many physicians believe that magnesium deficiency is responsible for the alcohol-withdrawal syndrome and often treat it with magnesium compounds.

After several attacks of delirium tremens, a very serious condition called "wet brain" may develop. This is a chronic or long-term condition, seldom curable, and often fatal. The alcoholic's thought processes are completely disrupted. All functions of his or her nervous system are impaired. The alcoholic who reaches this stage will either die or spend the rest of his or her life in an institution.

The Alcoholic

The alcoholic has several defense mechansims that partially deal with the guilt resulting from his or her drinking problem. He or she may appear extremely jovial, but the remorse felt will show itself in crying jags and serious periods of depression.

Most alcoholics have problems with employment and finances. Intoxication on the job is usually grounds for dismissal from any position. Once a person has been fired for drinking, it becomes very difficult for him or her to find another job. Failing to earn a living, alcoholics may go on spending sprees, making extravagant investments and purchases. As their self-esteem sinks lower, alcoholics may try to establish a social position. This often results in them buying drinks for total strangers or the whole bar.

Many of alcoholism's effects on marriage are the result of financial strains. Money problems always place a strain on a marriage. But, when these problems are the direct result of the excessive drinking of one spouse, the other spouse is likely to be highly resentful. Other marriage problems arise from the loss of companionship. In some cases, the abuse of family members while the alcoholic is under the effects of alcohol becomes a major problem.

The alcoholic's family tends to become socially isolated. Members no longer bring friends home because they fear embarrassment by the alcoholic's

● ASK YOURSELF

1. What are the "warning signals" of beginning alcoholism? Can you see any of these in yourself, a friend, or family member?
2. What is the most important symptom of alcoholism? Have you, a friend, or family member ever experienced this symptom? What does this imply about the person involved with alcoholism?
3. What are some of the possible outcomes of alcoholism?

actions. This fear is an especially painful problem for children who can never be free of the fear that an alcoholic parent may be at home and intoxicated at any time of the day or evening.

Another problem in the alcoholic's marriage is jealousy. This is one of the many cause-and-effect dilemmas of alcoholism. Some alcoholics give a spouse's infidelity as a reason for their drinking problem, while others recognize that their drinking problem has ruined the marriage and driven a spouse into an extramarital relationship.

The alcoholic often suffers a loss of sexual drive. As the sexual relationship in the marriage deteriorates and intercourse becomes less frequent, the alcoholic tends to blame the deterioration on anything but the real cause — alcohol-induced reduction of sexual drive. Very often, the alcoholic's spouse is then accused of having extramarital love affairs. And it is this suspicion and jealousy that can lead to the eventual end of the marriage.

The female alcoholic Until recently, most of our knowledge about alcohol has been based on studies of small clinical populations. These studies are not representative of larger populations of untreated alcoholics. Separating myth from half-truth from fact is no easy task. But, a number of statements frequently made about females and alcohol consumption are not valid in light of current North American research data. There is no easy way of counting the number of alcoholics. But, several approaches have been used to estimate rates of alcoholism among males and females: (1) Self-reports of heavy drinking and alcohol-related problems; (2) estimates of the proportion of male and female liver cirrhosis deaths that are alcohol-related; (3) hospital admissions for alcohol-related conditions; and, (4) estimates of heavy drinking based on body size of males and females.

The sex ratio for alcoholism is 1:1 (male: female) This is a myth. The fact is that the best current estimate of the sex ratio for alcoholism is 3 males for every 1 female. At times a greater number of women alcoholics has appeared because women are more likely than men to report consumption of alcohol and alcohol-problems accurately. This has caused an underestimation of male alcoholics.

Alcoholism is increasing at a faster rate among women than among men Despite popular media reports, there is no evidence women are becoming alcoholics at a faster rate than men. Alcohol consumption has increased since World War II, but the sex ratio for adult drinking, and apparently alcoholism, has not varied significantly over the past 35 years.

Alcoholism develops more quickly in women than in men The interval between the onset of heavy drinking and admission for treatment appears to be shorter for women than men. However, women are more likely than men to seek help for health problems, and to do so at an earlier stage. Proper studies to prove the first statement would require long-term detailed drinking histories.

Many housewives are hidden alcoholics This notion stems from early Freudian concepts of female inadequacy. Freud felt that women engage in as much deviant behavior as men but cleverly conceal it. The research does not support this idea, but the stereotype of the devious female persists.

There are reports that women alcoholics suffer a far greater social stigma from their drinking than do men. As a result, they and their families make every effort to conceal their drinking. Protected and able to drink secretly at home, these housewives have become our "hidden alcoholics." Actually, housewives are underrepresented among women alcoholics. Other studies indicate that women are no more likely than men to drink secretly. It is also not at all certain that the spouses of women alcoholics protect them. Frequently the spouse is also an alcoholic. Recent research on the stigma of being an alcoholic suggests that men and women alcoholics are equally stigmatized.

Some evidence suggests that married women who work develop more alcohol-related problems. A working woman can find it very difficult in the male-oriented business world. Working women who are heads of households may also find that children double or triple the pressures on them.

THE TREATMENT OF ALCOHOLISM

Most chronic alcoholics do not voluntarily stop drinking. Even if an alcoholic could stop, he or she would risk serious or even fatal withdrawl symptoms. An alcoholic must, therefore, have intensive medical treatment during the sobering-up ("drying-out") period. He or she may require hospitalization during this time. Such drugs as minor tranquilizers, insulin, thiamine, magnesium compounds, and caffeine may be used in the treatment. Once alcoholics have passed through the more serious physical parts of the withdrawal period, many therapists feel that, generally, they should seek continuing treatment the rest of their lives.

Before an alcoholic person goes for help and while being treated, there are some things family members can do. Al-Anon (a group that helps the families of alcoholics understand and cope with alcoholism) makes the following suggestions:

1. Don't regard alcoholism as a family disgrace. Recovery from alcoholism is the same as recovery from any other disease.

2. Don't nag, preach, or lecture to the alcoholic. Chances are they have already told themselves everything you can tell them. You may only increase their need to lie or force them to make promises that they cannot possibly keep.

3. Guard against taking on a "holier-than-thou" or martyrlike attitude. Be careful—it is easy to project this attitude without even saying a word.

4. Don't use "if you loved me" appeals. Alcoholic drinking is compulsive and cannot be controlled by willpower. This is like saying, "If you loved me, you would not have a heart attack."

5. Avoid any threat—unless you are definitely prepared to carry it out. There are times when you must take action to protect yourself or the children.

6. Don't hide alcoholic beverages found in the

The Awareness Process

As with any treatement program, alcoholics must become aware of their drinking behavior. (It is better to use the phrase "uncontrolled drinker" instead of "problem drinker" and "alcoholic.") Uncontrolled drinkers are individuals who refuse to stop drinking regardless of the environmental and personal consequences. These people are unable to control their emotional impulses or direct their energies toward constructive goals.

Uncontrolled drinkers are not fully aware of how much they are drinking. They have been told they are drinking too much but do not know what this really means. The awareness process will help them recognize their drinking behavior and accept responsibility for it.

Step 1. The first step is to have them review their drinking habits and establish the quality, quantity, circumstances, and financial cost of their drinking behavior.

Step 2. The second step is to have them keep an accurate record of each day's drinking—quality, quantity, circumstances, and the cost. The drinker is then asked to evaluate the financial and personal costs of his or her drinking behavior. This process develops an awareness of the drinking behavior and records the costs and motivations for drinking.

Step 3. The third step consists of a study of the structure and function of the drinking behavior. What aspect of the drinking does the individual like? What part does he or she get something from? What part is not liked? What are the negative effects?

Step 4. The fourth step is to detail the "solutions" that alcohol is providing the individual and some mutually acceptable alternatives.

Step 5. The fifth step is to develop alternatives to the real problems faced, such as handling anger, anxiety, and awkwardness.

Step 6. Finally, the individual becomes fully aware of the extent of his or her use and abuse of alcohol. The cessation of the use of alcohol as a means of coping lets the individual use his or her own resources. When individuals reach this step of awareness, they begin to shift their focus from drinking to themselves, other people, and the world. Such individual awareness is the unstated but understood goal of most psychoactive drug abuse treatment.

home or dispose of them. This only forces the alcoholic into depression or to find new ways of obtaining alcohol.

7. Don't let alcoholics persuade you to drink with them on the grounds that it will make them drink less. This lets them put off getting help because they feel you really condone their drinking.

8. Don't be jealous of the people the alcoholic turns to for help. You may feel left out when the alcoholic turns to others for help in staying sober. You wouldn't be jealous of a doctor treating him or her for that heart attack, would you?

9. Don't expect an immediate 100 percent recovery. In any illness there is a convalescence period and, often, relapses during times of tension.

10. Don't try to protect the recovering alcoholic from drinking situations. She or he must learn to say No gracefully and meaningfully. If you warn people against serving drinks,

● ASK YOURSELF

1. How can family members help an alcoholic to seek treatment?
2. How would you find a local AA meeting? Attend an AA meeting, and report on it. (It is also interesting to attend AA meetings from surrounding areas and compare them.)
3. How effective are the different types of alcohol treatment programs in your area, town, or city?

you will stir up old feelings of resentment and inadequacy.

11. Don't do for alcoholics what they can do for themselves. You cannot take medicine for someone who is sick. Don't remove the problem before the alcoholic can face it, solve it, or suffer the consequences.
12. Offer love, support, and understanding—but not excuses.

Psychotherapy

Since alcoholism is at least partly the result of emotional illness, it is understandable that one approach to its treatment is psychotherapy. The success of psychotherapy depends on how well the therapist understands the alcoholic personality. It is very difficult for someone who has never been an alcoholic to understand what it means to be one. Group psychotherapy is becoming increasingly important, because alcoholics really do understand each other.

Aversion Therapy

This type of therapy involves the use of drugs or other methods that make people sick if they drink alcohol. These can be administered in two ways. One is by a daily dosage of a drug such as Antabuse (disulfiram), which causes unpleasant bodily reactions if any alcohol—even a small amount—is consumed. Breathing becomes difficult, the heart pounds, and nausea and vomiting occur. As long as a person is taking Antabuse, he or she is not likely to drink. This drug is sometimes taken for months or years. For Antabuse to be successful, the patient must want to stop drinking; otherwise he or she will simply stop taking the drug.

Another type of aversion therapy that sometimes works is to give alcoholics a drink of alcohol along with a drug that makes them sick. After several of these treatments, they may develop a conditioned reflex so that alcohol alone makes them sick.

Alcoholics Anonymous (AA)

One of the most successful approaches to the treatment of alcoholism has been that of Alcoholics Anonymous. This is in reality a "multimodular" program—that is, it supports the alcoholics in all aspects of their lives. AA actually becomes "a way of life." It is believed that AA has the greatest recovery rate of any program. Seventy-five percent of those attending more than one meeting recover from their drinking. These people *want* to stop drinking and have the self-discipline to stay with the program. Or, they develop the self-discipline through association with other people in AA.

Alcoholics Anonymous is an organization whose only purpose is to help its members stay sober. Today, almost every city has regularly meeting AA groups ranging in size from a handful of members to over a hundred. A large city might have groups meeting every night of the week. There are even special groups for teen-age alcoholics and for spouses and children of alcoholics.

An evening's program usually consists of several members telling informally how miserable their lives were during their drinking years and how they have changed since joining AA. New members often find that these admitted alcoholics have had experiences similar to theirs. They can identify with older members who "speak their language." As they tell of their past experiences, the older members are helped too. The stories serve as a constant reminder of the unhappiness of their drinking periods and help to prevent a return to drinking.

Alcoholics Anonymous does not claim to cure the alcoholic; rather it helps him or her stop drinking and regain sobriety. AA emphasizes that an alcoholic is always an alcoholic, even when not drinking; if one starts drinking again, he or she would still drink in an alcoholic manner. For this reason, members always begin their personal stories by stating "I am an alcoholic."

There are many cases where members of AA decided, after years of sobriety, to return to social drinking. These attempts are never successful. Alcoholics Anonymous can only help those who have a strong desire to stop drinking forever and are *aware* that their drinking behavior is the basis of their personal, social, or economic problems.

THE ACCEPTABLE USE
OF ALCOHOLIC BEVERAGES

Some people argue that any consumption of alcoholic beverages is improper. But the majority of Americans find no medical, moral, legal, or religious reason for not making moderate use of alcoholic beverages. We shall therefore offer some suggestions that may help a person avoid drinking problems.

Even those people who fully approve of drinking and themselves drink regularly usually disapprove of certain types of drinking behavior — for example, drunken driving or such antisocial behavior as physical or verbal violence. Almost all those who approve of drinking do feel that there are times and places where drinking is not appropriate. Any time when a person needs his or her fullest mental facilities, such as when driving, flying, or operating machinery, is obviously a poor time to drink. For many employers, drinking or being drunk on the job is grounds for immediate dismissal. It is very poor policy to drink for courage, for example, in preparation for a job interview or sales conference. This is using alcohol as a crutch and is a step in the direction of alcoholism.

There is, of course, no set answer for the question of how much to drink. While drinking is acceptable, getting drunk is definitely frowned upon. The person who drinks is expected to drink in moderation, without serious impairment of physical or mental functions. Social drinkers learn how to drink: They drink slowly and pace their drinking in order not to build up a high blood-alcohol level and become drunk. If it takes an hour to oxidize the alcohol from one drink, a person who spaces four drinks over the span of a four-hour party will not become drunk.

The host or hostess of a party at which drinks are served should feel a certain responsibility for the amount of alcohol the guests drink. They must ask themselves how they would feel if someone was involved in a fatal accident after leaving their home. Non-alcoholic drinks should be available for those who prefer them, and the person who prefers not to drink should not be pressured or ridiculed. No pressure should be put on any guest to drink more than he or she really wants. If guests want to stop at one drink, they should be allowed to. During the last hour or so of a party, coffee should be served. This

"If you need a drink to be social, that's not social drinking." In other words, you may have a drinking problem.

serves several purposes. Coffee does not counteract alcohol, but the caffeine may help overcome the drowsiness that can be as much a cause of accidents as intoxication. The time spent drinking coffee serves as a "sobering-up" period, as well. Finally, the serving of coffee is accepted by most guests as a signal that the party is about to end. Those who are obviously in no condition to drive home should be strongly encouraged to stay overnight, take a taxi, or share a ride. They should never be permitted to drive.

● ASK YOURSELF

1. How can drinkers control their drinking so as not to become drunk during a party?
2. Name some occasions when drinking would be inappropriate or antisocial behavior.
3. What are the responsibilities of a host or hostess of a party? How would you structure a party to fulfill these responsibilities?

IN REVIEW

1. The most prevalently used and potentially dangerous mood-modifying drug is alcohol. The effects of alcohol are due to some persons' body chemistry and others' lack of self-control or sense of responsibility.
2. Alcoholic beverages are abundant in "empty" calories. The best quantitative measure of alcohol intake in a healthy person is the blood-alcohol concentration.
   a. Low blood-alcohol levels reduce a person's level of judgment. Sight is the first sense affected by alcohol. Hearing is also affected. Alcohol also interferes with both the storage and retrieval of information. Alcohol causes the most serious organic damage to the liver. Cirrhosis of the liver is six times more common among alcoholics than among the general public.
3. The consumption of alcoholic beverages by women during pregnancy can cause significant fetal development problems; this is termed the Fetal Alcohol Syndrome (FAS).
4. Alcoholism is America's major "hidden" health problem.
   a. Serving a child a nonalcoholic drink that simulates the appearance of an adult alcoholic cocktail is the first step in the making of an alcoholic.
   b. With alcohol, as with any mood-modifying drug, there are levels of use.
      (1) A moderate social drinker enjoys the company of others.
      (2) Heavy drinkers drink daily or heavily at times. They have no problems resulting from their alcohol use.
      (3) Problem drinkers are heavy drinkers with alcohol-related problems.
      (4) An alcoholic is a person whose drinking interferes with a major aspect of life on a continuing basis.
   c. The causes of alcoholism have never been clearly determined. All drinkers should be aware of the warning signals of impending alcoholism. The most important symptom of alcoholism is loss of control.
   d. Alcoholism and drug addiction are similar processes. The major differences are the length of time and the dosage required for development of physical dependency.
   e. Most alcoholics have problems with employment, finances, marriage, and family. A number of statements frequently made about female alcoholics are not valid.
5. Most chronic alcoholics do not voluntarily stop drinking. Alcoholics, generally, should seek continuing treatment the rest of their lives.
   a. Since alcoholism is at least partly the result of emotional illness, one possible treatment is psychotherapy.
   b. Aversion therapy uses drugs or other methods that make people sick if they drink alcohol.
   c. The most successful approach to the treatment of alcoholism is that of Alcoholics Anonymous (AA).
6. Most Americans use alcoholic beverages in moderation. There are times and places where drinking is not appropriate. Social drinkers learn to drink slowly and pace their drinking in order not to build up a high blood-alcohol level and become drunk. Hosts and hostesses should feel a certain responsibility for the amount of alcohol their guests drink; they should never allow a drunk guest to drive home.

CHAPTER 7
SMOKING

"WARNING: Smoking may be harmful to your health." This is probably the greatest understatement of all time. Smoking *is* harmful to individuals and to society in many ways. In the surgeon general's 1979 report on smoking and health the then secretary of Health, Education, and Welfare, Joseph Califano, wrote: *"Smoking is the largest cause of death in America,"* and "smoking is public health enemy number one in America."

Tobacco is an $8-billion-a-year industry, and every smoker is expected to pay his or her share. The average smoker experimented with cigarettes between the ages of 9 and 11, was smoking fairly regularly by 14, and was "hooked" by 17. Over a lifetime, the average smoker has spent $8,000 on cigarettes and given it to the tobacco industry. The advertising industry has "bought" and "sold" them to some brand, just like a slave on the auction block 150 years ago.

In 1970, Congress passed the Public Health Cigarette Smoking Act, which prohibited the broadcasting of cigarette commercials on TV or radio after January 2, 1971. Cigarette advertising has two purposes. First, it helps develop brand loyalties by emphasizing minor differences between basically similar tobacco products. This goal is accomplished by creating an image of the product and of the type of people who choose a particular brand. Second, cigarette advertising encourages young people to take up smoking in the first place, which is the reason the government has sought to control the cigarette advertising.

An array of cigarette ads designed to appeal to women. The ads all appeared in "women's" magazines.

The cigarette advertising industry is immense and powerful. From 1970 on, the cigarette companies have strengthened their campaigns in the printed media. As anyone can notice in magazines, newspapers, and billboards, a large percentage of all ads are cigarette advertisements. The industry is able to spend between 90 and 120 percent more money promoting cigarettes than agencies such as the American Cancer Society and the American

● ASK YOURSELF

1. Ask some smokers if they read the warning on cigarette packages. What do they think about it? If you are a smoker, ask these questions of yourself.
2. Have smokers calculate the amount of money they have spent on cigarettes so far in their smoking lives. Then have them calculate how much more they will spend on cigarettes during the rest of their lives, taking into account their reduced life expectancies.
3. Quiz smokers on which brands they smoke. Find out just how loyal they are to specific brands. Does this loyalty warrant the amount of money manufacturers spend on establishing brand loyalties? Give the reasons for your answer.

Heart Association can spend on anticigarette campaigns.

THE EXTENT OF SMOKING

The use of tobacco by Americans has increased tremendously since the turn of the century. In 1900 the consumption rate was less than 50 cigarettes per year per person (both smokers and non-smokers, 18 years of age and over). Today smokers start younger and smoke far more cigarettes. The bulletin "1978 Cancer Facts and Figures" (published by the American Cancer Society) explains that there are 53.3 million smokers in America, 45.6 million over 21 years of age and 7.7 million between 12 and 20. The total consumption of cigarettes is about 617 billion, or 15,530 cigarettes per smoker per year (2 packs a day for a year, or $40 \times 365 = 14,600$ cigarettes).

Since 1970 the number of adult smokers, especially men, had declined. From 1964 to 1975, the percentage of adult males who smoked dropped from 52 to 39 (a decline of 13 percent), while percentage of adult female who smoke dropped from 34 to 29 (a decline of only 5 percent). During this same period, teen-age girls (12 to 20 years of age) increased their smoking from 5 percent to 27 percent, while males within the same age group stayed

at 30 percent. Contributing to this rise in smoking by women is aggressive advertising in magazines.

THE EFFECTS OF SMOKING

Tobacco contains about 4,000 known chemical compounds, including nicotine. Some of the substances found in tobacco remain in the ashes of a burned cigarette. Other chemicals are greatly changed during the burning process. Moreover, additional compounds are produced during combustion, and it is some of these materials that are of great concern to scientists and physicians. An analysis of the cigarette smoke that enters the human body has been the primary aim of many studies.

Cancer-causing Substances

At least fifteen and possibly more compounds found in cigarette smoke are known to be *carcinogens*—cancer-causing substances. In addition to those known carcinogens, cigarette smoke also yields substances that have not yet been tested to determine their cancer-causing properties. Also present are *hydrocarbons*—chemicals closely related to the chemicals in gasoline. As explained earlier in the section on solvents, the long-term effects of hydrocarbon inhalation may cause death.

Evidence has accumulated that explains the processes and chemicals in tobacco smoke that cause the many diseases associated with smoking. The amounts of carcinogenic chemicals in tobacco are very small; in some cases they are measured in fractions of micrograms. But the normal functioning of the cell is a very delicate process. Constant irritation over long periods of time allows these carcinogens to change normal cells into cancerous cells. The chances of cancer for nonsmokers and smokers, verified by scientific investigations, are described and summarized in Table 7.1, Expected and Actual Death Rates for Cigarette Smokers. This table also shows the increased death rates from cancer expected for heavy smokers in any one year.

The most common type of cancer found in smokers is lung cancer. More people die of lung cancer than of any other type—the majority of them being male smokers. The number of victims has risen sharply during the past thirty years. Lung

TABLE 7.1 EXPECTED VS. ACTUAL DEATH RATES FOR CIGARETTE SMOKERS*

Underlying cause of death	Expected number of deaths in the general population	Actual number of smokers dying in general population	Increased ratio of smoker deaths
Cancer of lung	170.3	1,833	10.8 to 1
Bronchitis and emphysema	89.5	546	6.1 to 1
Cancer of larynx	14.0	75	5.4 to 1
Oral cancer	37.0	152	4.1 to 1
Cancer of esophagus	33.7	113	3.4 to 1
Stomach and duodenal ulcers	105.1	294	2.8 to 1
Other circulatory diseases	254.0	649	2.6 to 1
Cirrhosis of liver	169.2	379	2.2 to 1
Cancer of bladder	111.6	216	1.9 to 1
Coronary artery disease	6,430.7	11,177	1.7 to 1
Other heart diseases	526.0	868	1.7 to 1
Hypertensive heart disease	409.2	631	1.5 to 1
General arteriosclerosis	210.7	310	1.5 to 1
Cancer of kidney	79.0	120	1.5 to 1
All causes of death†	15,653.9	26,223	1.7 to 1

*This table shows the expected and actual deaths for smokers of cigarettes only and the ratios of such deaths to expected deaths in the general public.
† Includes all other causes of death as well as those listed above.
Source: Adapted from U.S. Department of Health, Education, and Welfare, *The Health Consequences of Smoking* (Washington, D.C.: 1973).

cancer now exceeds automobile accidents as a significant cause of death.

Cardiovascular Diseases

Recently, studies of large groups of people have shown that cigarette smokers are more likely to die of certain cardiovascular disorders than nonsmokers. These diseases of the heart and blood vessels are the most common causes of death in our population. A cause-and-effect association has theoretically been established between cigarette smoking and the incidence of coronary attacks in humans, especially among men between 35 and 55 years of age. The risk of death for male cigarette smokers in relation to nonsmokers is greater in middle age than in old age. Statistics indicate that smokers are often struck down with disease when they should be most active and enjoying life. They also imply that men who stop smoking have a lower death rate from coronary diseases than those

who continue to smoke. Heart disease is discussed in detail in Chapter 20.

Respiratory Diseases

Smoking is linked to the development and progression of respiratory diseases, such as bronchitis and emphysema. Air pollution and respiratory infections, as well as smoking, cause and aggravate chronic bronchitis and emphysema. Any pollutant, condition, or infectious agent that can cause permanent damage to the respiratory system can be linked with these diseases. However, smoking causes an increased irritation above and beyond the pollutants and irritants commonly encountered.

All of the effects of cigarette smoke on the tissues of the body are damaging. The actual role of cigarettes in the production of diseases is great because of a combination of harmful factors. Any one of them could be responsible for damage, but together they are deadly. As explained in the Jan-

● ASK YOURSELF

1. What is the most common disease linked to smoking? Have you ever known someone who had this disease? If so, interview this person about the disease, if possible.
2. What is the meaning of the term *carcinogenic*. How does a carcinogen affect a cell? The American Cancer Society publishes a pamphlet listing the many carcinogens we are exposed to every day in our environment. How does smoking add to this exposure.?
3. At what period of life is a smoker most prone to a heart attack?

Is it fair to force your baby to smoke cigarettes?

uary 12, 1979, *Morbidity and Mortality Weekly Report* of the U.S. Center for Disease Control, "Overall, current cigarette smokers have an approximately 70 percent greater chance of dying from disease than nonsmokers." A few of these diseases are discussed in detail in Chapters 20 through 22.

Smoking and Fetal Development

It is very disturbing that the greatest number of women who smoke (38.6 percent of all adult women) are between the ages of 25 and 44, which includes the critical childbearing years (25 to 34). Women who smoke during a pregnancy affect two lives: their own and that of the developing child.

While the bloodstream of a pregnant woman is completely separate from that of her developing baby, certain components in her blood can pass into the blood of the fetus. When the pregnant woman smokes, harmful gases and poisonous chemicals in the smoke cross the placenta into the baby's blood. One of the gases found in smoke is *carbon monoxide* (CO). This chemical replaces oxygen in the red blood cells, greatly reducing the amount of oxygen available to the fetus. Nicotine further reduces the oxygen available to a fetus by narrowing the blood vessels in the placenta and the developing baby. This lack of oxygen reduces the fetus's growth rate. Thus, the unborn babies of mothers who smoke during pregnancy do not develop as fast and are more likely to be undersized at birth. These smaller babies also have a greater chance of dying soon after birth. Studies have also shown that women who smoke during pregnancy produce more still-

births and spontaneous abortions (miscarriages) than nonsmoking women.

Smoking and Children

Infants have very small lungs and lung airways. Breathing smoke-filled air causes these tiny airways to contract and become even smaller. This can impair or block breathing. Also, infants and children breathe much faster than adults; consequently, they breathe in more polluted (smoke-filled) air. This may lead to pneumonia and bronchitis. Parents with lung ailments may spread their infections to their children. Acute respiratory problems are more common in children if their parents smoke.

Parental smoking is a major factor in motivating youngsters to smoke. Few parents want their children to take up a habit that destroys lungs and shortens lives.

THE SMOKING HABIT

All evidence indicates that there are two types of dependent smokers. One group consists of social smokers and their dependence is mostly psychosocial; these people smoke for social reasons. Stopping smoking is relatively easy for them. The other group consists of those who are physically dependent, or "addicted" to the nicotine in tobacco. These people have a difficult time overcoming their

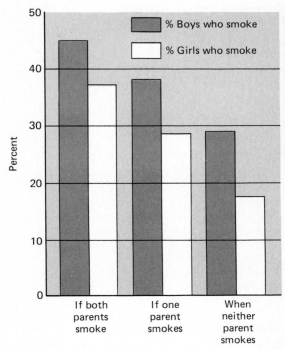

Percent

| % Boys who smoke |
| % Girls who smoke |

If both parents smoke / If one parent smokes / When neither parent smokes

Smoking habits of teenagers and their parents. Most teenage smokers come from homes where one or both parents smoke. A strong relationship exists between parents' and youngsters' smoking habits. Parents' smoking habits influence the age at which children take up smoking more than they influence whether the children will be smokers or nonsmokers. According to many authorities, the most effective way to cut down smoking among young people is to decrease it among their parents. (From American Cancer Society, Profile 1970.)

dependence, and they experience withdrawal symptoms.

Is Tobacco a Dangerous Drug?

Tobacco is not legally classified as a dangerous drug. For the most part, this is because it does not cause the drastic mood modifications or behavior changes found among those who abuse more potent drugs. However, smokers do exhibit mood modifications when they have to do without cigarettes. These modifications are not dangerous enough to society to warrant legal controls; but tobacco should be classified as a "socially acceptable"

A very young smoker. His smoking habit seems well established.

mood-modifying, psychoactive drug. Also, although the immediate effects of tobacco are mild, the overall, long-range physical damage to the body is drastic. There is enough damage, in fact, to classify smoking as a form of drug abuse.

Starting to Smoke

Most smokers begin smoking before they are 20 years of age. Young people begin puffing on cigarettes because of curiosity, rebelliousness, pressures from friends, or the desire to appear more

● **ASK YOURSELF**

1. What chemicals in tobacco smoke are the most damaging to a developing child? Ask some mothers who smoke how much their children weighed at birth. Were these weights less than average?

2. Have any of these women experienced miscarriages or stillbirths?

TABLE 7.2 HOW MUCH DO YOU ACTUALLY SMOKE?

Cigarettes smoked per day		Puffs* per day	Cigarettes smoked per year	Puffs per year
(½ pack)	10	80	3,612	28,896
(1 pack)	20	160	7,224	57,792
(1½ packs)	30	240	10,836	86,688
(2 packs)	40	320	14,448	115,584

*Assuming that a person takes 8 puffs from one cigarette.

grown-up and sophisticated. Often girls begin to smoke because their first boyfriends smoke. Most individuals past 20 start smoking during times of stress or unusual disruption in their lives (college pressures, divorce, deaths of loved ones). People must *learn how to smoke,* which takes time and practice. Smokers must learn how to adjust their depth of inhalation in order to inhale a mixture of smoke and outside air that will not make them sick.

Continuing the Habit

Nicotine Newly acquired smoking behavior sets smokers up for the chemical motivator: *nicotine.* Besides being poisonous, nicotine is a powerful *reward* chemical for humans and other animals. People enjoy the feelings nicotine produces. Nicotine reaches the brain in about 7½ seconds after inhaling. The pleasurable effects of nicotine last for 30 to 40 minutes. This is why regular smokers commonly consume at least one cigarette every half hour. It is not uncommon for a smoker to wake up during the night to "have a cigarette." Smokers strive to reach a personally satisfying level of nicotine in their daily smoking routine. When smoking reduced-nicotine cigarettes, they compensate by either inhaling deeper or smoking more cigarettes. Each puff on a cigarette equals one "dose" of nicotine. Average smokers take at least eight puffs from each cigarette they smoke. Thus, one pack of twenty cigarettes contains at least 160 puffs. If a one-pack-a-day smoker smokes 7,224 cigarettes a year, he or she receives at least 57,792 doses of nicotine per year.

Psychological reinforcement The act of smoking regularly is enough to establish it as a major part of a smoker's daily routine. Because smoking is socially acceptable, it is possible to smoke in almost any social situation or during almost any physical activity. When associated with pleasurable activities, the positive reinforcement makes the act of smoking even more pleasurable. Often smokers link together two pleasurable events, such as a satisfying meal or finishing work, and a cigarette.

Smoking can also be an escape mechanism. It supplies a convenient way to escape tension, anxiety, embarrassment, boredom, a break in the daily routine, or fatigue.

No other substance provides so many kinds of positive reinforcement, is so cheaply available, and can be used in so many different settings. The cigarette advertising agencies have much to work with in enticing the public to use cigarettes.

BREAKING THE SMOKING HABIT

The ability to stop is related to the forces that led to the smoking habit in the first place. Most important are the number of cigarettes smoked per day and the number of years of smoking. The ability to stop smoking has consistently been found to be highest among those who started late in life and whose average cigarette consumption is low.

Of all the substances in tobacco smoke, only nicotine has effects that produce the dependence associated with smoking. This statement is based on established scientific facts concerning the properties of nicotine, the descriptions of symptoms given by those who smoke (or try to stop smoking), and comparisons made with other drug abuses.

Steps Toward Reducing the Habit

If you are unable or unwilling to quit, at least try to reduce your consumption of cigarettes as a means of decreasing the harmful effects of smoking. The National Clearinghouse on Smoking and Health recommends the following five steps to reduce ciga-

These two people obviously enjoy smoking.

rette consumption to a level where smokers may be able to break the habit altogether.*

1. *Choose a cigarette with low tar and nicotine.* The U.S. Department of Health and Human Services publishes a list of the tar and nicotine content of cigarettes every six months. See how your brand compares and find out how much you can reduce your tar and nicotine intake by switching to another brand. Will such a switch result in your smoking more? Probably not. Most smokers who make such a change either continue to smoke at their previous rate or even smoke less. One possible reason for the lower rate is that nicotine and tar contribute to the taste of a cigarette. Lower tar and nicotine cigarettes will not seem as flavorful to a smoker who is used to high tar and nicotine concentrations. This can result in finding smoking less enjoyable.

2. *Don't smoke your cigarette all the way down.* No matter what cigarette you smoke, the most tar and nicotine is found in the last few puffs. The sooner you put your cigarette out, the lower your dose of harmful ingredients. This same fact also points up the added risk of extra-long cigarettes. Their extra puffs are really extra perils for you.

3. *Take fewer draws on each cigarette.* With practice, some people find they can substantially cut their actual smoking time without really missing it.

4. *Reduce your inhaling.* Easier said than done? Perhaps. But remember, it is the smoke that enters your lungs that does most of the dam-

* *If You Must Smoke,* Public Service Publication No. 1786 (Washington, D.C.).

● **ASK YOURSELF**

1. What are the two types of smokers? Interview some smokers and see if they sort themselves out into these two types.
2. After your interviewing, do you believe there is such a thing as a *nicotine addict?*
3. Ask some ex-smokers to describe their withdrawal symptoms when they were giving up smoking.
4. How does nicotine help make smoking part of a person's life style?

age; it causes lung cancer and creates the cardiovascular changes that can bring on heart attacks.

5. *Smoke fewer cigarettes per day.* Pick a time of day when you promise yourself not to smoke. It may be before breakfast, or while driving to school or work, or after a certain hour each evening. It's always eaiser to *postpone* a cigarette if you know you will be having one later. Maybe you're a pack-a-day smoker. Buy cigarettes one pack at a time. Try buying your next pack an hour later each day. It may also help to carry your cigarettes in a different pocket. Or, at work, keep them in a drawer of your desk or in your locker — any place where you aren't able to reach for one automatically. The trick is to change the habits you have developed over the years. Make a habit of asking yourself, "Do I really want *this* cigarette?" before you light up. You may be surprised how many cigarettes you have been smoking that you don't really want.

Five-Day Plan

You must have strong motives to stop smoking. In the spaces below, write down your reasons for quitting.

1. _____
2. _____
3. _____

Day 1. *Keep a record*   Keep a *smoking diary* for the next five days of every cigarette you smoke. In the form below, write down *what you were doing* when you smoked (situation) and the *intensity* of each smoking urge. Use the following intensity scale for your diary: I want a cigarette:

    1. A little (not really at all)
    2. Somewhat (perhaps)
    3. Moderately (a vague desire)
    4. A lot (a need)
    5. Badly (a craving)

Smoking Diary:

Situation Hour	Day 1 A.M.	P.M.	Day 2 A.M.	P.M.	Day 3 A.M.	P.M.	Day 4 A.M.	P.M.	Day 5 A.M.	P.M.
12										
1										
2										
3										
4										
5										
6										
7										
8										
9										
10										
11										

Day 2. *Know your habit*   In your smoking diary you will see a pattern of situations that "trigger" your urge to smoke. Some will be stressful triggers, others will be relaxed triggers. Write your triggers in the following form:

1. _____
2. _____
3. _____
4. _____
5. _____

Day 3. *Do something instead*   You smoke when you are under stress, and you smoke when you are relaxing. This step gets you to do other things at either time.

Step 1: Learn to relax. When you are feeling stress, instead of smoking:

1. Concentrate your thoughts on something pleasant
2. Tense and relax your muscles
3. Breathe deeply

Step 2: Do something else. Stay active: Take a walk, find a hobby, putter in the garden.

Day 4. *Learn to quit permanently*   Improve your chances of becoming a permanent ex-smoker by predicting things and situations that will tempt you to smoke.

Step 1: List what future situations might tempt you to smoke:

1. _____
2. _____
3. _____

Step 2: List the things you plan to do in the future when you are tempted to smoke.

1. _____
2. _____
3. _____

Day 5. *Sign a contract with yourself*   Now you understand your smoking habit through your Smoker's Diary and you have the necessary skills to become a permanent nonsmoker.

Step 1: List your rewards. Write down all the benefits from not smoking.

1. _____
2. _____
3. _____

Step 2: Now sign a contract with yourself, a personal commitment to yourself that you will never smoke again.

My personal contract: I, (your name) _____ will quit smoking forever as of (time) _____, (day and date) _____. I will be saving $_____ per day by not smoking, and this money will be saved toward (personal reward) _____.

Source: Smoking Withdrawal Clinics, American Cancer Society.

Methods of Breaking the Habit

To stop smoking altogether, people must first stop ignoring the health dangers of smoking and accept smoking as a personal problem they must conquer. The habit can be broken in a number of ways. The most successful methods fall into the following categories: (1) Individual medical care provided by physicians or psychologists; (2) self-help programs based on books, magazines, lectures, and pamphlets (these services are available from such organizations as the American Cancer Society, American Heart Association, and others); and (3) withdrawal or "smoker's" clinics, conducted by hospitals or medical and health organizations.

Research has shown that the best way to quit smoking is to stop all at once. But facts and statistics will not lead smokers to quit; they need a program to follow. The first three days seem to be the hardest period to get through. Consequently, a number of "five-day plans" to stop smoking have

Students participating in a "Smoke Out." Smoking withdrawal programs were first organized by the American Cancer Society.

been very successful. The following five-day plan was designed by Dr. J. Wayne McFarland and has been highly successful.

Five-day plan to stop smoking Through the use of will power, all smokers can bring their habits under control. The purpose of this program is to help smokers overcome the craving for a cigarette as fast as possible. Some people can quit in three days, the majority do it in five, while others will need to follow this program for ten days. Any smoker who stays with the program for ten days will make it.

Smokers should never think about *when* they are going to stop smoking. They should just get up one morning and say to themselves, "I *choose* not to smoke!" They should keep repeating this decision throughout the day. From the first time they open their eyes in the morning until the last thought at night, they should think: "I choose not to smoke." The mind has great power over the body; how we think and use our willpower has an immediate effect upon our body's craving for a cigarette. Smokers should never forget this fact. This strong, posi-

tive decision exerts an immediate effect on a smoker's physical craving to smoke. Many feel the urge to smoke weakening as soon as *they* have made the decision not to smoke.

Through the correct use of will power, smokers not only weaken the craving for a cigarette, but also bring the habitual part of smoking under control. Whenever smokers strongly crave a cigarette, they should watch the second hand of their watches as it sweeps around the dial. Regardless of how strong the urge to smoke, they can certainly keep from smoking for a *mere* sixty seconds. With one minute gone, they can now hold on for one more minute. As the third minute elapses, they will usually find that the craving has peaked and is weakening. As minutes turn into hours they will need all the will-power they have. The following suggestions will help smokers over the peak urges and put them on the way to becoming *ex-smokers.*

1. Water on the outside   This is the time to give yourself some luxury. Take a warm bath two or three times a day for fifteen to twenty minutes at a time. Just relax and enjoy it. If you have times when you feel you cannot stand it any longer, jump into a warm shower. It is pretty hard to smoke in a shower.

Have you ever rubbed your arm hard and noticed how stimulated it feels when you are done? Here is another relaxing procedure: Get up in the morning a few minutes early. In a warm bathroom fill the washbasin with cool water. Dip a washcloth into the cool water, wring it out thoroughly. Now, firmly rub your arm until the skin begins to glow and a pink color appears. The pink color is due to increased peripheral blood circulation. It may require considerable rubbing before your skin turns pink. Your peripheral blood vessels have become very sluggish due to poor circulation from smoking.

On only the second morning you will notice that your blood vessels are more easily dilated by rubbing. Rub both arms. Also, using progressively cooler water each morning will increase the tonic effect of the rubbing. On the third morning add the chest. On the fourth the legs, so that your entire body is now stimulated by cold rubbing. This cold vigorous rubbing will make you feel wide awake and stimulated without triggering the craving for a morning cigarette.

● ASK YOURSELF

---

1. List at least five *personal* reasons for smoking.
2. List at least five *personal* reasons for *not* smoking.
3. If you cannot stop smoking today, what can you do to reduce the amount that you smoke?

---

2. Water on the inside   Drink six to eight glasses of water between meals. Keep a record if necessary. The more liquids you drink, the quicker the nicotine leaches out of your body. Don't drink any alcoholic beverages. Keep to fruit juices. The brain is approximately 75 percent water. The nervous system does not function properly without adequate fluids. Besides removing the nicotine from your system these fluids allow your nervous system to work better.

3. The importance of rest   Obtain sufficient sleep at night, especially during these five days. Some smokers habitually stay up till the national anthem closes the TV station. Plan to retire earlier than usual. You are going "all out" to conserve your nervous energy and give those nerves a good rest.

4. No sitting around after meals   *Get outside, and walk.* Do not sit down in front of the TV after dinner. This is the time of day you want that cigarette most. If you can't go out, find something to do. Work at your favorite hobby, or do the dishes.

5. Be careful what you drink   Do not drink alcohol, tea, coffee, or cola. Avoid all depressants and stimulants. You want to build up your reserves as quickly as possible. For a cold drink, try milk or buttermilk. For a hot beverage, try warm water with some lemon in it or a cereal drink.

Many ex-smokers have made excellent progress until a fateful afternoon or evening when they drop into their favorite bar or go to a cocktail party. They take a drink and one cigarette because "What is a drink without a cigarette?" Half a pack later, they are still trying to figure out what made a shambles of their willpower.

6. Eat the best possible way   Heavy smokers often like their food highly spiced. They also eat too much red meat, gravy, fried foods, and rich desserts. Tobacco deadens taste buds, and smoking makes the tongue crave heavy, highly spiced, or sweet food. Substitute fish or fowl for red meat. Dispense with rich pastries and desserts. Give your body and nerves the best possible chance to weather the storm of nicotine withdrawal.

Eat mostly fruits, grains, vegetables, and nuts at regular meals, and little or nothing in between. Eat a generous breakfast and lunch and a light dinner. Reduce all your servings by one third. For your oral cravings, eat sugarless hard candies and chew sugarless gum. Snack only on vegetable sticks.

During withdrawal from nicotine the body needs extra amounts of B complex vitamins. Put one or two tablespoons of bran on your sugarless, hot, cooked cereal in the morning. Or stir a tablespoon of brewer's yeast into a glass of tomato juice, hold your nose, and drink it down.

7. Avoid special tablets, pills, or aids   While these may help, they have not been overly successful. Preparations containing *lobeline,* a drug with actions similar to nicotine, are sold over-the-counter at drugstores. Smokers take lobeline products in ever-decreasing amounts after they have stopped smoking.

Recently a new product known as *Nicorette* gum has been introduced as an aid in stopping smoking. It contains nicotine, but in smaller amounts than cigarettes. The idea is to withdraw the person from cigarettes and then, at a later date, withdraw them from the Nicorette gum.

Now for a word of warning: Don't forget that just below the surface of your willpower lies a once well-established addiction. It is ready, without warning, to unleash a craving for a cigarette. Keep your guard up. Do not become careless in your eating, drinking, working, or sleep habits. Your job, now that you are an ex-smoker, is to establish the habit of not smoking just as firmly as you once established the habit of smoking.

BENEFITS TO EX-SMOKERS

Little is known of the relationship between not smoking and normal health. The main areas of study have so far been confined to the relationships between smoking and disease. These studies are just beginning to show which body changes in response to cigarettes are transitory and which are permanent.

In a study based on twelve years of experience in conducting smoking withdrawal clinics. Dr. Borje E. V. Elrup, clinical associate professor of medicine, Cornell Medical Center, New York, has been able to demonstrate the following anatomical and physiological changes in ex-smokers.

How to Quit Smoking and Not Gain Weight

When you stop smoking, three things begin to happen immediately:

1. Your body's metabolism begins to work more efficiently. Thus, you utilize food more efficiently.
2. For years your taste buds have been deadened by nicotine. Now your food tastes better, and you are tempted to eat more.
3. For years you have been used to having something in your mouth—a cigarette. You are tempted to indulge in between-meal snacks.

Here is a list of ten tips to help you stay healthy and keep your weight down after you quit smoking:

1. Eat a good breakfast.
2. Eliminate between-meal snacks.
3. Eat one normal helping at a meal, then get up from the table. If you still are hungry, tell yourself you will eat more in 15 or 20 minutes—you won't be as hungry then.
4. Remove as many empty calories as possible from your diet.
   a. Eliminate or cut down on all fats and oils (Crisco, Mazola oil, margarine, butter, and salad dressings).
   b. Reduce your intake of sugar by eliminating:
      1. Desserts
      2. Jams and jellies
      3. Sweetened cereals or sugar on breakfast cereals
      4. Soft drinks or fruit drinks (Use natural, unsweetened fruit juices.)
      5. Canned fruit (Use fresh fruit.)
   c. Use *unrefined* cereals and grains by substituting:
      1. Brown rice for white
      2. Whole-wheat bread for white
      3. Cooked cereals (with fruit if you want) for dry (*See* Chapter 8.)
   d. Avoid alcoholic beverages.
5. Drastically reduce your intake of animal fats by:
   a. Eliminating red meat whenever possible (Substitute fish and skinless fowl.)
   b. Eliminating butter fat (Use nonfat milk or buttermilk and no cheese except skim-milk cottage cheese.)
   c. Using spare amounts of spread such as avocado, diet margarine, or peanut butter in place of butter or regular margarine
6. Eat a light fruit or vegetable supper or none at all.
7. Exercise regularly
8. Get adequate rest
9. Have regular checkups
10. Use the stress-management techniques recommended in Chapter 1

Digestive and eating patterns start to change even during withdrawal from cigarettes. Soon after a person stops smoking, intestinal motility decreases, often causing constipation for a short time. The absorption of food is greater, the appetite is better, and both the taste of food and the sense of smell are improved—all contributing to a weight gain.

Patterns of circulation also change. Ex-smokers are less tired. They often arise earlier in the morning and are more alert during the day. Skin circulation improves. The complexion of the face can be seen to change for the better, even during the process of stopping. Increased circulation helps to slow the pulse rate, reduce blood pressure, and increase heart efficiency—both at rest and after exercise.

Responses from the respiratory tract show a decrease in breathing rate, as well as an increase in maximal breathing capacity, and a better exchange of oxygen between the lungs and the circulatory system. Such respiratory conditions as chronic bronchitis improve and coughing disappears during withdrawal. Emphysema patients are able to breathe more easily and many asthmatic conditions improve substantially. Also, if lung cancer does not develop within two years, the ex-smoker has no more of a chance of developing lung cancer than the nonsmoker.

RIGHTS OF THE NONSMOKER

Studies have shown that exposure to a "smoking environment" causes measurable body effects. These include increased heart rate, blood pressure, and amount of carbon monoxide in the blood. Other possible effects individuals may feel include eye and nose irritation, headache, sore throat, cough, hoarseness, nausea, and dizziness. Out of regard for nonsmokers, air carriers have agreed to set aside nonsmoking areas. The American Medical Association has asked member physicians to keep people from smoking in their waiting rooms. In 1971 the Interstate Commerce Commission issued a regulation requiring separate seating on all interstate buses for smokers and nonsmokers. Since this time hotels have set aside nonsmoking floors. States

● ASK YOURSELF

1. What does a smoking withdrawal program try to establish in ex-smokers? Interview some ex-smokers and see if this is so.
2. Make a list of aids to help people quit smoking.
3. What are the benefits of quitting smoking? Can you think of others besides those listed in this chapter?
4. How would you go about enforcing this statement: *Make smoking socially unacceptable in your home?*
5. How can you help enforce the rights of nonsmokers?

and cities have outlawed smoking in public buildings and on public transportation. And many restaurants have nonsmoking areas.

Smoking in the presence of a nonsmoker should be considered "an act of aggression." Cigarette smokers in a crowded, ill-ventilated room or automobile can raise the level of carbon monoxide to a dangerous point. Experiments show that in a small room a smoker can raise the level of carbon monoxide to 50 parts per million. At this level, after an hour and a half, a nonsmoker can have trouble discriminating time intervals and visual and auditory cues. The right of smokers to enjoy their habit is frequently cited in opposition to antismoking regulations. However, the rights of nonsmokers to a clean, smoke-free environment must also be recognized. Nonsmokers definitely should feel free to discourage smoking in their presence, especially in their homes.

Although quitting smoking is seldom easy, the effort required may be handsomely rewarded. For example, two-pack-a-day smokers will realize savings of $475 per year (average cost of cigarettes is $.65 per pack). Those who stop will have a cleaner mouth and will require less professional cleaning work by a dentist. Smokers who stop, especially women smokers, will keep their own teeth longer. They will not run the risk of burned clothing, upholstery, or furniture. Added years of good health and fewer chances of early death are, of course, the most important gains. If you smoke—quit now. If you don't smoke—*don't start.*

IN REVIEW

1. Tobacco is the largest single cause of death in the United States today. Tobacco is an $8-billion-a-year industry. Cigarette advertising has two purposes: to develop brand loyalties and to encourage young people to smoke.
2. Smoking has increased tremendously since the turn of the century. Since 1970, however the proportion of adult smokers has declined.
3. Tobacco smoke contains over 4,000 known chemical compounds. Many of these cause damage to the human body.
   a. At least 15 compounds found in cigarette smoke are known to cause cancer. The most common type of cancer linked with smoking is lung cancer.
   b. Cigarette smokers are much more likely to die of heart attacks than nonsmokers. This is especially true of men between 35 and 55 years of age. Also, men who stop smoking have a lower death rate from coronary diseases than those who continue to smoke.
   c. Smoking is linked to the development and progression of respiratory diseases. All of the effects of cigarette smoke on the tissues of the body are damaging.
   d. Women who smoke during a pregnancy are damaging two lives: their own and their developing child's. Babies have a greater chance of dying at birth from being undersized or premature.
   e. Parental smoking is a major factor in motivating children to smoke. Smoking in the home also harms growing children in many ways.
4. There are two basic types of dependent smokers: the psychosocially dependent and the physically addicted. The first type experiences very little discomfort when stopping smoking. The other type experiences withdrawal symptoms when stopping.
   a. Tobacco is a socially acceptable, legal, mood-modifying drug. There is enough long-term damage to classify smoking as a form of drug abuse.
   b. Most smokers begin to smoke before they are 20 years of age. Often girls begin to smoke because their first boyfriends smoke.
   c. People continue to smoke because of social reasons and because they become addicted to nicotine. Nicotine is a powerful reward chemical in humans and animals. A one-pack-a-day smoker smokes 7,224 cigarettes a year, which yield 57,792 doses of nicotine.
   d. The act of smoking regularly establishes it as a major habit and part of a person's daily routine. No other substance provides so many kinds of positive reinforcement, is so cheaply available, and can be used in so many different settings as tobacco.
5. The ability to stop smoking is related to the forces that led to the smoking habit in the first place.
   a. Smokers who are unable to quit should at least try to reduce their consumption of cigarettes. The five steps in the chapter will help smokers reduce their smoking to a level where they may be able to break the habit altogether.
   b. Smokers who want to stop must first recognize the health dangers of smoking and accept their habit as a personal problem they must conquer.
   c. Research has shown that the best way to stop smoking is to stop all at once.

Five-day plans to stop smoking help overcome cravings for cigarettes and strengthen willpower.

6. Ex-smokers enjoy a number of benefits, especially improved physiological functioning.

7. The exposure to cigarette smoke causes measurable effects in the body. Smoking in the presence of a nonsmoker is a form of aggression.

SECTION

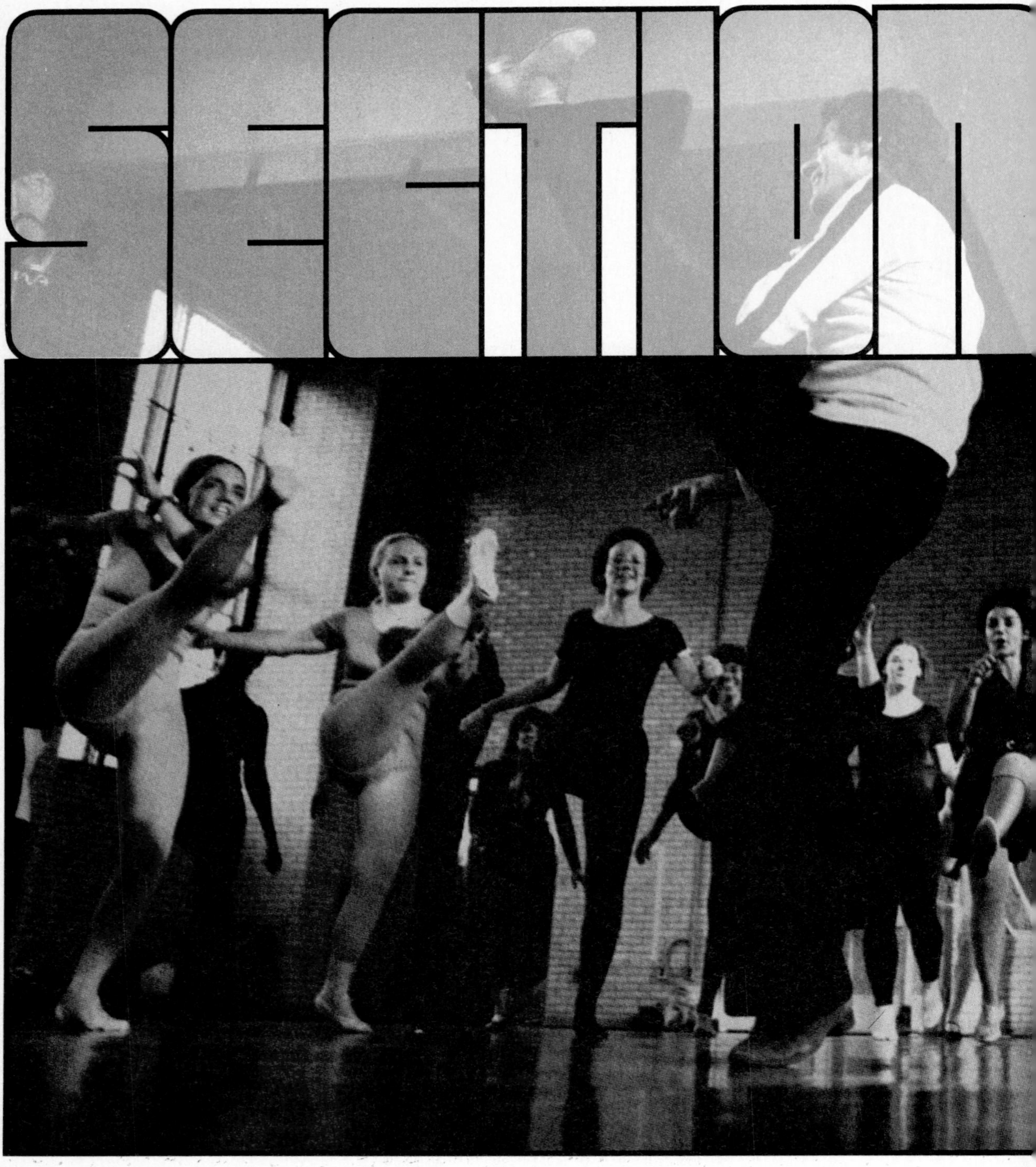

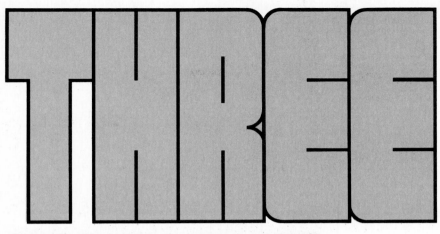

THREE

GOOD HEALTH AND THE MARKETPLACE

Everyone should be able to select proper foods, health services, health professionals, and personal services, such as weight control and fitness programs. Proper selection of these services is one of our responsibilities in maintaining health for ourselves and our families.

The modern supermarket contains thousands of products, ranging from foods, to "nonfoods," which contain no nutrients, to over-the-counter drugs and cosmetics. Everyone should learn to make intelligent selections when purchasing any product.

The ability to maintain a proper weight requires the proper selection of food and participation in a regular schedule of exercise throughout life. Exercise should be done at a level that produces total fitness. Exercise at an improper level produces only fatigue or physical damage to the body. Instruction in a proper fitness program is imperative for good health.

The purchase of health services is one of the major expenses in each of our lives. These services should be selected very carefully. Many health services are dispensed by inadequately trained individuals, quacks, and charlatans. Everyone should be able to select the proper health services or product they need and know who is qualified to dispense that service or product.

CHAPTER 8
NUTRIENTS AND NUTRITION

A food can be defined as any substance—besides air and medicines—that provides for the body's growth, maintenance, repair, and reproduction. The substances in food that perform these functions are called *nutrients*. Food may be either natural or man-made, may come from plants or animals, and is the source of all nutrients. Nutrients are chemicals that:

1. Provide energy for body activities
2. Provide materials for growth and maintenance of body tissues
3. Provide substances that regulate body processes

WHAT IS FOOD?

Although the hundreds of substances we consume as food (bread, steak, salad, ice cream, and so on) show little similarity, the nutrients they provide fall into only six chemical classes.

Carbohydrates and Fiber

The carbohydrates consist of the simple carbohydrates (the sugars) and the complex carbohydrates (starch and cellulose). For the majority of people in the world today, the complex carbohydrate starch is the most important source of energy.

Fiber Fiber consists of the complex carbohydrate *cellulose*. Cellulose is present in plant foods. It is *not* converted into energy in humans, because we lack the enzymes necessary for its digestion. Fiber is usually referred to as the "bulk" or "roughage" in our diet. Fiber performs many functions in the digestive tract. It stimulates intestinal activity and increases the movement of undigested materials through the digestive tract. It reduces the absorption of cholesterol. It also removes water and bacterial toxins from the tract, and alters the types of bacteria that flourish in our large intestine.

Carbohydrates Carbohydrates consist of one or more simple sugar units. The simple sugars are: glucose (also known as *dextrose*); fructose, found in many fruits and honey; and galactose, found in milk. Examples of carbohydrates consisting of two simple sugar units connected together (compound sugars) are: sucrose, which is table sugar (cane and beet sugars are identical); maltose, produced by germinating grains; and lactose, found only in milk.

Carbohydrates consisting of long chains of simple sugar units connected together are called *starches*.

Simple sugars require no digestion; they are ready to be absorbed into the blood. Double sugars, such as table sugar, are quickly digested into simple sugars. Starches are more slowly digested to simple sugars.

After simple sugars are absorbed into the body, they are converted by the liver into *glucose*. Glucose is the only carbohydrate that can be used by the body cells as a source of energy. The liver and muscles convert glucose into the starch called *glycogen*, which is then stored and kept available for rapid conversion back to glucose when extra muscular energy is needed.

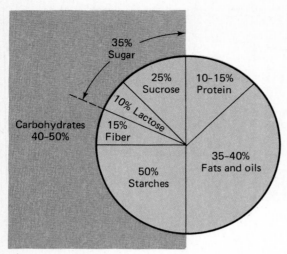

The current average American diet.

If digestion reduces all starches and compound sugars into simple sugars, it might seem to matter little whether we consume our carbohydrates as sugars or starches. Apparently it matters a lot. One problem is that foods that are high in sugars — candies and jams, for example — often lack any other nutrients. They represent "empty" calories — foods that contain only calories and no other nutrients.

Another disadvantage of sugars is that they are quickly absorbed into the bloodstream. The body responds by releasing a burst of insulin to lower the blood-sugar level. Much of the sugar is converted to body fat and within two or three hours you are hungry again. Not only does this eating pattern promote excess fatness, but it is a leading cause of one form of diabetes. Apparently, the frequent bursts of insulin eventually cause the body cells to become insulin resistant.

● ASK YOURSELF
━━━━━━━━━━━━━━━━━━━━━━━━━━━━━━━━

1. What are the functions of the different nutrients in the body?
2. What is the difference between *foods* and *nutrients?*
3. What is the major function of carbohydrates within the body?
4. What is fiber? What functions does fiber perform for the body?
5. Why should simple sugar foods be restricted in the diet?

The complex carbohydrates (starches) are digested more slowly and absorbed into the bloodstream gradually, over a longer period of time. High insulin levels are not required, and hunger does not return as soon.

Still another reason to avoid sugars is their contribution to tooth decay. Bacteria on the surface of the teeth convert sugars into decay-causing acids.

Fats and Oils

In addition to being a high calorie source, fats store the body's energy, form a part of the membranes of cells, act as carriers for the fat-soluble vitamins A, D, E, and K, and provide both insulation and protection for the body. Oils are fats that are liquid at room temperature.

A fat is made up of one glycerol molecule connected to three fatty-acid molecules. The human body is able to produce most of the fatty acids it needs through the conversion of carbohydrates. Fatty acids that the body needs but is unable to produce in sufficient amounts are called *essential fatty acids.* These must be obtained from our food. Fortunately, essential fatty acids are widely and abundantly distributed in foods such as meats, whole milk, cheese, nuts, olives, and fish. Such food products as butter, margarine, oils, and shortenings are almost pure fat.

Some fats are designated as "saturated" fats and others as "unsaturated" fats according to the amounts of hydrogen in the molecule — the more hydrogen, the more saturated the fat.

Proteins

Proteins form the soft tissues of the body. They make up the enzymes and many other chemicals produced by the cells. The building blocks of proteins are the *amino acids.* Proteins are made up of long chains of amino acids linked together. There are twenty naturally occurring amino acids, eight of which cannot be synthesized by the body. (These essential amino acids which must be derived from the proteins consumed in food are shown in Table 8.1.) Adults need eight essential amino acids, while children need nine. The other amino acids can be made from molecular pieces and other substances in the body.

Proteins containing all essential amino acids in significant amounts are called *complete proteins.* Proteins low in one or more of these amino acids are *incomplete.* Most animal proteins are complete,

TABLE 8.1 ESSENTIAL AMINO ACIDS

Essential amino acids	Dietary sources*	Problems when deficient	Problems when taken in excess
Phenylalanine Lysine Isoleucine Leucine Valine Methionine Threonine Tryptophan Histidine† (required by infants)	*Good sources:* 　Legumes (plants of the 　　pea family) 　Dairy products 　Red meat 　Fowl 　Fish 　Eggs *Adequate sources:* 　Rice 　Corn 　Wheat 　Soy *Poor sources:* 　Cassava 　Sweet potatoes	All amino acids must be in constant circulation. Reduced protein intake leads to retarded growth, senility, low energy levels, and two specific diseases: kwashiorkor and marasmus. Essential amino acids must be in the diet or death will result.	Excess protein intake can cause kidney damage and aggravate some chronic diseases.

*Plant sources need to be ingested together or within a short period of time to yield the proper balance of essential amino acids (*see* Table 8.2).
†Not always listed as an essential amino acid for adults; is essential to infants for proper growth.

while most plant proteins are incomplete. This is why it may be difficult for a vegetarian to obtain all the essential amino acids. Two incomplete proteins may be used to complement each other, if they are deficient in different amino acids. Table 8.2 shows how to supplement plant foods to insure obtaining all essential amino acids.

From the constitutent amino acids, the body makes the *enzymes* (chemical catalysts), hormones, secretions, and tissues it needs. These highly complex substances are formed by the literally millions of possible combinations of amino acids that the body can prepare.

In contrast to fats and carbohydrates, amino acids are not stored in the cells. Thus, a person needs a daily supply of protein in the diet. Table 8.3 on energy values compares the three basic energy-yielding food groups (carbohydrates, proteins, and fats). These are the only nutrients that can be converted into energy.

Minerals

Many mineral elements are found in the body. They may occur as simple compounds or be incorporated into very complex materials. Many of these elements (such as calcium, phosphorus, sodium, potassium, chlorine, magnesium, iron, sulfur, iodine, manganese, cobalt, copper, and zinc) perform essential functions in the body—they make up vital

● ASK YOURSELF

1. Explain the functions of fats and oils in the diet.
2. What is an essential fatty acid? Why are these needed by the body?
3. What is the difference between a "saturated" and an "unsaturated" fat?

parts of cells, bones, teeth, and the blood. Other mineral elements make up important parts of hormones and secretions.

As is true of all nutrients, specific patterns of deficiency and abundance exist in the distribution of minerals in the American diet. For example, phosphorus deficiency can occur when someone does not eat adequate amounts of fresh green vegetables. Men consume more calcium than women at all age levels, because of their greater total intake of food. Also, food processing with water washes away many minerals, especially magnesium, phosphorus, and calcium.

Vitamins

Vitamins are a group of important compounds that are found in very small proportions in food. They are needed in trace amounts for the proper functioning of the body. Vitamins function along with enzymes to carry out very specific, important chemical reactions in the body. Like enzymes, vi-

TABLE 8.2 COMBINATIONS OF PLANT FOODS TO ENSURE AMINO ACID INTAKE*

Plant food deficient in specific amino acids	Amino acids deficient	Plant food providing deficient amino acids
Vegetables: Potatoes, green peas, asparagus, kale, broccoli, spinach	Methionine & cystine	Rice, wheat, oats, rye, corn, & nuts (except peanuts)
Grains: Rye, wheat, rice, millet, & oats	Lysine	Soybeans, dried beans and peas, peanuts, & vegetables
Legumes: Dried peas, dried beans, frozen beans and peas, peanuts, & tofu.	Methionine & tryptophan	Rice, wheat, oats, corn & nuts (except peanuts)
Nuts (except peanuts)	Lysine	Soybeans, dried beans and peas, peanuts, lentils, & vegetables.
Seeds: Sesame & sunflower seeds	Lysine	Soybeans, dried beans and peas, peanuts, lentils, & vegetables.
Corn & wheat germ	Tryptophan	Soybeans, lima beans, & peas.

*Combining these specific foods will improve the biological quality of the protein being eaten.
   Good combinations to obtain proper essential amino acid intakes:
      Rice and beans or peas
      Peanut butter and bread
      Bread and cheese
      Rice and milk or yogurt
   Fair combinations to obtain proper essential amino acid intake:
      Bread and nuts (except peanuts)
      Wheat and rice
      Peanuts and potatoes

● **ASK YOURSELF**

1. What is the major function of proteins in the body?
2. Why do we need the essential amino acids daily?
3. Explain the difference between a complete and an incomplete protein.

tamins act by helping a reaction take place (in some cases, enzymes and vitamins make reactions possible that could not otherwise occur), but vitamins are neither changed nor incorporated into the products of the reaction. Because of this action, vitamins are called *coenzymes.*

Except for vitamins D and K, vitamins cannot be synthesized directly in the body; they must be

obtained from the diet. And even though vitamins D and K are synthesized within the body, the chemicals from which they are synthesized still must come from what we eat. Consequently, they also depend on a proper diet. As shown in Table 8.4, whether a vitamin will dissolve in water or in fat (or oil) is important. This can tell you the source of a vitamin, how it is absorbed into the body, and what happens to it inside the body. Water-soluble vitamins are not stored in the body and should be taken each day. In general, fat-soluble vitamins are stored within body tissues and can cause toxic effects in large overdoses. Excessively high intakes of fat-soluble vitamins, such as A and D, especially in infants, whose bodies are small, should be avoided. On the other hand, the absence of fat-solu-

TABLE 8.3 CALORIC VALUES OF ENERGY-YIELDING NUTRIENTS*

Type of nutrient	Kilocalories per gram	Kilocalories per pound
Carbohydrates	4	1,860
Fats	9	4,220
Proteins	4	1,860

*This table assumes a gram or pound of pure substance. Very few foods are pure and release the total number of kilocalories.

ble vitamins has serious consequences. An entire class of diseases, called *vitamin deficiency diseases* (*see* Table 8.4), can result from a lack of these vital chemicals in the diet.

A number of vitamins, such as B_6 (pyridoxine), C, and E, have become very controversial in the last few years. Powers have been ascribed to them that many nutritionists feel are unproven, while others insist large amounts of these vitamins are needed for good health. It will take years of continued research to completely substantiate many of the claims made for these vitamins.

Water

No material serves the body in as many vital functions as water. The importance of water to the body is so great that the loss of only 10 percent of the body's water can cause death. The body is over 50 percent water, and many of its tissues (such as blood) are as much as 90 percent water. Digestion, absorption, and the secretion of materials must take place in water. All chemical reactions of metabolism also require water. It provides the moisture in the lung cells that enables the membranes to exchange oxygen and carbon dioxide. It is important in distributing heat uniformly throughout the body. It transports many vital substances throughout the body. And it serves as a cushion for the brain and spinal cord.

How much water people require each day depends largely on the air temperature around them and the kind of physical activity they are engaged in. Water loss may range from 2½ quarts for a moderately active person to several times that much for a person working vigorously in the hot sun. The loss occurs primarily through the kidneys, lungs, digestive tract, and skin.

This water loss can be replenished by liquids

What is a calorie? A kilocalorie is the amount of heat needed to raise the temperature of 1 liter of water (1,000 milliliters) 1 degree Celcius (centigrade).

and foods of all kinds. All foods—even dry bread—contain some water. Some water is produced within the body through the metabolic breakdown of stored nutrients. Since there are variables both in water needed and water available from different sources, it is not possible to state the specific amount of water a person should drink each day. In general, a person should drink a little more water than is sufficient to satisfy thirst. The slight excess beyond thirst promotes good kidney health.

THE CHEMISTRY OF NUTRITION

Each cell of the human body converts the chemical energy of our food into the many forms of energy (chemical, electrical, mechanical, and heat) needed by our bodies to perform and grow.

TABLE 8.4 VITAMINS

Vitamin	Rich Sources	Properties	Function	Deficiency Symptoms
Fat-soluble vitamins				
Vitamin A	Cheese, deep green and orange vegetables, butter, eggs, milk, fish-liver oils; carotene in vegetables converted to vitamin A by liver	Lost through oxidation during long cooking in open kettle; overdose possible	Necessary for growth, tooth structure, night vision, healthy skin	Slow growth, poor teeth and gums, night blindness, dry skin and eyes (lack of tears)
Vitamin D	Beef, butter, eggs, milk, fish-liver oils; produced in the skin upon exposure to ultraviolet rays in sunlight; no plant source	One of the most stable vitamins; large doses may cause calcium deposit, poor bone growth in children, congenital defects	Necessary for metabolism of calcium and phosphorus; essential for normal bone and tooth development	Rickets; poor tooth and bond structure; soft bones
Vitamin E	Widely distributed in foods; abundant in vegetable oils and wheat germ	Lost through oxidation during long cooking in open kettle; overdose not known	Not definitely known for humans	Not definitely known for humans
Vitamin K	Eggs, liver, cabbage, spinach, tomatoes; produced by bacteria of intestine	Destroyed by light and alkali; absorption from intestine into blood depends on normal fat absorption	Necessary for blood clotting	Slow blood clotting; anemia (low oxygen-carrying capacity of blood)
Water-soluble vitamins*				
Vitamin B (thiamine)	Meat (especially pork) whole grains, liver, yeast, nuts, eggs, bran, soybeans, potatoes	Usually not destroyed by cooking, but can be destroyed by alkali—may dissolve in cooking water. Not stored in body, daily supply needed	Necessary for carbohydrate metabolism, normal nerve function; promotes growth	Beriberi; slow growth, poor nerve function, nervousness, fatigue, heart disease

Vitamin	Sources	Stability	Function	Deficiency symptoms
Vitamin B_2 (riboflavin)	Milk, cheese, liver, beef, eggs, fish	Not destroyed by cooking acid foods, unstable to light and alkali	Essential for metabolism in all cells	Fatigue; sore skin and lips, bloodshot eyes, anemia
Niacin (nicotinic acid)	Bran, eggs, yeast, liver, kidney, fish, whole wheat, potatoes, tomatoes; can be synthesized from amino acid tryptophan	Not destroyed by cooking, but may dissolve extensively in cooking water	Necessary for growth, metabolism, normal skin	Pellagra; sore mouth, skin rash, indigestion, diarrhea, headache, mental disturbances
Vitamin B_6† (pyridoxine)	Meat, liver, yeast, whole grains, fish, vegetables	Stable except to light	Functions in amino acid metabolism	Dermatitis, convulsions; deficiency rare
Vitamin B_{12} (cyanocobalamin)	Meat, liver, eggs, milk, yeast	Unstable to acid, alkali, light	Necessary for production of red blood cells growth	Pernicious anemia
Vitamin C‡ (ascorbic acid)	Citrus fruits, tomatoes, potatoes, cabbage, green peppers, broccoli	Least stable of the vitamins; destroyed by heat, alkali, air, dissolves in cooking water	Essential for cellular metabolism, necessary for teeth, gums, bones, blood vessels	Scurvy; poor teeth, weak bones, sore and bleeding gums, easy bruising, poor wound healing
Folacin (folic acid)	Liver, yeast, leafy vegetables such as asparagus, lettuce, and broccoli, and whole wheat products	Destroyed by cooking at high temperatures; keep leafy vegetables under refrigeration	Needed for synthesis of DNA and RNA; extremely important during pregnancy	Gastrointestinal disorders, diarrhea, and anemia; can cause anemia during pregnancy and childhood

* Several other water-soluble vitamins are believed essential to human nutrition, but are not as well understood as the above vitamins and their deficiency is less common.

† A female using an oral contraceptive pill should increase her intake of vitamin B_6 to 25 milligrams daily.

‡ Some controversy exists over the amount of vitamin C an individual should consume.

TABLE 8.5 KILOCALORIE EXPENDITURE PER KILOGRAM PER HOUR *

Physical activity	Calories per kilogram per hour	Physical activity	Calories per kilogram per hour
Bedmaking	3.0	Golf	1.5
Bicycling		Horseback riding	
moderate speed	2.5	walk	1.4
racing	7.6	trot	4.3
Boxing	11.4	gallop	6.7
Carpentry	2.3	Office work, standing	0.6
Cleaning windows	2.6	Painting furniture	1.5
Dishwashing	1.0	Playing cards	0.5
Dressing & undressing	0.7	Ping-Pong	4.4
Driving car	0.9	Reading aloud	0.4
Eating	0.4	Running	7.0
Exercise activities		Sewing	
Very light (sitting, with about 2 hours of standing & walking)	0.9	hand	0.4
Light (sitting, typing, moderate walking)	1.4	electric machine	0.4
		Singing in loud voice	0.8
Moderate (housework, gardening, little sitting)	3.1	Sitting quietly	0.4
Severe (no sitting, standing, walking, skateboarding, group games, dancing)	5.4	Skiing (moderate speed)	10.3
		Standing relaxed	0.5
Very severe (dual or individual sports, tennis, swimming, basketball, heavy work)	7.6	Swimming (2 mph)	7.9
		Tennis	5.0
Football (organized)	6.8	Typing	
Gardening, weeding	3.9	manual machine	1.0
		electric machine	0.5
		Walking	
		slowly (3 mph)	2.0
		rapidly (4 mph)	3.4
		at high speed	8.3

*Exclusive of sex, age, body build, resting metabolism, or food.

● **ASK YOURSELF**

1. What is a mineral? Why are minerals important in the body?
2. What is a vitamin? What is the function of vitamins in the body?
3. What is the difference between a water-soluble and an oil (or fat)-soluble vitamin? Do you take a vitamin supplement? Which vitamins?
4. What functions does water perform in the body?
5. How do we lose water?

The quantity of energy released from a given amount of food is measured in *large calories,* or *kilocalories*—abbreviated *kcal.* One kilocalorie is the amount of heat required to raise the tempera-ture of one liter of water (approximately a quart) by one degree Celsius (centigrade).

Total Energy Needs

The total need for calories is proportional to the amount of daily exercise plus one's resting metabolism. The number of calories used in exercise is directly proportional to total body weight. Table 8.5 shows some physical activities and the calories expended for each kilogram of body weight. To calculate your caloric use, simply multiply the numbers in the table by your weight in kilograms and then by the time you spend at the activity. For example, swimming uses 7.9 kcal per hour per kilogram of body weight. A 128-pound woman would use (7.9 × 58 kilograms × 1.0 hours) 458 calories per hour of swimming.

TABLE 8.6 KILOCALORIC USE BY 128-POUND WOMAN STUDENT DURING HYPOTHETICAL DAY

Physical activity	Calories per kilogram per hour	Hours spent in activity	Total calories used
Dressing & undressing	0.7	1.5	60.9
Eating meals & snacks	0.4	1.5	43.5
Sitting in class	0.4	4.0	92.8
Laboratory class	0.5	1.0	29.0
Walking slowly (3 mph)	2.0	1.0	116.0
Walking at high speed (5 + mph)	8.3	0.5	240.7
Driving an automobile	0.9	1.0	52.2
Writing	0.4	0.5	11.6
Typewriting	1.0	0.5	29.0
Ping-Pong game	4.4	0.5	127.6
Swimming (on school team)	7.9	3.0	1,374.6†
Housework	1.2	1.0	69.6
Sleep*	− 0.1	8.0	− 46.4
			2,247.5

*Calories used in sleep are less than in RM; subtract savings   −46.4

Calories used in the day and night   2,200

Calories used in RM   +1,320

Total calories used   3,520 kcal

† Without workout during swim-team practice caloric use   2,145 kcal

When calculating your daily energy usage, you must take into consideration your total energy needs for internal body functioning and the calories used by your body at rest [termed resting metabolism or RM] plus the calories you expend in your daily activities. Table 8.6 shows the caloric usage of a 128-pound woman student during a hypothetical day. By using these two tables you can calculate your approximate daily energy requirements and see if you are eating too much food or exercising too little.

DIETARY GUIDELINES

In 1972 the U.S. Senate established the Select Committee on Nutrition and Human Needs. This committee was to study the diets of Americans. In 1977 the committee published its *Dietary Goals for the United States,* which kicked up a whirlwind of conflicting opinions. This controversy led to the disbanding of the committee at the end of 1977. The information was turned over to the Department

● ASK YOURSELF

1. What is a kilocalorie?
2. What does it mean when someone says: "A piece of toast contains 250 calories?"
3. How do you calculate your caloric use?
4. Using Table 8.5 as a guide, calculate your caloric needs for one day. Now, calculate the number of calories contained in the food you ate for that day. Are you eating too much food or exercising too little?

of Agriculture (USDA) and the then Department of Health, Education, and Welfare (USDHEW), now the Department of Health and Human Services. In 1979 representatives of these two departments produced a far less controversial document entitled *Nutrition and Your Health: Dietary Guidelines for Americans*. This document consists of seven guidelines that are outlined in the feature below.

The following is a discussion of the original 1977 *Goals* and the current *Guidelines* to help people change their diets and establish more healthful eating habits.

Carbohydrates

The committee found that the average diet consisted of between 40 and 50 percent carbohydrates. These carbohydrates were about 30 percent simple sugars, 15 percent cellulose fiber, and 50 percent

---

Dietary Guidelines for Americans (and Suggestions for Food Sources)

1. *Eat a variety of foods daily.* In your everyday diet, include: fruits and vegetables; whole grains and enriched breads and cereals (unsweetened); milk and milk products (low fat or fat free); fish, poultry, and eggs (red meat in lesser amounts than in past); dried peas and beans.

2. *Maintain ideal weight.* Increase daily physical activity. Reduce calories by eating fewer fatty foods and sweets and less sugar. Drink only moderate amounts of alcohol. Lose weight gradually.

3. *Avoid too much saturated fat and cholesterol.* Choose low-fat protein sources such as lean meats, fish, poultry, low-fat or fat-free skim milk and dairy products, dry peas and beans. Use eggs and organ meats in moderation. Limit intake of fats on and in foods. Trim fat from red meats, skin poultry. Broil, bake, or boil—don't fry. Read food labels for fat content.

4. *Eat foods with adequate starch and fiber.* Substitute starches for fats and sugars. Select whole-grain breads and unsweetened cereals, fruits, vegetables, dried beans, peas, and nuts.

5. *Reduce sugar intake.* Use less sugar, syrup, and honey. Reduce concentrated sweets like candy, soft drinks, and cookies. Select fresh fruits or fruits canned in light syrup or their own juices (no sugar added). Study food labels for amounts of glucose, dextrose, maltose, lactose, fructose, syrups, "corn sweetener," and honey. These sugars should be avoided.

6. *Avoid salt.* Reduce salt in cooking. Add little or no salt at table. Limit salty foods like potato chips, pretzels, salted nuts, salted popcorn, cheese, pickled foods, and cured meats (luncheon meats). Read food labels for sodium or salt content, especially in processed and snack foods.

7. *If you drink alcohol, do so in moderation.* Limit the amount of alcohol (wine, beer, and liquors) to not more than one or two drinks per day. NOTE: The use of alcoholic beverages during pregnancy can result in Fetal Alcohol Syndrome—the development of birth defects and mental retardation in the fetus (*see* Chapter 6).

Adapted from USDA and USDHEW, *Nutrition and Your Health: Dietary Guidelines for Americans* (February 1980).

complex carbohydrates (starches). Most of the sugar consisted of highly refined table sugar (sucrose) from sugar beets and sugarcane, and lactose sugar found in milk.

The committee recommended that we increase our carbohydrate consumption to about 60 percent of our total calorie intake, especially complex carbohydrates and cellulose fiber from plant foods. We should greatly reduce our consumption of refined sugars.

Complex carbohydrates (starches) The best way to increase complex carbohydrates is to eat more cereals, nuts, vegetables, and fruits. Read the labels of commercial cereals and avoid those that contain sugar. Also, fruit and vegetable snacks increase complex carbohydrate intake without increasing refined sugar intake.

Fiber (cellulose) The current American diet consists mostly of low-fiber and nonfiber foods (*see* Table 8.7). We should increase our fiber intake to between 6 and 24 grams (0.2 to about 1 oz) per day. This is equivalent to one large bowl of high-fiber cereal, or an apple, or some other high-fiber food.

Fiber has been reported to help prevent a number of diseases, including cancer of the colon, atherosclerosis (by reducing the absorption of cholesterol), diverticular disease, irritable bowel syndrome, hemorrhoids, tooth decay, and certain forms of diabetes.

Simple sugars Sugars should be reduced to about 10 percent of our diet. Except for milk and dairy products, almost all sugar foods contain few other nutrients. Sugar's only real contribution to the diet is taste and calories.

TABLE 8.7 CURRENT CONSUMPTION OF FIBER AND NONFIBER FOODS

Food	Percent of total diet
Nonfiber foods:	85
Meat & eggs	23
Fats & oils	18
Refined cereals	17
Sugar	17
Milk	11
Fiber foods:	8
Fruits & vegetables	3
Legumes & nuts	3

What sugar we consume should come from low-fat or nonfat milk and dairy products, fruits, and vegetables. Reduce or eliminate candy, pastries, and sweetened cereals from your diet. Children's sugar intake should be watched. While a piece of pie or cake is only a small percentage of an adult's diet, it may contain almost all of a small child's needed daily calories.

Proteins

The study found that Americans consume an average of about 100 grams ($3\frac{1}{2}$ oz) of protein per day. The U.S. dietary goals recommended about 0.8 g of protein per kilogram (kg) of *ideal* body weight per day. The Food and Agriculture Organization of the World Health Organization recommends 0.57 g of protein per kg of body weight for a male and 0.52 g of protein per kg of body weight for females. For a 150-pound male, the recommended intake could range from 35.5 to 54.5 g, or between 1 (1.2) and 2 (1.9) oz of protein per day. Reducing the intake of protein from $3\frac{1}{2}$ oz per day to between

Sugar: How Sweet It Is

Americans currently obtain about 24 percent of their calories from sugar. Three percent comes from natural fruits and vegetables. Another 3 percent comes from dairy products (lactose sugar). The balance (18 percent) comes from refined sugar added to processed foods. In 1970 we consumed 122 pounds of sugar a year. By 1978 we were consuming 128 pounds of sugar a year. In the 1930s 75 percent of the sugar produced was purchased for home consumption. Today 75 percent of the sugar produced is used by the food and soft-drink industry and only 25 percent is used in the home. The next time you buy a soft drink, candy bar, or pastry, consider its sugar content.

Meat analog products. Plant proteins flavored and textured to simulate meat can be found in many supermarkets.

foods that have been textured and flavored to resemble meat. They are made from soybeans, wheat proteins, peanuts, and other nuts. Such foods have been used by vegetarians for years; they are now appearing in supermarkets as well as health food stores.

Fats and Oils

North Americans' high consumption of fats and oils is the major factor in their weight control problem. Fat consumption should be reduced from the current 40 percent of the diet to not more than 10 percent. Links have been established between fat consumption and heart attacks and cancer. Fats and oils heated to high temperatures during deep frying or broiling (especially with charcoal) are changed into what are known as *transfatty acids*. These transfatty acids have been shown to be carcinogenic. High intakes of saturated fats and cholesterol have been linked with heart attacks.

Polyunsaturated and saturated fat ratio Both saturated and polyunsaturated fats are needed for normal body functioning. But, an overabundance of saturated fats has been linked to heart problems. A healthy dietary balance must be established between saturated and polyunsaturated fats.

This balance, termed the *P/S ratio*, is the amount of polyunsaturated fat divided by the amount of saturated fat in a product. The higher the number, the more desirable the product. Currently, the U.S. populations P/S diet ratio is .44; it should be at least 1.00 for the general population and 1.50 for heart patients. Most food manufacturers do not list this number on their products. Table 8.8 lists the P/S ratios for some common oil-containing products.

1 and 2 oz is not a problem. The problem is the foods we use as sources for protein. Most of our proteins come from beef, pork, eggs, and whole milk or high-fat dairy products, such as cheeses. All of these foods contain high amounts of fats and oils. We should instead increase our consumption of plant proteins, and proteins from nonfat animal sources. As was discussed earlier, however, the major problem with plant proteins is that they are low in some essential amino acids. Thus, we must carefully combine plant foods to obtain an adequate supply of these essential amino acids.

Another way of obtaining adequate amounts of essential amino acids is to use *meat analogs,* or *textured vegetable protein* (TVP) foods. These are

---

Tips for Cooking with TVPs and Meat Analogs

Whenever using meat analogs or TVP products, *always* follow the directions on the package *exactly* as they are given. Whenever cooking with TVPs remember:

1. They do not need to cook as long as real meat products.
2. They will not make a broth for gravies.
3. Do not overuse when mixing with real meats.
4. Some have a very high salt content. Use less salt when cooking with them.
5. They contain no fat. You must add fat for frying or they will burn.

TABLE 8.8 THE P/S RATIOS OF SOME COMMON FOODS

Product	P/S ratio*
Butter and various margarines	
Butter	0.1
Nucoa, whipped	1.1
Fleischmann's	1.3
Parkay, soft	1.3
Golden-Glow, soft	1.4
Nucoa	1.5
Blue Bonnett, soft	1.7
Mazola	1.7
Imperial, diet	2.0
Fleischmann's, diet	2.2
Fleischmann's, soft	3.3
Chiffon, soft	4.0
Saffola, soft	4.2
Cooking oils	
Coconut oil	0.01
Olive oil	0.7
Peanut oil	1.6
Cottonseed oil	2.0
Soy oil	3.6
Corn oil	5.6
Safflower oil	9.0
Nuts	
Coconuts	0.01
Cashews	0.47
Peanuts	1.32
Brazil nuts	1.40
Pistachios	2.0
Hickory nuts	2.38
Almonds	2.5
Pecans	3.0
Beechnuts	4.13
Filberts (Hazelnuts)	7.4
Walnuts, black	9.17
Walnuts, English	10.5

*The higher the number the more desirable the product.

Cholesterol High intakes of cholesterol have been linked with heart problems (*see* Chapter 20). Thus, it is desirable to reduce the amount of cholesterol in the diet. We consume most of our cholesterol in red meats, eggs, and high-fat dairy products.

Currently most Americans consume between 500 and 1,000 mg of cholesterol per ml of blood per day. The recommended level is not more than 300 mg per ml of blood per day. This can be achieved by reducing the daily intake of high-cholesterol

● ASK YOURSELF

1. How does your diet compare with the average American diet?
2. How can you increase your consumption of complex carbohydrates and fiber while decreasing your consumption of simple sugars?
3. Do you eat enough fiber foods? If not, how can you increase your consumption of them?
4. How much red meat do you consume per day? Are you consuming too much protein and fat? If so, what can you do about it?
5. Plan a menu using only plant sources that would supply an adequate amount of protein.

foods and increasing the consumption of fish, fowl, plant protein, and low-fat or fat-free milk, cheese, or other dairy products.

Sodium

Another major dietary problem in the U.S. is the overconsumption of *sodium*. Sodium has been found to be one of the major causes of heart attacks and strokes. The average American consumes between 6 and 18 g of sodium a day. *One-half of a gram a day* of sodium is needed to sustain life. This amount occurs normally in vegetables and in the usual cooking of most foods. Any amount over 0.5 g a day is consumed solely for taste.

Most of the sodium consumed is in the form of table salt (sodium chloride). Some other products are very high in sodium. Nutritionists recommend cutting down on the amount of salt we use in cooking and not salting food before eating it. Read the

All these foods contain large amounts of sodium.

TABLE 8.9   THE U.S. FOUR FOOD-GROUP PLAN

Food group	Sample foods	Main nutrient contributed
Meat & meat substitutes	Beef, pork, lamb, fish, poultry, eggs, nuts, legumes	Protein, iron, riboflavin, niacin, thiamin
Milk & milk products	Milk, buttermilk, yogurt, cheese, cottage cheese, soy products, ice cream	Calcium, protein, riboflavin, thiamin, fats & oils
Fruits & vegetables	All fruits & vegetables	Vitamins A & C, carbohydrates, thiamin, iron, riboflavin
Grains (bread & cereal products)	All whole-grain and enriched flours & products	Carbohydrates, riboflavin, niacin, iron, thiamin

Source: E. N. Whitney and E. M. N. Hamilton, *Understanding Nutrition* (Los Angeles: West, 1981), pp. A100–A104.

● **ASK YOURSELF**

1. What is the P/S ratio? What are the five best margarines in your supermarket?
2. Name some ways in which you could reduce the number of high-cholesterol foods and amount of sodium in your diet.

labels of prepared foods and be on the lookout for salt or any chemical name that begins with "sodium."

Caffeine

A diet problem that is often overlooked is the enormous amount of caffeine in the American diet. Caffeine has been linked to hyperactivity in children, heart attacks, and fetal malformations when taken during pregnancy (*see* Chapter 5). Most people should reduce or eliminate caffeine from their diets by drinking noncaffeinated soft drinks, and less coffee, tea, and cocoa.

THE U.S. FOUR FOOD-GROUP PLAN

Food-group plans are designed to provide an easy way to eat adequately. They are simple, quick guidelines for diet planning. Such diets are more apt to be nutritionally adequate than those that are randomly chosen. Table 8.9 shows the four food

groups. Each of the food groups contains foods that are similar in nutritional content. The foods shown in the four food group table are used as indicator foods. Other foods that are similar in nutrient content can be used to replace these foods.

The four food-group plan was devised to ensure an adequate amount of all important nutrients in the diet. By following this plan, a person will consume all necessary carbohydrates, fats, and amino acids. The meat and milk group contributes amino acids and vitamin B_{12}; this important vitamin is found only in animal products. All the foods in the meat, milk, and cereal groups provide zinc. The importance of this metal is just becoming recognized.

A certain amount of each nutrient should be consumed daily for the plan to work. Table 8.10 shows the number of servings recommended.

Canadian Food Guide Plan Canada's Food Guide Plan is similar to the U.S. Four Food-Group Plan and was developed with the same intent. Table 8.11 shows the Canadian plan. Table 8.12 shows the number of servings recommended.

FOOD FACTS AND FALLACIES

Food supplies the nutrients for growth and replacement of worn or damaged cells, as well as for the manufacture of cellular products, such as enzymes

TABLE 8.10 SERVINGS IN THE U.S. FOUR FOOD-GROUP PLAN

Food group	Recommended number of servings for an adult	Serving size
Meat & meat substitutes	2	2–3 oz cooked meat, fish or chicken. 1 cup cooked legumes
Milk & milk products	2*	1 cup (8 oz) milk; 1–2 oz cheese
Fruits & vegetables	4†	½ cup fruit, vegetables, or juice.
Grains (bread & cereal products	4‡	1 slice bread; ½ cup cooked cereal; 1 cup ready-to-eat cereal.

*For children up to 9: 2–3 cups; for children 9 to 12: 3–4 cups; for teen-agers and pregnant women: 3–4 cups; for nursing mothers: 4 or more cups.
† One daily selection should be rich in vitamin C, and one selection every other day should be rich in vitamin A.
‡ Enriched or whole-grain products only.
Source: Canadian Ministry of Health and Welfare, *Canada's Food Guide-Handbook* (Ottawa, Ontario, 1977).

and hormones. The overall composition of the body is about 59 percent water, 18 percent protein, 18 percent fat, and 4.3 percent minerals. At any one time there is less than 1 percent carbohydrate in the makeup of the body.

These substances that make up the body are not distributed equally in all organs. For example, the percentage of water varies from 90 to 92 percent in blood plasma in 72 to 78 percent in muscle, 45 percent in bone, and only 5 percent in tooth enamel. Proteins are found most abundantly in muscle. Fat tends to concentrate in the adipose (fat) cells under the skin and around the intestines. Carbohydrates are found mainly in the liver, muscles, and blood. As for the minerals, calcium and phosphorus form the bones and teeth, sodium and chlorine are found

TABLE 8.11 CANADIAN FOOD GROUP PLAN

Nutrient	Milk & milk products	Bread & cereals	Fruits & vegetables	Meat & alternatives
Vitamin A	Vitamin A		Vitamin A	Vitamin A
Thiamin		Thiamin		Thiamin
Riboflavin	Riboflavin	Riboflavin		Riboflavin
Niacin		Niacin		Niacin
Folic acid			Folic acid	Folic acid
Vitamin C			Vitamin C	
Vitamin D	Vitamin D			
Calcium	Calcium			
Iron		Iron	Iron	Iron
Protein	Protein	Protein		Protein
Fat	Fat			Fat
Carbohydrates		Carbohydrates		Carbohydrates

Source: Canadian Ministry of Health and Welfare, *Canada's Food Guide-Handbook* (Ottawa, Ontario, 1977).

TABLE 8.12   SERVINGS IN CANADIAN FOOD PLAN

Food group	Recommended number of servings for an adult	Serving size
Meat & meat substitutes	2	60–90 g (2–3 oz) cooked lean meat, poultry, liver, or fish; 60 ml (4 tbsp) peanut butter; 250 ml (1 cup) cooked dried peas, beans, or lentils; 80–250 ml (⅓–1 cup) nuts or seeds; 60 g (2 oz) cheddar, processed, or cottage cheese; 2 eggs
Milk & Milk products	2*	250 ml (1 cup) milk, yogurt, or cottage cheese; 45 g (1½ oz) cheddar or processed cheese
Fruits & vegetables	4–5†	125 ml (½ cup) vegetables or fruits; 125 ml (½ cup) juice; 1 medium potato, carrot, tomato, peach, apple, orange, or banana
Bread & cereals	3–5‡	1 slice bread; 125–250 ml (½–1 cup) cooked or ready-to-eat cereal; 1 roll or muffin; 125–200 ml (½–¾ cup) cooked rice, macaroni, or spaghetti

*Children up to 11: 2–3 servings; adolescents: 3–4 servings; pregnant & nursing women: 3–4 servings.
† Include at least 2 vegetables. Choose a variety of vegetables and fruits.
‡ Whole-grain breads or enriched; whole-grain breads are preferred.
Source: Canadian Ministry of Health and Welfare, *Canadian Food Guide-Handbook* (Ottawa, Ontario, 1977).

● **ASK YOURSELF**

1. What is the U.S. Four Food-Group Plan? Try following this plan for one week. Is it hard or easy to follow?
2. Name some substitutes within each of the four groups.
3. Can this program be used with the *Dietary Guidelines?* How well do they fit together?

mainly in the body fluids (blood plasma and lymph), potassium is the main mineral in the muscles, iron is essential to red blood cells, and magnesium is distributed throughout the body.

These are the main minerals supplied to the body as food, but many other minerals are essential to the human body in proportionately smaller amounts. These minerals are termed *trace elements*, and they too must be ingested with our food. Other chemicals (vitamins) are needed in very small amounts for various functions of the body to take place.

Food Fads and Quacks

Food fads are food beliefs supported by very little scientific evidence. A *food faddist* is someone who relies upon belief and refuses to acknowledge fact, regardless of the evidence. Food faddists are sincere in their beliefs about food and are honestly concerned about their health. Somehow, they find it easier to believe various bizarre and spectacular claims than the more moderate but realistic statements of qualified authorities.

A *food quack,* by contrast, is a fraud who pretends to have skill, knowledge, or qualifications that he or she does not possess. Most food quackery exists because people are gullible and are looking for a "miracle" food like the "miracle" drugs developed in recent years. The motive for quackery is money.

Food faddism Although there are many forms of faddism, many proponents use words such as *organic, natural, health,* and *raw* to identify their foods.

The word *organic* is commonly misunderstood and misused. According to biological and chemical definitions, organic materials are complex combinations of chemicals, all of which contain the element carbon. Thus, all foods from plant or animal sources and most synthetic foods are organic. Although not official, the following definitions are generally accepted by nutritionists.

Organic matter, or humus, is used in the growing of plants we eat or feed to animals that furnish milk products or meat. Humus contains manures, plant composts, and other plant residues such as peat moss and sawdust. The major value of humus is that humus-rich soils absorb and hold water better and are easier to till than humus-depleted soils. Commercial fertilizers contain the same chemical nutrients but in simpler forms.

Organically grown food is supposedly grown without pesticides or artificial fertilizers and in humus-rich soils. *Organically raised animals* are fed on organically grown pasture and feed and are given no growth stimulants, antibiotics, or synthetic materials. *Organically processed food* is organically grown food that has not been treated with preservatives, hormones, antibiotics, or synthetic additives of any kind in its processing. Any unprocessed food is preferable to processed food; this is true whether the food is called organic or not.

There is no scientific evidence that plants grown with only organic fertilizers have greater nutritive value than our regular food produced by the usual methods. If manure is used as fertilizer, the organic compounds present must be broken down by bacteria into the same simple compounds present in

Some typical health foods. These organic and natural foods provide the same essential nutrients as conventional packaged products.

commercial fertilizers before they can be absorbed into the roots of the plant.

Unfortunately, organically grown food is not necessarily free of pesticides. Pesticides are easily airborne, and their chemical residue may remain in the soil for years after their use has been discontinued. Pesticide residues are often found in foods grown organically.

Although organically grown foods have no greater nutritive value, they often cost more than the same items produced and marketed by regular commercial methods. Their production methods are more costly, their storage life is shorter, and packaged food must be watched for contamination and spoilage. There is evidence that much more food is being sold as organic or natural than is being produced, indicating many unscrupulous marketers.

Natural foods are foods sold in the same form as they were when harvested. Fresh fruits and vegetables are natural, but canned or frozen must be processed. Natural foods may or may not be grown organically. Many natural foods advertised as organically grown and sold at higher prices may have come from large commercial producing operations.

Raw food faddists contend that cooking or heating natural foods destroys much of their nutritive value. While natural food advocates will cook their food, raw food advocates eat all food raw. They are also opposed to the pasteurization of milk and milk products. Pasteurization protects consumers from dangerous bacteria and does not destroy the nutri-

● **ASK YOURSELF**

1. What is *food faddism?* What are some of the more common food faddisms?
2. What does "organic" mean? How is organically grown food different from normally grown food?
3. Are unprocessed foods always better than processed foods? Explain your answer.

tive value of milk. Vegetables and other foods, if not overcooked, retain most of their nutritive value. In addition, cooking often improves taste and increases digestibility.

Most health food stores feature various types of raw or less processed sugar. Whether the sugar is "raw," "less processed," or refined makes no particular difference. The basic problem is that many people eat far too much of all forms of sugar, a dietary factor that contributes to dental decay, obesity, malnutrition, and heart disease.

Food quackery Food quackery is a big business. Half the money spent on quackery is spent in the area of nutrition. Over 10 million Americans have been seriously confused by food faddists and health food quacks. These unfortunate people are encouraged to follow expensive, complicated, and often unpleasant diets. Rather than being better fed as a result, they are actually more likely to suffer from nutritional deficiency than those who eat ordinary diets, following the simple rules of basic nutrition.

Food quackery products are sold in several ways. They are often featured in health food stores, sold by door-to-door salespeople, advertised for mail-order sale, and promoted in "health" lectures. Regardless of the sales approach used, the food quack uses scare tactics to frighten people into buying certain products. Almost all operators in this field rely on a few myths. While each of these myths may contain an element of truth, the conclusions drawn by the quack are not supported by scientific evidence. The following are some of these myths:

1. All diseases are due to faulty diet Of course, such deficiency diseases as scurvy are entirely the result of poor diet, and a person's resistance to many infections is lower when the diet is poor. But no known diet can protect a person from all infectious diseases or from cancer, as is claimed by certain quack nutrition products.

2. A particular product is indispensable Salespeople often represent their products as being the only source of a vital food substance and imply that maximum good health is possible only if their products are used. The truth is that every substance known to be important in nutrition is available from a variety of common grocery-store foods.

3. Soil depletion causes malnutrition A common story is that repeated cropping of the land has removed some substance, which is therefore lacking in the foods produced. Iodine is the only substance that becomes depleted with repeated cropping. Since people today obtain adequate iodine through diet and the use of iodized salt, iodine deficiency is rare. If any other mineral is lacking from the soil, this deficiency is reflected in a lowered *quantity* of produce, but the nutritional *quality* is not affected.

4. "Organic" or "natural" foods These are the key words in the health food business. The claim is often made that foods grown with commercial fertilizers are inferior to those grown with natural fertilizer (manure). But a plant absorbs only simple inorganic nutrients from the soil. The organic compounds in both manure and commercial fertilizers must be broken down by bacteria into inorganic compounds before the plant roots can absorb them.

A related claim is that synthetically produced vitamins are inferior to naturally occurring vitamins. This statment is usually made by salespeople of high-priced food supplement products to indicate the superiority of their products over lower-priced products. Actually, the synthetic vitamins are chemically identical to the naturally occurring vitamins, are absorbed in the same manner, and function in the body in exactly the same way.

The very word *chemical* is often used in a derogatory manner by salespeople who apparently do not know or choose to ignore the fact that all food is nothing but a mixture of chemicals. They deplore the use of chemical food additives such as antioxidants, coloring agents, mold inhibitors, and numerous other additives important to modern food processing. In fairness, it must be pointed out that scientific debate continues over possible unknown effects of food additives. Caution must be exercised in allowing claims of proponents or opponents to replace substantial information.

5. Myths about vitamins There have been claims of unusual therapeutic effects for a host of conditions through the use of huge doses of vitamins E, B, and C. No claims have been more spectacular than those made for vitamin E. The list of ailments claimed to be relieved by vitamin E is staggering. It includes most noninfectious diseases such as sterility, muscular weakness, cancer, ulcers, and shortness of breath. To date, extensive tests have failed to demonstrate therapeutic benefit from supplemental vitamin E.

However, an ample supply of vitamin E is found in substantial amounts in some foods. The chief sources of vitamin E in the diet are wheat-germ oil and other vegetable oils and vegetables, especially lettuce. If you eat these foods, you should not need to supplement the vitamin E in your diet.

6. Myths about food processing Some critics exaggerate the loss of food value through modern food-processing methods. Although some loss definitely does occur, it is minor in comparison with the benefits we receive from modern food technology. Today's processing methods are often less destructive to vitamins than were the methods used in the past. Highly nutritious "processed" fruits, vegetables, and meats are now available throughout the year, rather than just during limited seasons. The vitamin loss seldom exceeds 25 percent and is generally much less than that. At the same time, this is not meant to discourage the consumption of fresh foods. When fresh food is available and of good quality, people who can afford to buy it should do so.

7. Advantages of "raw" sugar. Most health food stores feature various types of raw or less processed sugar. Two fallacies are involved here. The first is that the nutritional value of less refined sugar is not signficantly higher than that of pure sugar. The second is that the sugar content is not much lower.

The basic problem is that many people in the United States eat far too much sugar, a dietary factor that contributes to dental decay, obesity, malnutrition, blood-sugar disorders, and possible heart disease. It makes no difference whether the sugar is "raw" or refined. In many of the cereal products

● ASK YOURSELF
1. What is food quackery?
2. What are some common food myths?
3. What are some of the serious consequences of food quackery?

that are highly promoted on children's TV programs, the most abundant ingredient is sugar. Even the more "natural" granola-type cereals often have an extremely high sugar content (*see* feature on page 000). It is believed that many of today's children are acquiring such a strong desire for sugar that their lifelong eating habits may be altered. One thing to look for in a market or health food store is *fructose sugar*. This is the sugar contained in fruits and vegetables. Fructose is twice as sweet as sucrose (beet and cane sugar). Thus, you can use half as much for the same amount of sweetness. This can help to reduce your caloric intake if you *must* have sugar.

One of the most serious consequences of food quackery is that it frequently interferes with useful, scientific evelation of nutrition, food processing, and preserving methods. There is presently a good deal of concern about food additives and their known and unknown effects—but this is a different concern from food quackery. (See Chapter 11 for a more detailed discussion of quackery.)

NUTRITION LABELING

The labels of foods processed in one state and shipped to another are required by law to state what nutrients the foods contain. As shown in Table 8.14, food labels must list the ingredients, what nutrients the food in the package provides, and in what quantities. Uniform labeling enables consumers to shop economically and to compare the nutritional values of different foods. One problem with this labeling is that foods that are sold in the same state where they are processed are exempt from such labeling requirements.

WHAT'S IN YOUR FOOD?

One major problem with most foods is their relatively short storage life. In the past, people have used many methods to try to preserve perishable

TABLE 8.13 STANDARD FORMAT FOR NUTRITIONAL LABELING

NUTRITION INFORMATION PER SERVING	Comments
SERVING SIZE: ONE OUNCE (ABOUT 1 CUP) CEREAL ALONE AND IN COMBINATION WITH ½ CUP VITAMIN D FORTIFIED WHOLE MILK.	Defines the size of a serving or portion
SERVINGS PER CONTAINER: 8	Number of servings or portions in container

	CEREAL		Comments
	1 OZ.	WITH ½ CUP WHOLE MILK	Breakdown of calories, protein, carbohydrates, and fat. Total fat and cholesterol content is optional.
CALORIES	110	180	
PROTEIN	2 g	6 g	
CARBOHYDRATES	25 g	31 g	
FAT	0 g	4 g	

PERCENTAGE OF U.S. RECOMMENDED DAILY ALLOWANCE (U.S. RDA)

	CEREAL		Comments
	1 OZ.	WITH ½ CUP WHOLE MILK	Statement of protein, vitamin, and mineral content in percentages of the U.S. Recommended Daily Allowances for each.
PROTEIN	4	15	
VITAMIN A	25	30	
VITAMIN C	25	25	
THIAMIN	25	25	
RIBOFLAVIN	25	35	
NIACIN	25	25	
CALCIUM	*	15	
IRON	10	10	
VITAMIN D	10	25	
VITAMIN B_6	25	25	
FOLIC ACID	25	25	
PHOSPHORUS	*	10	
MAGNESIUM	*	4	

CONTAINS LESS THAN 2 PERCENT OF THE U.S. RDA OF THESE NUTRIENTS.

INGREDIENTS: MILLED CORN, SUGAR, SALT, MALT FLAVORING, SODIUM ASCORBATE (C), VITAMIN A PALMITATE, NIACINAMIDE, ASCORBIC ACID (C), REDUCED IRON, PYRIDOXINE HYDROCHLORIDE (B_6), THIAM HYDROCHLORIDE (B_1), RIBOFLAVIN (B_2) FOLIC ACID AND VITAMIN D_2. BHA and BHT ADDED TO PRESERVE PRODUCT FRESHNESS.

Ingredients in descending order of predominance, except when blends of fats or oils are used. Fat and oil ingredients must be listed by common or usual names, such as lard, coconut oil, beef oil, or soybean oil shortening blend, instead of more generalized names such as animal fat, vegetable oil, or shortening.

AT ALTITUDES OVER 3500 FEET.
Stir 1/4 cup flour into mix. Mix as
directed using 1-1/2 cups water and 2
egg whites. Bake at 375° until done
(20-30 minutes). Makes 3 layers if 8"
round pans are used, makes 2 layers if
9" are used.
FOR ONE LAYER OR 12 CUP CAKES.
Use 1/2 package (2 cups) mix, 2/3 cup
water and 1 egg white. Mix and bake
as directed.

For nutrition information write to
Food and Nutrition Center, Procter
& Gamble, P.O. Box 2, Dept. D-5,
Cincinnati, Ohio 45299.

INGREDIENTS

Sugar and dextrose enriched flour
(bleached), vegetable shortening, wheat
starch, leavening, propylene glycol
monesters, iodized salt, soy protein
isolate, artificial flavoring, dried whey
solids, soy lecithin, vegetable gum,
ascorbic acids as a freshness preserver.

The additives in packaged foods (additives are underlined).

foods—the most common being drying and salting,
or adding salt, sugar, or spices.

Additives

There are two ways in which we classify chemicals
that are in our food as a result of food-processing
technology. *Additives* are substances in foods that
are deliberately added to foods for specific reasons.
Residues are substances that remain in foods as a
result of agricultural processes or environmental
contamination. By conservative estimates, at least
10,000 additives or residues may occur in most
common food.

Convenience foods are made possible by the use
of food additives that improve the appearance, texture, taste, nutritional value of food and prolong its
shelf life. Over half of the foods sold contain additives.

● ASK YOURSELF

1. What items should be listed on a food label?
2. Check the food labels in your market and see if
   they list nutrients in the proper order.
3. Go to a health food store and check the nutritional labels on the products. Do they conform
   to the law?

Hazards posed by additives Official policy governing the use of additives varies widely. While
some countries have strict regulations, the United
States has allowed wide proliferation of new additives, while at the same time it has prevented them
from being adequately evaluated. Many are loosely
accorded "generally recognized as safe" (GRAS)
status, on the basis of long-established use without
"evidence of harm." Relatively few food additives
have been accepted for specific and limited uses on
the basis of sound investigation.

It is acceptable for food processors to give food
a reasonable shelf life, but not to "embalm" food
that can no longer be considered fresh.

The most disturbing element in the use of additives is that they are not always fully investigated
before being used in foods. This is particularly true
of substances in use prior to the Food Additives
Amendment to the federal Food, Drug, and Cosmetic Act, and also of those included as GRAS. In
some cases, the use of a particular additive, with
great food industry resistance, has been discontinued only after the fact—after possible physical
damage to unwary and trusting consumers.

Before purchasing processed foods, read the
ingredients list on the label. The ingredients are
listed in descending order of abundance—the first
item listed is the most abundant. If there seem to be
more chemicals than food, you might consider buying another product. Remember that some of the
ingredients with chemical-sounding names, such as
ascorbic acid, are actually important vitamins or
other nutrients. Some foods, such as ice cream, can
still be sold without listing ingredients, although
many chemicals may be included.

Antioxidants Some foods, particularly unsaturated
fatty-acid-containing foods, tend to oxidize (combine with oxygen). Frozen peaches become brown

---

● ASK YOURSELF

---

1. What is a food *additive?*
2. What is a food *residue?*
3. What hazards do the additives in our food pose?
4. What is the function of antioxidants? Emulsifiers? Artificial sweeteners? Coloring agents?

---

and unattractive. Some cake mixes become useless unless the shortening in them is kept fresh. Antioxidants are used to minimize these problems.

Acids Baking powder or cream of tartar (tartaric acid) reacts with baking soda and produces carbon dioxide to leaven bread and cake, making it light. Certain other acids (phosphoric, citric, malic) are used to counteract the excessive sweetness of many soft drinks.

Emulsifiers Emulsifiers break up fats and oils into very small particles. They are used in bakery goods to improve uniformity of texture, fineness, and softness; in ice cream to control particle size (which results in smoother ice cream); in salad dressing to prevent the oil and vinegar from separating.

Artificial sweetners Substitute sweeteners (sweet-tasting compounds with no food value) have a long history. They have long been used by diabetics. Saccharin, for example, is over 300 times as sweet as table sugar. Although possibly safe within recommended doses and as used by diabetics, the use of saccharin is restricted or prohibited in some countries because it is a nonfood (contains no food value) and may cause cancer when consumed in large quantities.

Coloring agents Coloring agents have been used in food since the early 1900s. Many scientists question the safety of a number of federally certified food-coloring agents, particularly the sodium nitrite used in meats and the "rainbow" of coloring agents used to increase the eye-appeal of beverages, cereals, meats, desserts, fruits, and vegetables.

Sodium nitrite is used as a preservative, flavoring agent, and coloring fixative in hot dogs, bacon, ham, luncheon meats, smoked fish, and related products. It is added to over 7 billion pounds of such food products each year to give them the pink color everyone expects. The FDA limits the use of nitrites to 200 parts per million in food; this amount is not toxic. Many food manufacturers have been cited for overstepping these limits up to 3,000 parts per million. The amount of sodium nitrite can be toxic. This is the only material added to food during processing that is known to cause death in humans. Further, it is converted within the intestine to cancer-causing nitrosamines.

Four million pounds of dyes are added to commercial foods every year. Many dyes are complex chemicals that are not found in nature. Some in current use have never been adequately tested for harmful effects. Over the years numerous dyes have been banned after it was proven that they caused cancer, birth defects, long-term mutagenic harm, or serious allergies.

These coloring agents, along with artificial sweeteners, in recent studies by Dr. Ben F. Feingold, chief emeritus of the department of allergy at the Kaiser-Permanente Medical Center in San Francisco, have been suggested as the cause of hyperkinetic behavior in some schoolchildren. When these youngsters were allowed to eat foods containing these materials, they showed the classic symptoms of a hyperkinetic condition. After being put on a diet free from such additives they returned to normal. Dr. Feingold has asked the federal government and the food-processing industry to identify more clearly the coloring and sweetening agents used in foods. A listing of dyes (and other additives as well) is not required on so-called *standard food items* such as ice cream, cheese, and butter. On other food products the coloring agents need only be identified as "artificial coloring," or "certified color added."

Chemical bleaching Years ago, flour for white bread was aged, stored, and turned periodically to bleach it. Today, chemical bleaching agents are added to white flour for making bread. The baking industry states that the sole purpose of bleaching is consumer demand for white bread. One or a combination of chemicals may be used to bleach flour. A flour treated by these chemicals must be marked "bleached" on the label. The complete actions of many of these chemicals are unknown.

Enrichment chemicals Some chemicals are added to white flour in order to "enrich" it. These chemi-

Food Fortifying

The FDA does not approve of labeling that claims that the addition of nutrients makes a food nutritionally equivalent to a food which it resembles and for which it is a substitute. Food should be fortified by the addition of nutrients under the following guidelines:

1. To overcome a nutritional deficiency, such as adding iodine to salt to prevent goiter
2. To restore nutrients lost in storage, handling, or processing (All nutrients lost should be restored.)
3. To supply other nutrients in proportion to the calories contained
4. To fortify foods that substitute for or resemble traditional foods, so that they will be equal nutritionally to the traditional food
5. To meet nutritional standards required for the food, such as in the enriching of bleached, white bread (Not all the nutrients removed in processing are restored; only one guideline needs to be followed.)

cals are identical to the vitamins and minerals that have been removed in processing. Vitamins and minerals are often added to certain other food products to replace *some* of those that have been lost during processing or to make up for an inherent deficiency in the food.

White flour is enriched not only to restore nutrients lost in processing but also to supply the general public with nutrients commonly deficient in their diets.

Other additives *Pectin* is a thickening agent that is added to certain fruits in order to give a consistent and desirable thickness to jams and jellies. Wieners, like other sausages, contain *flavoring agents.* Canned shredded coconut contains a *humectant* to keep it moist. Table salt, powdered sugar, and malted milk powder all contain *anticaking* agents. Salt and sugar are still used as preservative and flavoring agents.

NUTRITION AND HEALTH

Good health is much more than the absence of disease; it is also physical and intellectual vigor, vitality, and freedom from emotional and functional illnesses of all kinds. The level of general public health in the United States has been improving through the years, but consistent health problems do trouble certain segments of our population. Diet, the customary amounts and types of food and drink taken by a person from day to day, and nutrition, the relationship between the needs of our bodies and the food we consume, are important factors in good health. In fact, to a certain extent, food is the basis of good health. How adequate our diet and nutrition are in contributing to good health may be measured in one or more of the following ways.

Nutritional Levels

Adequate nutrition is attained when individuals are eating diets that enable them to grow, mature, reproduce, and function in a healthy and normal manner. Insufficient nutrition occurs when the diet is inadequate for an individual to maintain health. It may result from one of two things: undernutrition or malnutrition. The effects of these two may appear separately or together.

Undernutrition This is an insufficiency of calories in the diet, and is usually caused by lack of food. When famine strikes, those most severely hit are the very young, the old, and those in the lower socioeconomic groups. The United Nations Food and Agricultural Organization (FAO) estimates that 10 to 15 percent of the world's population is undernourished.

The most obvious symptoms of continued calorie deficiency are the conditions of underweight and

Kwashiorkor is caused by a severe protein deficiency. It is one of the most common diseases in the world today.

TABLE 8.14 THE CHARACTERISTICS OF GOOD VS. BAD NUTRITION

Healthy	Unhealthy
Well-developed body	Body may be under-sized, or show poor development
Average weight for height	Thin (more than 10 percent underweight) or overweight (fat or flabby)
Muscles firm	Muscles small and flabby or underdeveloped
Skin firm and of a healthy color	Skin loose, sallow, waxy, or off color
Membranes of eyelids and mouth reddish pink	Membranes of eyelids and mouth pale
Eyes clear	Eyes reddened or puffy; dark hollows or circles below
Full of life	Irritable, overactive, fatigues easily, listless, fails to concentrate
General health excellent	Susceptible to infections, lacks endurance and vigor

becomes severely waterlogged, and death commonly occurs from heart failure.

Pregnant women suffering from severe undernourishment may have longer periods of labor at childbirth, creating hazards to both themselves and the child. Since the mother's production of milk is often affected, infant mortality increases sharply.

Malnutrition Malnutrition is a type of selective starvation. It is the absence of some of the needed nutrients in the diet and is responsible for the deficiency diseases that affect human beings.

Kwashiorkor is the severe deficiency of protein; it is the most widespread and most serious deficiency disease in the world today. Other common deficiencies include the following four: Vitamin A deficiency can cause bone, skin, and vision problems. *Xerophthalmia*, a vitamin-A-deficiency disease, can impair night vision. *Scurvy* is caused by a deficiency of vitamin C, and is not as common today as it was in the past. Iodine deficiency can be

starvation. Undernourishment causes the body to use its own fat protein and other tissues; first the body loses weight, and then growth and development are stunted. In serious cases of starvation, the metabolic rate is reduced, the pulse is slowed and weakened, the blood pressure is reduced, muscle tone is decreased, the skin becomes less elastic, mental processes are dulled, and the person is easily fatigued. In cases of extreme starvation the body caused by an insufficient supply in the diet or the body's inability to use iodine. An iodine deficiency can cause the thyroid gland to enlarge in an attempt to produce sufficient thyroxin. This is called *goiter*. Since hemoglobin in red blood cells contains iron, an iron deficiency can reduce the hemoglobin concentration and cause *iron-deficiency anemia*.

Other nutritional disorders Nutrition plays an important role in such metabolic diseases as gout, diabetes, and obesity. Acne, eczema, dermatitis, and other skin diseases often have nutritional origins. The skin is affected by many nutritional deficiencies, especially the lack of vitamins A and C and protein. Well-nourished skin is better able to resist skin infections. High-quality nutrition also helps to counteract both physical and emotional stresses.

Inadequate protein in children's diets can affect their intellectual development. Lysine, one of the

● **ASK YOURSELF**

1. What is meant by "undernutrition"?
2. How does malnutrition differ from undernutrition?

amino acids, plays an important part in supplying adequate protein, which, according to some authorities, may determine how well the memory operates.

The human body is like any other natural system. It must maintain a balance between the food it takes in and the energy it expends. Adequate nutrition and exercise are needed for optimum health. Often parents assume that a child is healthy because they have no standard of what a healthy child should be. Table 8.14 contrasts the general characteristics of healthy, well-nourished individuals with those of undernourished or poorly nourished ones.

IN REVIEW

1. Food is any substance—besides air and medicines—that provides for the body's growth, maintenance, repair, and reproduction. The substances within food that perform these functions are termed nutrients.
   a. Carbohydates consist of the complex carbohydrates (starches), cellulose fibers, and the simple and compound sugars. The complex carbohydrates are important for the production of energy. Fiber is useful because of the functions it performs in the digestive tract. Simple sugars supply ready energy.
   b. Fats and oils are very high in caloric content. They form part of the membranes of cells, carry fat-soluble vitamins into the body, and insulate and protect the body.
   c. Proteins form the soft tissues of the body; they make up enzymes and other chemicals produced by cells. Proteins are long chains of amino acids, of which there are twenty. Eight are essential for adults, while growing infants and children need nine. Proteins containing all essential amino acids are biologically complete, while those missing one or more are incomplete. Amino acids cannot be stored in the body and must be supplied daily in the diet.
   d. Many minerals are found in the body; they perform many essential functions in the bones and blood.
   e. Vitamins are a group of important compounds that are found in very small portions in food. Along with enzymes, vitamins to carry out very specific, important chemical reactions in the body.
   f. No material serves the body in as many vital functions as water. How much water people require each day depends largely on the temperature of the air, and the kind of physical activity they engage in.
2. Each cell of the body converts food energy into many other forms of energy. The amount of energy released is measured in kilocalories (kcal). Proteins and

carbohydates contain 1,860 kilocalories per pound. Fats and oils contain 4,220 kilocalories per pound.

   a. The body's total caloric needs depend upon metabolism and physical activity.

3. In 1977 and 1979 the Select Committee on Nutrition and Human Needs and the USDA and USDHEW published goals and guidelines to help Americans establish more healthful eating habits.

   a. The committee recommended that we increase our consumption of complex carbohydrates, high-fiber foods, and reduce our consumption of simple sugars.

   b. The committee recommended changes in the sources of our protein. We should reduce our intake of beef, pork, eggs, and whole milk and high-fat dairy products, and increase our consumption of low-fat foods such as seafood, poultry and nonfat dairy products. Low-fat protein can also be obtained from meat analogs and textured vegetable proteins.

   c. The high consumption of fats and oils in the United States is a major cause of overweight. Many links exist between high-fat diets and the incidence of heart attacks and cancer.

   d. Another major dietary problem in the United States is the overconsumption of sodium. Any amout of sodium over $\frac{1}{2}$ gram a day is unnecessary. Salt should not be used in cooking or put in cooked food; foods high in sodium should be avoided.

   e. There is also too much caffeine in the U.S. diet. Americans can reduce or eliminate caffeine in their diets by drinking noncaffeinated soft drinks, and less coffee, tea, and cocoa.

4. Food-group plans provide an easy way to adequate nutrition. The U.S. Four Food-Group Plan divides foods into four groups similar in nutritional content. The Canadian Food Guide Plan is similar to the U.S. plan and was developed with the same intent.

5. Food supplies all the nutrients for growth and for the replacement of worn or damaged cells, as well as the manufacture of cellular products.

   a. Food fads are food beliefs inadequately supported by scientific facts. A food faddist is someone who relies upon belief and refuses to acknowledge fact. Food quacks pretend to have skills, knowledge, or qualifications that they do not possess.

6. Foods processed in one state and shipped to another are required by law to state on their labels what nutrients they contain. Foods sold in the same state where they are processed, however, are exempt from these labeling requirements.

7. Most foods have a short storage life; therefore many chemicals have to be added to our foods to perserve them.

   a. Additives are substances in foods that are deliberately added for specific reasons.

   b. Residues are substances that remain in foods as a result of processing or contamination.

8. Food is the basis of good health.

   a. Adequate nutrition enables individuals to grow, mature, reproduce, and function normally.

   b. Insufficient nutrition occurs in two ways.

      (1) Undernutrition results from an insufficiency of calories in the diet.

      (2) Malnutrition is caused by the absence of needed nutrients.

   c. Many other nutritional problems arise from faulty nutrition.

CHAPTER 9
DIET AND WEIGHT CONTROL

The American preoccupation with problems of overweight brought on by excessive eating is a modern paradox. In a world plagued by too little food, we spend too much time with our fears of being overweight. While such concern might be considered a compliment to our knowledge of what proper weight might be, it is at the same time a frank commentary on many people's failure to maintain satisfactory weight.

Appetite and hunger are controlled by a small area in the brain called the "appetite center," or *appestat*. This area is composed of two sets of nuclei (brain control centers): One nucleus determines our perception of hunger, and the other determines our perception of feeling satisfied. We regulate our need to eat by comparing the perceptions we receive from each. Thus, the appestat works something like a thermostat controlling the temperature of a room.

These nuclei in turn are regulated by our emotions, body chemistry, and inheritance. All of these factors influence our ability to control our weight. Most of us succeed in accomplishing things we want to accomplish. We do things that interest or motivate us. If we are to maintain a desirable weight, we must *want* to. Our motivations for losing weight or maintaining a desirable weight include the following:

1. *A desire to look attractive.* Whether a person likes it or not, he or she must admit that clothing styles are directed toward slender figures. Of course, larger sizes are made for those who need them, but these are not the fashion ideal. Few people are not influenced by the fashion market. Then again, it's always pleasant to fit into a standard-size theater or lecture-hall seat or to take no more than our third of the car seat. The overweight person faces such dilemmas daily.

2. *Longer life.* According to studies, a man 45 years of age, of medium height and frame and weighing 170 pounds, can expect to live 2 to 4 years less than a similar man weighing 150 pounds. A man 45 years of age, of medium height and frame, weighing 200 pounds, can expect to live 4 to 6 years less than a similar man weighing 150 pounds. For every pound of added fat an additional two-thirds of a mile of blood vessels are required to keep this pound of fat alive.

   Excess fat complicates all surgery and increases surgical risks. The same is true in the delivery of a child. It takes extra body effort to carry body weight that is not needed. Thus, the overweight person is more often tired.

3. *Fewer diseases.* Cardiovascular diseases, diabetes, gallbladder disease, cirrhosis of the liver, certain forms of cancer, and arthritis occur more often or can be more serious in overweight people than in those of desirable weight.

Overweight persons are less agile, have more balancing problems, move more slowly, and have more physical accidents than persons of normal weight.

4. *Fewer painful conditions.* Overweight is a factor in such common conditions as varicose veins, high blood pressure, gout, pulmonary emphysema, nephritis, and toxemia in pregnancy.

A HEALTHY DIET

A healthy diet must meet certain basic requirements. It should provide sufficient amounts of all the nutrients required for good health; and it should not provide excessive amounts of any one nutrient. Some excess amounts of nutrients can be stored in the body; others can be eliminated. Still other nutrients can be toxic (poisonous) in high concentrations in the body.

A diet should include a variety of textures to maintain good intestinal tone and to provide enough water. Several guidelines can help you select a good diet. Some were developed by research agencies in government and private industry. Others were provided by educational institutions to help in the teaching of sound nutrition.

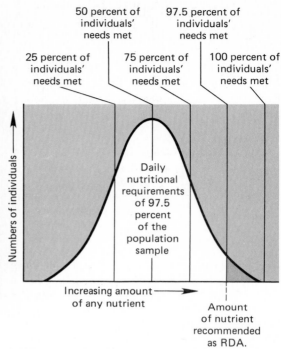

The basis of RDAs. The area under the bell-shaped curve represents the degree to which the nutritional requirements of the population sampled were met. International committees use this curve in establishing the recommended daily allowances of any nutrient.

Recommended Daily Dietary Allowances (RDAs)

Because of metabolic variations among people, it is impossible to predict the precise nutritional needs for everyone. Consequently, nutritional recommendations are based on the needs of normal, healthy individuals, statistically placed upon a "bell-shaped curve" as shown in the illustration.

The amount of each nutrient needed by the greatest number of individuals is shown by the middle or "mean" of the bell-shaped curve. It is not practical to recommend amounts of each nutrient needed by 100 percent of the population. But, the RDAs are adequate for 97.5 percent of the people who were sampled when constructing the curve. As shown by the curve, a few normal individuals (about 2.5 percent) require more nutrients than can be practically recommended in a daily allowance.

Also, the recommendations for adults are often inadequate for pregnant women, breast-feeding women, the elderly, and people suffering from diseases.

Because of the size of the population sample used, and the variability of individuals within the population, RDAs are only a guide for the design or evaluation of adequate diets. Always keep this in mind when a recommendation for daily allowance is made. The RDAs are shown in Table 9.1.

Since some people need more, some less, than the recommended amounts of certain nutrients, it must not be automatically assumed that certain food practices are poor or that people are malnourished simply because the recommendations are not being completely met. Also, the RDAs are ample enough to provide some margin of safety for each nutrient in order to take care of variations among different individuals. For most people in this country, the recommended allowances can be considered adequate for maintaining good health and preventing nutritional diseases.

In addition to defining sufficient amounts of each nutrient, we should take a look at the matter of excess amounts of certain nutrients in the body. Fat-soluble vitamins can be stored in the body, and can be toxic when take in excess. The RDAs for such vitamins should be adequate for most individuals. Excess water-soluble vitamins are excreted by the body, but if excess amounts are consumed without enough water to carry them through the kidneys, they can cause permanent kidney damage. Also, excess amounts of fatty and starchy foods produce overweight individuals, while high-protein diets can cause kidney damage. All the nutrients that one consumes should be balanced to promote good health.

Calculations of Dietary Needs

There are no definite requirements for carbohydrate intake, and the amount of carbohydrates in the diet can be varied considerably. Carbohydrate-rich foods provide an economical energy source and furnish some minerals and vitamins.

Some fat or oil should always be present in the diet. The body needs *essential fatty acids*. Most fat-soluble vitamins, especially A and D, gain entrance to the body through fat-rich foods. Whenever greatly reducing fat intake in a low-fat diet, you should supplement your diet with vitamin A.

The amount of protein consumed should be liberally above the recommended daily allowances listed on the RDA table. On the other hand, super-abundant protein intake provides no advantage and just adds weight. In general, it is unwise to limit the intake of protein over long periods. Each day an individual should eat a minimum of about 0.8 grams of protein per kilogram (2.2 pounds) of body weight to ensure the best nutritional conditions. It is important the protein come from a wide variety of sources.

Children need about twice the minimum RDA of protein to allow a reasonable margin for growth. More calories, proteins, minerals, and vitamins are required during the body's growth period. Development for normal individuals will vary considerably depending on differences in height and weight.

Caloric needs during pregnancy If a woman has established good eating habits and is well nourished when she becomes pregnant, she has little cause for concern. She needs only to increase her intake of some foods she is accustomed to eating. Her nutri-

Eating during pregnancy. Mothers should increase their food intake by 150 calories a day during the first trimester (3 months) and 350 calories a day during the second and third trimesters.

TABLE 9.1 RECOMMENDED DAILY DIETARY ALLOWANCES*

	Age (years)	Weight		Height		Protein (g)	Fat-soluble vitamins		
		(kg)	(lb)	(cm)	(in)		Vitamin A (μg RE)†	Vitamin D (μg)‡	Vitamin E (mg α-TE)§
Infants	0.0–0.5	6	13	60	24	kg × 2.2	420	10	3
	0.5–1.0	9	20	71	28	kg × 2.0	400	10	4
Children	1–3	13	29	90	35	23	400	10	5
	4–6	20	44	112	44	30	500	10	6
	7–10	28	62	132	52	34	700	10	7
Males	11–14	45	99	157	62	45	1,000	10	8
	15–18	66	145	176	69	56	1,000	10	10
	19–22	70	154	177	70	56	1,000	7.5	10
	23–50	70	154	178	70	56	1,000	5	10
	51+	70	154	178	70	56	1,000	5	10
Females	11–14	46	101	157	62	46	800	10	8
	15–18	55	120	163	64	46	800	10	8
	19–22	55	120	163	64	44	800	7.5	8
	23–50	55	120	163	64	44	800	5	8
	51(+)	55	120	163	64	44	800	5	8
Pregnant						+30	+200	−5	+2
Lactating						+20	+400	+5	+3

*The allowances are intended to provide for individual variations among most normal persons as they live in the United States under usual environmental stresses. Diets should be based on a variety of common foods in order to provide other nutrients for which human requirements have been less well defined.

† Retinol equivalents. 1 retinol equivalent = 1 μg retinol or 6 μg β carotene.

‡ As cholecalciferol. 10 μg cholecalciferol = 400 IU of vitamin D.

§ α-tocopherol equivalents. 1 mg d-α tocopherol ≡ 1 α-TE.

‖ 1 NE (niacin equivalent) is equal to 1 mg of niacin or 60 mg of dietary tryptophan.

¶ The folacin allowances refer to dietary sources as determined by *Lactobacillus casei* assay after treatment with enzymes (conjugases) to make polyglutamyl forms of the vitamin available to the test organism.

IU stands for International Unit.
1 kilogram (kg) = 2.2 pounds (lbs)   1 gram (g) = 1,000 milligrams (mg)
1 kilogram (kg) = 1,000 grams (g)   1 milligram (mg) = 1,000 micrograms (μg)
1 kilocalorie (kcal) = 1,000 calories

● **ASK YOURSELF**

1. What factors regular appetite?
2. How strong are emotional factors in regulating appetite? How can we alter our emotions to control our appetite?
3. Discuss the reasons for losing weight or maintaining a desirable weight. How many of these reasons guide you? Should you loose some weight?
4. Explain some of the adverse effects of fat on the body.

tive needs are best met by a simple, wholesome diet, based upon milk (and dairy products), eggs, meat, legumes, whole grains, fruits, and vegetables.

From the first through the fourth month of pregnancy, nutrient needs for the daily growth of the fetus are small. The mother should eat just what any woman normally would consume to preserve or build up health and vitality. However, nausea during this time may cause a woman to eat smaller meals at shorter intervals. If vomiting is severe or prolonged, she should notify her physician. Vitamin-mineral supplements should be taken only if prescribed by her physician.

After the fourth month, the National Research Council recommends increased allowances for almost all essential nutrients. These recommendations are shown in the RDA table. By the sixth month, the fetus is gaining about 10 grams of weight daily. About half of the total weight increase of the fetus occurs in the last two months. Therefore, it is very important that a pregnant woman eat foods unusually rich in all nutrients during this two-month period.

Water-soluble vitamins							Minerals					
Vita-min C (mg)	Thia-min (mg)	Ribo-flavin (mg)	Niacin‖ (mg NE)	Vita-min B$_6$ (mg)	Fola-cin¶ (μg)	Vitamin B$_{12}$ (μg)	Cal-cium (mg)	Phos-phorus (mg)	Mag-nesium (mg)	Iron (mg)	Zinc (mg)	Iodine (μg)
35	0.3	0.4	6	0.3	30	0.5**	360	240	50	10	3	40
35	0.5	0.6	8	0.6	45	1.5	540	360	70	15	5	50
45	0.7	0.8	9	0.9	100	2.0	800	800	150	15	10	70
45	0.9	1.0	11	1.3	200	2.5	800	800	200	10	10	90
45	1.2	1.4	16	1.6	300	3.0	800	800	250	10	10	120
50	1.4	1.6	18	1.8	400	3.0	1,200	1,200	350	18	15	150
60	1.4	1.7	18	2.0	400	3.0	1,200	1,200	400	18	15	150
60	1.5	1.7	19	2.2	400	3.0	800	800	350	10	15	150
60	1.4	1.6	18	2.2	400	3.0	800	800	350	10	15	150
60	1.2	1.4	16	2.2	400	3.0	800	800	350	10	15	150
50	1.1	1.3	15	1.8	400	3.0	1,200	1,200	300	18	15	150
60	1.1	1.3	14	2.0	400	3.0	1,200	1,200	300	18	15	150
60	1.1	1.3	14	2.0	400	3.0	800	800	300	18	15	150
60	1.0	1.2	13	2.0	400	3.0	800	800	300	18	15	150
60	1.0	1.2	13	2.0	400	3.0	800	800	300	10	15	150
+20	+0.4	+0.3	+2	+0.6	+400	+1.0	+400	+400	+150	††	+5	+25
+40	+0.5	+0.5	+5	+0.5	+100	+1.0	+400	+400	+150	††	+10	+50

**The recommended dietary allowance for vitamin B$_{12}$ in infants is based on average concentration of the vitamin in human milk. The allowances after weaning are based on energy intake (as recommended by the American Academy of Pediatrics) and considera-tion of other factors, such as intestinal absorption.

†† The increased requirement during pregnancy cannot be met by the iron content of habitual American diets nor by the existing iron stores of many women; therefore the use of 30–60 mg of supplemental iron is recommended. Iron needs during lactation are not substantially different from those of nonpregnant women, but continued supplementation of the mother for 2–3 months after parturition is advisable in order to replenish stories depleted by pregnancy.

Source: Food and Nutrition Board, *Recommended Dietary Allowances*, 9th ed. revised (Washington, D.C.: National Academy of Sciences–National Research Council, 1980).

Any caloric increase during pregnancy must be closely observed. An excessive weight gain during pregnancy is undesirable and may lead to complica-tions at delivery. Also, some women may find that this excess weight gained during pregnancy may never be lost. A normal and desirable weight gain during the 40 weeks of pregnancy is 20 to 25 pounds. The expectant mother must greatly curtail her consumption of sweet, starchy, or fatty foods. She should eat foods that yield large amounts of protein, vitamins, and minerals in proportion to their caloric value. A woman who wants to control her weight during pregnancy should increase her physical activity and not cut down on the essential nutrients she is eating (*see* Chapter 10).

Teenage mothers are responsible for about 15 percent of all the births in the United States. At the present time, about 40 percent of all brides in the United States are between the ages of 15 and 18 years. Yet studies have shown that in this country, teenagers are more likely to be undernourished than any other segment of the population. There-fore they are poor prospects for a healthy preg-nancy.

● ASK YOURSELF

1. What are the basic requirements of a healthy diet?
2. What are the Recommended Daily Allowances? How were they established?
3. Using the above information, calculate your die-tary needs for one day. Are you currently eating a healthy diet? What should you change to make this a healthy diet?

Pregnant Women Should Not Drink Caffeine

The U.S. Food and Drug Administration has cautioned pregnant women about consuming caffeine. Pregnant women should not drink coffee, tea, and other caffeine-containing products because caffeine may well cause birth defects. Experiments in 1979 and 1980 showed that even modest caffeine consumption (equivalent to two cups of coffee a day) by pregnant rats can retard bone growth in developing rat fetuses. Heavy caffeine consumption (equivalent to 12 to 24 cups of coffee a day) resulted in rat offspring with missing toes.

A large-scale study of the effect of caffeine consumption on human babies was started early in 1980. It is being conducted by the soft-drink and coffee industries and will take from two to four years to complete.

"Cola" and "Dr. Pepper" drinks are made from cola beans, which contain caffeine. In fact, *truth-in-advertising laws* require such drinks to contain certain levels of caffeine. These laws are in the process of being changed at this time.

Until proved otherwise, it would be a good idea for pregnant women to drink caffeine-free soft, drinks, decaffeinated coffee, milk, or just water. (*See* Table 9.7 for a list of common beverages that contain caffeine.)

At birth, a child is already more than 9 months old in terms of nutrition and development. A child's nutrition can be no better than that of the mother.

Producing milk and reducing weight. Mothers need to take in 80 to 95 calories a day for every 100 ml of milk they produce.

Both for themselves and their children, it is important that pregnant women eat a balance diet.

Caloric needs during breastfeeding The nutritive requirements for nursing mothers are higher than those for pregnant women. As shown in the RDA table, all nutrients need to be increased for the lactating (milk-producing) mother.

Only about 11 pounds of weight is lost at delivery in the form of the child, placenta, membranes, and fluid. An additional $4\frac{1}{2}$ to $5\frac{1}{2}$ pounds of weight is also lost during the next several weeks. The remaining 10 or so pounds added as stored fat will be lost by the women when breastfeeding. If a woman does not breastfeed, this weight and any other excess weight is likely to remain permanently. As a consequence, many women grow heavier with each child. Other women, who exercise regularly during and after their pregnancies, retain slender figures.

While breastfeeding, a woman should eat an extra 500 calories per day. She should make sure that at least 85 of these calories, or about 20 grams, are protein. Since milk is very rich in calcium and phosphorus, a breastfeeding mother should increase her intake of milk by one pint to one quart to meet the demands of her child. The vitamin content of the mother's milk is greatly dependent upon her daily intake of vitamins, especially water-soluble vitamins. She should increase her intake of vitamins by about 50 percent. Before giving birth, the wise mother will find nutritional information to plan

Eating as you age. As people age, their food requirements decrease. In order to maintain an ideal weight, the caloric intake must be reduced by 5 percent per decade between the ages of 23 and 50, by 8 percent per decade between the ages of 50 and 70, and 10 percent per decade after age 75. However, older people must be very careful about what nutrients they reduce in their diet.

how she is going to meet the increased needs of breastfeeding.

Caloric needs during aging Many adults do not realize that their energy requirements decline as they grow older and continue their life-long eating habits. Many grow obese in their old age. A reduction in caloric intake can prevent such obesity. As people get older, their rates of basal metabolism (BMR) and their level of physical activity decline. Because the exact degree of reduction in basal metabolism and activity is impossible to predict for each individual, tables that show desirable weights for adults serve only as a reference.

The body's ability to utilize proteins efficiently also decreases as the individual ages. Therefore, it is suggested that the protein allowance remain the same for all ages. The same suggestion is made for minerals and vitamins.

In order to cut down the number of calories consumed, there must be a reduction in the intake of such common foods as fats, sweets, and especially alcoholic beverages. This reduction is important, since too much caloric intake from these sources will interfere with the proper utilization of proteins. Consequently, the best advice is to cut down on fats, sweets, and alcohol as you grow older.

● **ASK YOURSELF**

1. How can a woman best meet her nutritional needs during pregnancy?
2. What is a normal and desirable weight gain during pregnancy? Interview some mothers and see how much weight they gained during their pregnancies.
3. What are the values of breastfeeding to the mother and the child? How many of the above women breastfed their children? Were they the slender ones of the group?

TABLE 9.2   EXAMPLES OF DAILY ENERGY EXPENDITURES OF MATURE WOMEN AND MEN IN LIGHT OCCUPATIONS

Activity Category	Man, 70 kg [154 lb]			Woman, 58 kg [127.6 lb]	
	Time (hr)	Rate (kcal/min)	Total (kcal)	Rate (kcal/min)	Total (kcal)
Sleeping, reclining	8	1.0–1.2	540	0.9–1.1	440
Very light Seated and standing activities, painting trades, auto and truck driving, laboratory work, typing, playing musical instruments, sewing, ironing	12	up to 2.5	1,300	up to 2.0	900
Light Walking on level, 2.5–3 mph, tailoring, pressing, garage work, electrical trades, carpentry, restaurant trades, cannery workers, washing clothes, shopping with light load, golf, sailing, table tennis, volleyball	3	2.5–4.9	600	2.0–3.9	450
Moderate Walking 3.5–4 mph, plastering, weeding and hoeing, loading and stacking bales, scrubbing floors, shopping with heavy load, cycling, skiing, tennis, dancing	1	5.0–7.4	300	4.0–5.9	240
Heavy Walking with load uphill, tree felling, work with pick and shovel, basketball, swimming, climbing, football	0	7.5–12.0		6.0–10.0	
Total	24		2,740		2,030

Source: Food and Nutrition Board, *Recommended Dietary Allowances,* 9th ed., revised (Washington, D.C.: National Academy of Sciences–National Research Council, 1980), p. 24.

● ASK YOURSELF

1. What are your energy needs? How much energy do you expend each day?
2. Do you eat enough food to maintain an adequate energy level?
3. Do you exercise enough to balance your intake of food?
4. Outline a diet and exercise program that would enable you to maintain an ideal weight.

RDAs for energy The healthy body needs energy for metabolic processes, physical activity, growth (including pregnancy), lactation (breast feeding), and to maintain body temperature. In contrast to other nutrients for which RDA recommendations are made, the energy allowance is established at a level thought to be needed for good health of an *average* person within a given activity category. Table 9.2 outlines the different activity levels. From this you can judge your activity level category.

TABLE 9.3 MEAN HEIGHT AND WEIGHTS AND RECOMMENDED ENERGY INTAKE*

Category	Age (years)	Weight (kg)	Weight (lb)	Height (cm)	Height (in.)	Energy Needs (with range) (kcal)	
Infants	0.0–0.5	6	13	60	24	kg × 115	(95–145)
	0.5–1.0	9	20	71	28	kg × 105	(80–135)
Children	1–3	13	29	90	35	1,300	(900–1,800)
	4–6	20	44	112	44	1,700	(1,300–2,300)
	7–10	28	62	132	52	2,400	(1,650–3,300)
Males	11–14	45	99	157	62	2,700	(2,000–3,700)
	15–18	66	145	176	69	2,800	(2,100–3,900)
	19–22	70	154	177	70	2,900	(2,500–3,300)
	23–50	70	154	178	70	2,700	(2,300–3,100)
	51–75	70	154	178	70	2,400	(2,000–2,800)
	76 +	70	154	178	70	2,050	(1,650–2,450)
Females	11–14	46	101	157	62	2,200	(1,500–3,000)
	15–18	55	120	163	64	2,100	(1,200–3,000)
	19–22	55	120	163	64	2,100	(1,700–2,500)
	23–50	55	120	163	64	2,000	(1,600–2,400)
	51–75	55	120	163	64	1,800	(1,400–2,200)
	76 +	55	120	163	64	1,600	(1,200–2,000)
Pregnancy						+300	
Lactation						+500	

*The data in this table have been assembled from the observed median heights and weights of children, together with desirable weights for adults. The mean heights of men (70 in.) and women (64 in.) between the ages of 18 and 34 years are as surveyed in the U.S. population (HEW/NCHS data).

The energy allowances for the young adults are for men and women doing light work. The allowances for the two older age groups represent mean energy needs over these age spans, allowing for a 2 percent decrease in resting metabolic rate per decade and a reduction in activity of 200 kcal/day for men and women between 51 and 75 years, 500 kcal for men over 75 years, and 400 kcal for women over 75 years. The customary range of daily energy output is shown in parentheses for adults and is based on a variation in energy needs of ±400 kcal at any one age, emphasizing the wide range of energy intakes appropriate for any group of people.

Energy allowances for children through age 18 are based on median energy intakes of children of these ages followed in longitudinal growth studies. The values in parentheses are 10th and 90th percentiles of energy intake, to indicate the range of energy consumption among children of these ages.

Source: Food and Nutrition Board, Recommended Dietary Allowances, 9th ed., revised (Washington, D.C.: National Academy of Sciences–National Research Council, 1980), p. 23.

Many individuals in the United States are overweight and may require less food energy than is recommended because they have very sedentary work and leisure patterns. For individuals who are already obese, caloric intake should be reduced below the suggested levels, and exercise activity increased for a sound weight control program. Adults or children who gain excessive amounts of body fat while consuming appropriate amounts of food should increase their physical activity until the desired weight is achieved. It cannot be overemphasized that the maintenance of a desirable body weight throughout adult life depends upon achieving a balance between energy intake and energy output. Table 9.3 lists the suggested energy (caloric) intake for differing groups of individuals.

TABLE 9.4 PERCENTILES FOR WEIGHT AND HEIGHT FOR CHILDREN AND TEENAGERS, 0–18 YEARS OF AGE*

	Males						Females					
	Weight (lbs/oz)†			Height (inches)§			Weight (lbs/oz)†			Height (inches)§		
Age	5%‖	50%	95%	5%	50%	95%	5%	50%	95%	5%	50%	95%
Months												
1	6/15	9/07	11/13	20.2	21.8	23.4	6/08	8/12	10/13	19.7	21.4	22.8
3	9/12	13/03	16/03	22.7	24.4	26.2	9/03	11/14	14/13	22.2	23.8	25.4
6	13/10	17/04	20/13	25.4	27.1	28.9	12/12	15/14	19/03	24.7	26.4	28.1
9	16/09	20/03	20/01	27.2	28.9	30.8	15/06	18/13	22/12	26.4	28.2	30.0
12	18/06	22/05	26/06	28.7	30.4	32.5	17/04	20/16	24/12	27.9	29.7	31.6
18	25/04	25/04	29/09	31.0	33.0	35.2	19/10	23/13	28/08	30.4	32.4	34.4
Years												
2	23/01	27/02	34/02	33.0	34.7	37.8	21/14	25/15	31/02	32.6	34.7	37.4
3	26/08	32/03	39/02	35.6	38.0	40.8	25/09	31/00	37/14	35.3	37.6	40.2
4	30/00	36/11	44/10	38.3	41.2	44.0	28/13	35/02	43/13	38.0	40.6	43.3
5	33/10	41/01	50/13	40.8	44.0	46.8	32/00	38/14	49/12	40.4	43.4	46.2
6	37/04	45/08	57/15	43.1	46.4	49.4	35/05	42/15	56/10	42.6	45.8	49.1
7	41/00	50/04	66/04	45.2	48.7	51.9	38/15	48/01	65/05	44.7	48.2	51.8
8	44/14	55/11	75/15	47.2	50.8	54.3	43/03	54/10	76/06	46.8	50.6	54.5
9	48/15	61/14	87/01	49.2	52.9	56.7	48/10	62/10	89/07	48.8	52.9	57.2
10	53/08	69/03	99/10	51.1	55.0	59.2	53/09	71/10	103/12	51.0	55.3	59.8
11	58/15	77/11	113/04	53.0	57.3	62.0	59/15	81/05	118/13	53.4	57.9	62.5
12	65/09	87/08	127/13	55.0	59.9	64.9	67/02	91/06	133/13	55.9	60.6	65.1
13	74/00	98/14	143/01	57.2	62.6	67.9	75/02	101/07	148/01	58.1	62.8	67.2
14	84/01	111/11	158/11	59.5	65.2	70.7	83/01	110/10	160/12	59.5	64.2	68.5
15	94/13	124/12	174/01	62.1	67.6	72.8	90/03	118/02	171/02	60.2	64.7	69.1
16	105/00	136/10	188/06	64.4	69.4	74.2	95/08	122/15	178/03	60.6	65.0	69.3
17	113/05	145/14	200/14	66.0	70.5	74.9	98/07	124/11	181/07	61.1	65.2	69.4
18	118/12	151/09	210/11	66.3	70.7	75.0	99/09	124/09	181/07	61.4	65.4	69.4

* Data in this table have been used to derive weight and height reference points. It is not intended that they necessarily be considered standards for normal growth and development. Data pertaining to infants 2–18 months of age are taken from longitudinal growth studies at Fels Research Institute. Ages are exact, and infants were measured in the reclining position. The measurements were based on some 867 children followed longitudinally at the institute between 1929 and 1975. Data pertaining to children between 2 and 18 years of age were collected between 1962 and 1974 by the National Center for Health Statistics and involved some 20,000 individuals comprising nationally representative samples in three studies conducted between 1960 and 1974. In the studies children were measured in the standing position with no upward pressure exerted on the mastoid process.
† Weights were converted from kilograms to pounds and ounces using 1 kilogram equalling 2.2 pounds. Sixteen ounces to one pound.
‡ Heights were converted from centimeters to inches using 0.5 cm equalling 1 inch.
‖ Desirable height and weight is expressed in percentiles. The first number represents the average of 5 percent of the children studied; the second, 50 percent (median) of the children studied; and the third number, 95 percent. This allows for an excess of obese children (differences between first and third numbers) being part of the studies. Thus, the median (50 percentile) figures are a good measure of the desirable weight and height for nonobese children.
Source: Adapted from Food and Nutrition Board, *Recommended Dietary Allowances,* 9th ed., revised (Washington, D.C.: National Academy of Sciences—National Research Council, 1980), pp. 20–21.

DETERMINING DESIRABLE WEIGHT

The ideal weight for an individual is very difficult to establish. Scientists have attempted for years to set criteria for ideal height-weight relationships. So far, the criteria used are, at best, averages for large groups of people. Also, healthier women are giving birth to larger and healthier babies that show accel-

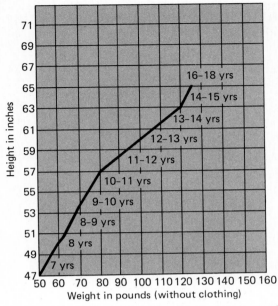

Average height and weight for girls 7 to 18 years of age. Based on the Iowa Growth Standards.

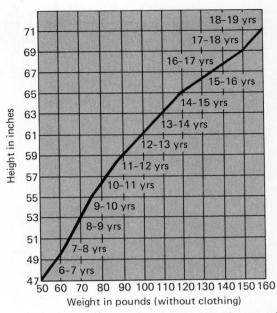

Average height and weight for boys 6 to 19 years of age. Based on the Iowa Growth Standards.

erated growth patterns, which lessens the value of these criteria.

Children and Teenagers

Physicians use two criteria to determine ideal height-weight relationships for growing children and teen-agers: The first is *bone-growth rate*, which is established by the use of X-ray examination of growing individuals. The second is the average height and weight of the parents. Both criteria are unusable by the general public. Growth rates for children can be assessed by using the diagrams based on the Iowa Growth Standards, which have been developed over the years by the University of Iowa. These diagrams are accurate for assessing normal growth and height-weight relationships for individuals up to 18 or 19 years of age.

Adults

There are a number of ways to establish idea weights for adults. The height and weight figures that have long been used have limitations. Two sets of figures are in use today. The first ignores age and

classifies individuals as having either small, medium, or large body frames. A good way to judge your body frame is shown in the following series of photographs. This set of figures based essentially on the average weight at age 23 for women and age 27 for men, the ideal being to maintain these weights throughout the life span.

The second set of tables is based upon records from insurance companies. The limitation of these tables is that the average weights they represent are not necessarily the most desirable. The most accurate insurance tables are based upon the measurements of 5 million people aged 15 to 69 years as shown in the following two graphs.

Individual tests The absence of any single effective method for measuring body fat has led to the use of many techniques other than the height-weight tables. Take the simple "mirror" test, for example. People can tell if they are greatly overweight just by looking at their reflected images. Body fat can also be determined by skinfold measurements. A physician measures skinfolds from several parts of the body, using a constant-pressure caliper, and then compares his findings with standard measure-

TABLE 9.5 SUGGESTED DESIRABLE WEIGHTS FOR HEIGHTS AND RANGES FOR ADULT MALES AND FEMALES

Height*		Weight†							
		Men				Women			
in.	cm	lb		kg		lb		kg	
58	147		–		–	102	(92–119)	46	(42–54)
60	152		–		–	107	(96–125)	49	(44–57)
62	158	123	(112–141)	56	(51–64)	113	(102–131)	51	(46–59)
64	163	130	(118–148)	59	(54–67)	120	(108–138)	55	(49–63)
66	168	136	(124–156)	62	(56–71)	128	(114–146)	58	(52–66)
68	173	145	(132–166)	66	(60–75)	136	(122–154)	62	(55–70)
70	178	154	(140–174)	70	(64–79)	144	(130–163)	65	(59–74)
72	183	162	(148–184)	74	(67–84)	152	(138–173)	69	(63–79)
74	188	171	(156–194)	78	(71–88)		–		–
76	193	181	(164–204)	82	(74–93)		–		–

*Without shoes.
†Without clothes. Average weight ranges in parentheses.
Source: Food and Nutrition Board, *Recommended Dietary Allowances*, 9th ed., revised (Washington, D.C.: National Academy of Sciences–National Research Council, 1980), p. 22.

ments for obesity. A variation of this method is called the "pinch" test. Pinch the underside of your upper arm. If the fold between our fingers is more than an inch thick, you have excess body fat. Generally, the skinfold thickness of body fat is unrelated to height. Another measure of obesity is the "ruler" test. Lie on your back and lay a yardstick from your chest to your abdomen. If you are obese, the yardstick will slant up toward your feet; if your weight is normal, the yardstick will remain flat or slant downward.

The "perfect 36" is another interesting way to determine ideal weight. Subtract your waist measurement, in inches, from your height, in inches. Values of 36 to 40 are considered to be normal. Values of 35 to 25 shows slight obesity. Values of less than 25 represent definite obesity.

● **ASK YOURSELF**

1. How would you judge the ideal weight of a growing child or teen-ager? Check with some children and see how they compare with these tables.
2. What is the "pinch" test?
3. What is the "ruler" test?
4. Work out the "perfect 36" calculation and determine your weight status.

All body measurements have obvious limitations. However, if you fail a number of these tests and your weight falls outside the height-weight tables, you can be sure that you need to lose weight.

OBESE OR OVERWEIGHT?

There is a difference between being *obese* and being *overweight*. *Obesity* is defined as an "excessive deposition of fat beyond what is considered normal for a given age, sex, and build." Weight, on the other hand, is defined as "a quality of heaviness." Overweight, then, is simply "overheaviness" without any regard to fatness. *Overweight* can be defined as any weight in excess of that recommended for a given person. Many people who are obese use the term overweight because emotionally it makes them feel better. The first thing you must do to reach an ideal weight is to be honest with yourself. Call excess fat what it is—*obesity*. This is the first step in weight control.

Causes of Obesity

The accumulation of excess fat that leads to eventual obesity is caused by an intake of excess calories, a deficiency of caloric expenditure because of lack of exercise, or a combination of both factors.

Each individual must develop eating habits that maintain ideal weight.

Certain factors affect your eating habits. Some contribute to overweight and obesity more than others.

1. *Home environment.* Some people come from homes where meals are rich and excessive eating is common. Others have become accustomed to too little exercise because of available transportation, lack of participation in sports, modern conveniences, and laziness.

2. *Poverty.* Some families, because of limited finances, buy or prepare foods that appease the appetite. Such foods are usually rich in fats and oils (fried foods) or sugar (candy, desserts).

3. *Occupation.* Housewives who are often around foods may become habitual samplers and nibblers. This is also true of people who work with food or beverages in restaurants, bars, fast-food outlets, bakeries, and candy shops.

4. *Emotional factors.* Some people find eating their only pleasure. Others use food as their *coping mechanism.* Anytime anything unpleasant happens to them, they respond to it by eating.

5. *Age and disease.* As people grow older, their need for calories declines. They often find it difficult to change long-established eating habits. Physical disabilities may reduce a person's activities and make a change in eating habits necessary.

6. *Inheritance.* A tendency toward obesity runs in some families. Studies show that the fatness of children relates closely to whether both parents, one parent, or neither parent is fat. The weight of children adopted from birth shows no correlation with that of their adoptive parents. Heredity and not environment, therefore, is often responsible for overweight.

7. *Inactivity.* When physical activity declines, the appetite increases. Thus, as you do less, you eat more. Mild increases in physical activity are very good for reducing appetite. Increased appetite and reduced activity means a weight gain. Persons who are suddenly immobilized tend to gain weight.

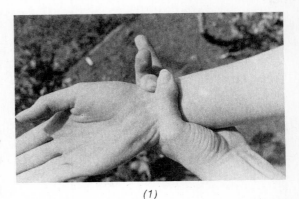

(1)

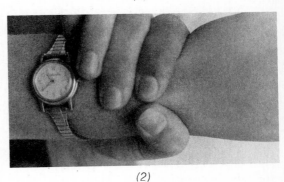

(2)

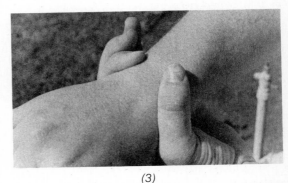

(3)

How to determine if you have a small, medium, or large frame (bone structure). (1) Small frames: Your index finger and thumb will overlap when you grasp your wrist. (2) Medium frames: Index finger and thumb just touch. (3) Large frames: Index finger and thumb do not touch.

Regardless of the underlying causes of obesity, the basic problem is that many of us simply take in more calories than we need for our total activities.

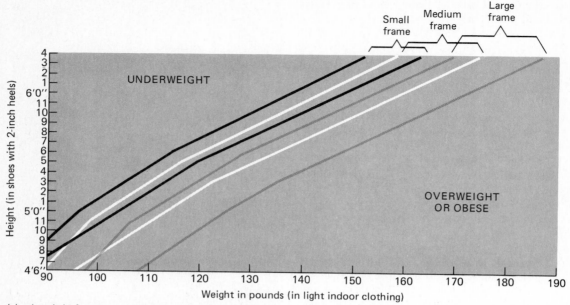

Ideal weight for women 25 years of age and over. Courtesy of the Metropolitan Life Insurance Co., How to Control Your Weight *(New York, 1960).*

● **ASK YOURSELF**

1. What is the different between *obesity* and *overweight?* Do you fit into either of these categories? If so, what do you think caused this?
2. What factors influence an individual's eating habits? Do any of these factors influence you?
3. How many calories does each pound of human body fat represent?

Unused calories from any source are stored in the body as fat. Each pound of human body fat represents about 3,500 stored calories. We either use the calories from the food we eat or store them. The more stored calories, the greater the obesity.

REDUCING YOUR WEIGHT

The first thing "heavy" people must do is to admit that their obesity in all likelihood is caused by overeating. If you eat the amount required by your activity level, you should be able to maintain your ideal weight. Obese people must go one step farther.

They must reduce their food intake to a point below their energy output. As the body uses more energy than it takes in, it will lose fat and thus weight. This is the simple basis of most effective reducing programs.

How can obese persons reverse the usual process? They can follow one of several methods. First of all, they can cut their food intake through a reducing diet. Second, they can use up their stored calories by exercising regularly. Best of all, they can do both. This combination reduces weight, improves muscle tone, and promotes emotional well-being.

Diet

As long as your caloric intake is equal to your caloric use, your body weight should not change beyond normal day-to-day fluctuations—usually a difference of about 2 pounds. As soon as your caloric use is more than your caloric intake, you begin to lose body fat. For every 3,500 calories you use up and do not replace by eating, you reduce your weight by one pound. It doesn't matter whether you

Obese persons are often very sedentary. To maintain normal weight, they must balance their food intake with their physical activity.

use up 700 excess calories every day for 5 days or 10 calories every day for 350 days. The end result in both instances will be the same; the loss of one pound of fat. Tables 9.6 and 9.7 show the caloric content of some common foods and beverages.

Of the many dietary aids constantly presented to the public, some have value, many are harmless but useless, and a few are useless and harmful. New "miracle" diets that purport to solve all weight problems often appear. The diversity of ideas on diets makes it almost impossible to evaluate each diet individually. Instead, we will present a few criteria that can help you judge new diets.

1. *The diet must reduce your caloric intake.* To determine whether the diet actu-

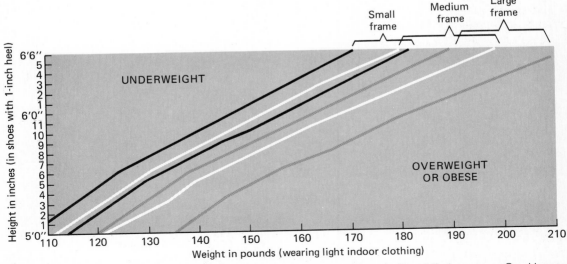

Ideal weight for men 25 years of age and over. Courtesy of the Metropolitan Life Insurance Co., How to Control Your Weight *(New York, 1960).*

TABLE 9.6 THE CALORIC CONTENT OF SOME COMMON FOODS

Common foods	Kcal	Common foods	Kcal
Breads		**Fish**	
Bagel (1)	165	Abalone (4″ × 5″)	118
Baking powder biscuit (1)	134	Bass (3″ × 3″ × 2″)	
Croutons (21–23 cubes)	62	baked	248
French or Vienna (1 lb loaf)	1,315	fried	274
Nut (1 slice)	146	Crab cocktail (⅓ cup)	67
Pumpernickel (1 slice)	95	Halibut, smoked (4″ × 2″)	219
Raisin (1 slice)	81	Oysters	
Rolls		canned (1 Tbsp)	8
French (1)	112	fried (4″ × 2″)	86
Parker House (1)	53	Salmon	
Rye (1 lb loaf)	1,100	baked (4″ × 5″)	540
White (1 lb loaf)	1,225	canned (½ cup)	206
Whole wheat (1 lb loaf)	1,095	smoked (3″ × 2″)	170
Zwieback (1 slice)	32	Shrimp	
		cocktail (½ cup)	79
Cakes		cooked or canned (¼ cup)	28
Angelfood (whole cake)	1,645	Prawns, fried (1)	65
Cupcake (2½″ diameter)		Swordfish, broiled (2″ × 3″)	181
without icing	90	Trout, steamed (1)	222
chocolate icing	130	Tuna, canned (¾ cup)	244
Devil's food (2-layer, iced)	3,755		
Pound (loaf)	2,430	**Fruits**	
Sponge (whole cake)	2,345	Apple (3″ diameter)	124
Yellow (whole cake)		Apple juice (1 cup)	121
2-layer, without icing	3,160	Apricots (3 to 4)	68
2-layer, chocolate icing	4,390	Apricots, dried (2 to 3)	96
		Avocado (1 medium)	485
Cookies		Banana (1 medium)	85
Brownies (1)	95	Cantaloupe (½ medium)	46
Chocolate chip (1)	50	Cherries, maraschino (5)	21
Commercial (read label)		Cherries, sweet (1 cup)	85
		Cider, sweet (1 cup)	100
Cereals (read label)		Cranapple juice (1 cup)	96
		Grapefruit (½ medium)	58
Cheese		Orange (4″ diameter)	72
Brick, processed (1″ × 2″ × ⅛″)	180	Peach (1 medium)	45
Blue (1 oz)	110	Pear (1 medium)	58
Cheddar (1 oz)	110	Prunes (2 to 3 medium)	21
Cottage (1 cup)		Strawberries (1 cup)	52
creamed	260		
uncreamed	170	**Meats**	
Creamed (1 oz)	100	Beef	
Limburger (1 oz)	100	rib roast (4″ × 4″)	257
Monterey Jack (1 oz)	100	pot roast (3″ × 3″)	198
Roquefort (1 oz)	110	steak, Porterhouse (3″ × 3″)	293
Swiss (1 oz)	110	steak, sirloin (3″ × 3″)	242
		Pork	
Cream		bacon, Canadian (1 slice)	122
Regular or whipped (1 oz)	110	bacon, crisp (1 slice)	149
Half and Half (1 cup)	320	chop (3″ × 5″)	179
Sour (1 cup)	485	ham, baked (4″ × 2½″)	279
Sour, imitation (1 cup)	440	spareribs (5″ × 1½″)	173
Coffee-mate (1 pkt)	17		

TABLE 9.6 (continued)

Common foods	Kcal
Poultry	
chicken (4" × 4")	
light meat, roasted	170
dark meat, roasted	186
turkey (4" × 4")	
light meat, roasted	205
dark meat, roasted	183
Milk	
Whole (3.5% fat) (1 cup)	160
Low-fat (2% fat) (1 cup)	150
Nonfat (skim) (1 cup)	90
Snacks	
Almonds (10)	60
Cashews (4 to 5)	58
Unsalted peanuts (20)	114
Pecans (8 halves)	70
Potato chips (5 chips)	54
Cheese tidbits (15)	8
Green olives (2)	13
Oysters, raw (3)	50
Herring, raw (1" × ½")	25
Sour pickles (1 slice)	30
Sweet gherkin (1)	10
Clam dip (3 tsp)	15
Popcorn, popped	
plain (1 cup)	25
with oil & salt (1 cup)	40
Pretzels	
thin, twisted (1)	25
stick, small (10 sticks)	10

Source: USDA, "Nutritive Value of Foods," *Home & Garden Bulletin, 72* (Washington, D.C.: 1977).

and high levels of protein delay hunger pangs for a longer time. They will last longer than diets containing only carbohydrates. It is easier to stay with a diet that stays with you.

4. *A good diet can be adapted readily from family meals and can be obtained in public eating places.* If the diet imposes extra preparation or cost, you are less likely to follow it for any length of time.

5. *The diet should be reasonable in cost.* It does this best by making use of seasonal foods and staple items.

6. *The diet should be realistic.* You should be able to adhere to it for the length of time needed to achieve and maintain the desired weight loss. The extremely obese should not lose more than 1½ pounds a week. Less obese people can safely lose more in a shorter period of time. "Crash diets" that limit your selection of food such as cottage cheese and peaches or steak, eggs, and tomatoes, are nutritionally inadequate. They are also monotonous, and their psychological appeal lasts only a short time. While people often lose weight on crash diets, they do not permanently modify their eating habits as necessary to keep the weight off.

7. *The diet must develop new eating habits.* Most important of all—if you are to achieve long-term weight control, the diet must wean you away from your former eating habits. You must acquire a new set of eating habits that you can enjoy for a lifetime.

ally does this, compare its caloric value with a reasonable estimate of your caloric needs.

2. *The diet should be adequate in all nutrients except calories.* Although this criterion is difficult to check, it should attempt to include servings from each of the major food groups. As the number of calories in a diet drops, it is increasingly difficult to obtain adequate vitamins and minerals. If the intake is less than 1,200 calories a day, the diet should be supplemented with vitamin pills and minerals.

3. *The diet should be filling and have staying power.* Diets containing some fats or oils

Maintaining an Ideal Weight

Weight-loss plans seldom address the problem of keeping weight off for long periods of time. This requires readjustment of the regular diet and eating patterns. Whatever is done to lose weight must be incorporated into a person's total life plan. The following are suggestions for losing weight and then maintaining an ideal weight for the rest of your life.

Graph your progress Since most people gain weight gradually, they should ideally lose it in the same way. Because gradual reducing programs can

TABLE 9.7 THE CALORIC CONTENT OF SOME COMMON ALCOHOLIC AND CARBONATED DRINKS

Alcoholic beverages	Calories	Carbohydrates (grams)	Protein (grams)
Ale (8 oz)	98	8.9	1.1
Beer (8 oz)	114	10.6	1.4
Light beer (8 oz)	64	1.8	0.6
Brandy or cognac (1 oz)	73		
Creme de menthe (cordial)	67	6.0	7.0
Gin, rum, vodka, whiskey			
80 proof (1 oz)	70		
86 proof (1 oz)	70		
90 proof (1 oz)	70		
94 proof (1 oz)	80		
100 proof (1 oz)	80		
Daiquiri (cocktail)	122	5.2	1.0
Eggnog (4 oz)	335	18.0	3.9
Highball (8 oz)	166	24.0	
Manhattan (3¾ oz)	164	7.9	trace
Martini (3¾ oz)	140	0.3	0.1
Old-fashioned (4 oz)	179	3.5	
Champagne (4 oz)	84	3.0	0.2
Port (3½ oz)	158	14.0	0.2
Sauterne (3½ oz)	84	4.0	0.2
Sherry (2 oz)	84	4.8	0.2
Vermouth, dry (3½ oz)	105	1.0	
Vermouth, sweet (3½ oz)	167	12.0	

Carbonated drinks	Calorie	Carbohydrate (grams)	Protein (grams)	Caffeine (milligrams)
Club soda (8 oz)	0			
Coca-cola (8 oz)	96	24.0		28
Diet Rite Cola (8 oz)	0–1	0–1		20
Pepsi-Cola (8 oz)	105	27.0		26
Diet Pepsi (8 oz)	0–1	0–1		24
Royal Crown Cola (8 oz)	105	27.0		20
Sugar Free RC (8 oz)	0–1	0–1		20
Shasta Cola (8 oz)	160	42.0		18
Shasta Diet Cola (8 oz)	0–1	0–1		18
Fresca (8 oz.)	2	0–1		
Ginger Ale (8 oz.)	80–90	20–34		
Mixers				
collins (8 oz)	130	21.0		
sour (8 oz.)	85–130	22–34		
vodka mix (8 oz)	130	34.0		
Mountain Dew (8 oz)	125	31		36
Orange (8 oz)	120–195	30–49		
Root beer (8 oz)	100–170	27–43		
Sprite (8 oz)	96	24.0		
Tab (8 oz)	0–1	trace		32
Teem (8 oz)	100	26.0		
Tonic water (8 oz)	84–115	20–26		
Wink (8 oz)	105	28.0		

Source: USDA, "Nutritive Value of Foods," *Home & Garden Bulletin,* 72 (Washington, D.C.: 1977).

be discouraging, people often lose sight of the initial goal. A good psychological "crutch" is a simple graph set up at the beginning of your reducing program. Plan your weight loss according to how many pounds you need to lose. Aim for 1 to 2 pounds per week and construct your graph accordingly. Weigh yourself each day at the same time; morning is perhaps best. Put a dot on the graph at your weight point. The line of dots will tell you how close you are (or how far you have strayed from) your weight-loss plan on any given day. The graph will serve as a daily reminder of your progress. The following graph shows a drop of a pound a week to 185 pounds. When you reach your goal, the graph will show whether you are gaining weight back. Used regularly, the graph will remind you of your progress in maintaining your ideal weight.

Exercise regularly A program of regular exercise is essential to help you lose weight. Besides using up calories, exercise improves muscle tone and produces a general feeling of well-being. Do not look upon physical exercise as a substitute for controlling your caloric intake. You must reduce that also.

Changing your behavior Small changes in your behavior can help you reduce your intake of food. Behavioral changes can make you more aware of when and why you eat, and can help you to eat at times when you are more likely to use up what you eat.

Do not prepare surplus food. Most people prepare more food than they actually need. Judge the amount needed by your activity level. Then put any surplus food in the refrigerator. In time you will be able to gauge the correct amount of food needed. If after eating, you are still hungry tell yourself to wait 15 or 20 minutes before eating any more; by that time you will probably not be hungry.

Learn to eat more slowly. Put your fork down between each bite. You will taste your food, enjoy it more, and also have fewer digestive problems. Eat foods like grapefruit that takes time to consume and make you feel as if you have eaten.

When you eat has a great deal to do with how your body uses food. You should never eat *just anything*. Also, do not eat unless you are going to be physically active within two hours after eating.

TABLE 9.8 A 1,200-CALORIE-PER-DAY DIET

Breakfast
Fruit: 1 medium serving, fresh, frozen, or canned
Egg: 1, poached or boiled
Toast: 1 slice with 1 teaspoon butter or margarine

or

Cereal: ½ cup with ¼ cup milk, no sugar
Coffee or tea: no cream or sugar

Midmorning snack
Nonfat milk or buttermilk: 1 glass

Luncheon
Meat or cheese: 1 3–oz portion
Vegetable: 1 medium serving; may be raw, such as a lettuce and tomato salad, or cooked; use lemon or vinegar for seasoning instead of butter or salad dressings
Fruit: 1 medium serving, fresh or unsweetened canned
Bread: 1 slice
Butter or margarine: 1 teaspoon or 1 pat
Tea or coffee: no cream or sugar

Midafternoon
Iced tea, lemonade, or a soft drink

Dinner
Bouillon or consomme or vegetable juice cocktail: 1 serving
Meat: 1 3–oz portion
Potato or potato substitute: 1 small serving of mashed or baked potato, steamed rice, corn, lima beans, or macaroni; or 1 slice bread
Vegetable: 1 serving, raw, as a salad, or cooked; one vegetable a day should be green and leafy
Butter or margarine: 1 teaspoon, for potato
Fruit: 1 medium serving, fresh or unsweetened canned
Tea or coffee: no cream or sugar

Evening or bedtime
Nonfat milk, buttermilk, soft drink, or glass of beer
Crackers or pretzels: 2

Some weight-reduction experts suggest that you can lose weight by eating only two meals a day: breakfast and a dinner at no later than 2 P.M. When you achieve the weight you would like to maintain,

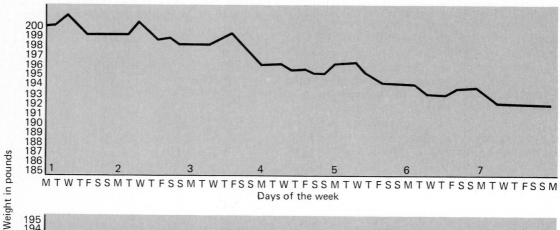

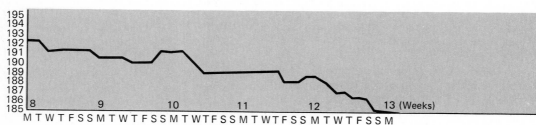

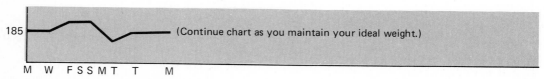

Weight reduction and maintenance graph.

● ASK YOURSELF

1. What is the first thing overweight or obese persons must do to start losing weight?
2. What must all weight-reducing diets contain to provide adequate nutrition?
3. What must a person do to keep weight off for long periods of time?
4. How can meals be spaced to help weight loss?

go back to three meals. But, always keep the third meal early in the evening—say, between 5 and 6 P.M. If you do not eat by 6 P.M., DO NOT EAT THAT EVENING. This will ensure your eating a good breakfast in the morning. Also, forget eating traditional foods at specific meals. Combine some fat with protein so that the meal will stay with you. Pizza for breakfast will last throughout the day because of its fat and protein content and contains enough calories for a day's physical activity. With a little imagination you can begin to eat some very interesting breakfasts.

Here is a little saying to remember:

Eat like a king at breakfast.
Eat like a prince at lunch.
Eat like a pauper at dinner.

Anorexia Nervosa

Like few other diseases, *anorexia nervosa* affects an individual's psychological well-being, family, and therapist. It can produce severe physical problems including death. The medical literature concerning anorexia nervosa is noticeably deficient prior to the 1960s. From 1954 to 1964, twelve cases were treated in the teaching hospitals of Washington University School of Medicine, and fifteen cases from 1964 to 1974; but, in the five years from 1974 to 1979, thirty-five cases of anorexia nervosa were treated.

What is this illness that appears to be increasing at an alarming rate? The following criteria are used in diagnosing the disease.

1. Usually individuals are under 25 years of age.
2. They have lost 25 percent or more of their total body weight.
3. Their attitudes toward food and body weight are distorted. Their refusal to eat food overrides their hunger and reasoning ability. They receive no pleasure from food. They take pleasure in losing weight, desire a thin body image, and hoard or handle food in unusual ways.
4. No physical condition or medical illness can account for the anorexia or the weight loss.
5. Individuals show at least two of the following signs:
   a. Bulimia (increased sensations of hunger)
   b. Lanugo (a growth of very fine body hair similar to the hair that covers newborn infants)
   c. Amenorrhea (stoppage of the menstrual flow). This condition is also common in female athletes and during normal weight loss until the body adjusts to the change in eating habits.
   d. Hyperactivity (periods of extreme overactivity)
   e. Bradycardia (slow heart rate—ofren less than 60 beats per minute)
   f. Vomiting after eating food

The anoretic is usually a teen-age girl who is a good student, is obviously intelligent, and shows no other signs of emotional disorder. Although almost a skeleton, she often participates in competitive sports. During treatment she still wants to run up the steps in the hospital and jog in the shower. She demonstrates a stubborn feeling that nobody can get her to eat "normally" or attain what most people would consider a normal weight. While anoretics persistently believe that they look fat, most of them seem to see others in proper perspective.

Reducing Don'ts

In addition to the specific reducing dangers already pointed out, several general points should be made:

1. *Don't blame obesity problems on heredity.* Although some people do inherit a predisposition toward becoming fat, this in no way implies that they must become fat. Persons with such a tendency simply need to watch their food intake more closely than others. Some who have no such natural tendency allow personal problems or family and cultural eating patterns to dictate against their better judgment. Be careful not to misconstrue faulty home food practices as hereditary problems. *Learn to take full responsibility for what you become.*

● **ASK YOURSELF**

1. When is underweight a problem??
2. Look up information on the disease anorexia nervosa. If possible, interview a doctor who treats this disease. How common is it in your neighborhood, town, or city?

2. *Don't be tempted by "crash diets."* Any diet plan that claims to take off any number of pounds in a short period of time should be carefully scrutinized. It may be a case of misleading advertising, or it may require taking drugs that cause organic damage to the body. Seeing a physician before going on any reducing diet should give one ample protection against attempting to correct overnight a weight condition created over many years.

3. *Don't follow fad diets.* Some fad diets are not palatable, others are monotonous, and many of them turn out to be expensive. Balanced nutrition calls for a variety of foods. A diet purporting to succeed with a single food could well result in malnutrition. Good nutrition is not a luxury but a necessity. Don't substitute a so-called perfect food for the wide variety of appealing and nutritious foods available at reasonable cost in our grocery stores today.

4. *Don't rely on mechanical devices.* In their desperation or gullibility, people are willing to invest in any device recommended for weight loss. Some of the more common devices are:

    Vibrators. They may be stimulating, but it's impossible to vibrate fat away.

    Massaging devices. They may be relaxing, but fat cannot be massaged off.

    Spot reducers. Fat deposits are genetically determined; it is absolutely impossible to mechanically control fat loss from any specific part of the body.

    Reducing creams. Although they are soothing, they are worthless.

The Problem of Underweight

The condition of underweight does not depend on weight alone, but on the presence or absence of

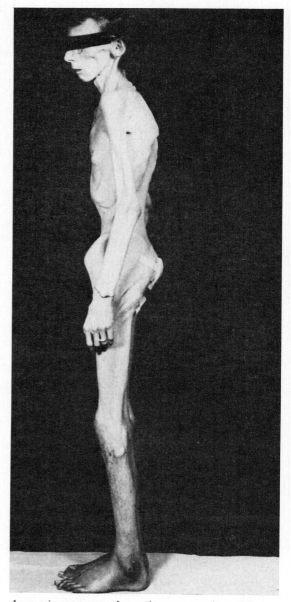

Anorexia nervosa. Anoretic persons lose 25 percent or more of their total body weight.

symptoms of malnutrition. A person can eat a balanced diet, in moderation, and maintain excellent health even though he or she remains slender. But symptoms such as lack of endurance, easy fatigue,

Recommendations for Gaining Weight

Here are some recommendations for people who need to *gain* weight:

1. *Don't skip any meals.* In fact you may want to eat five or six small meals instead of three.
2. *Drink a glass of orange juice before breakfast and treat yourself to a nightcap before bedtime* (malt drink, eggnog, cookies, ice cream).
3. *Add more rest or sleep to your daily routine.* Sleep an extra hour or take a daytime nap to conserve energy and calories.
4. *Eat on schedule every day, and be consistent.*
5. *Eat out if you can afford it.* The tendency is to eat more, and restaurant food is usually more fattening.
6. *Curb your smoking if it interferes with your food intake.*

frequent infections, intermittent diarrhea, or sores on the skin or mucuous membranes may indicate undernutrition — that is, insufficient food intake for normal body function. Severe cases can include deficiency diseases, injury to vital organs (especially heart and kidneys), and even death. Underweight can be caused by emotional state (nervousness, worry, anxiety), diseases, malnutrition (poor selection of foods), hormonal disorders, or unrealistic dieting.

The treatment of underweight is not always just a simple matter of starting to eat more. Underweight persons should consult a physician in case an infectious or glandular disease or other contributing condition may be involved. If the problem is an emotional one, as is often the case, individuals should try to correct the underlying condition, or seek therapeutic intervention. Increased exercise may help relieve nervous tension, leading to improved diet.

If the problem appears to be purely dietary, a good method for gaining weight is to increase the intake of food from the basic food groups; the individual can take either larger helpings or more helpings in a balanced food plan. There is a danger, however, in greatly increasing the intake of high-calorie foods such as butter, eggs, and cream or cream products (malted milks, milk shakes, ice-cream). These may increase the risk of premature heart disease.

Information on Food and Nutrition

Where can you obtain reliable information on foods and nutrition? You may check with your personal physician or nurse or write the Council on Drugs and Nutrition of the American Medical Association. Occasionally there are articles dealing with nutrition in the *American Journal of Nursing, the American Journal of Public Health, Science News Letter* and *Medical World News: Nutrition Reviews* is a professional nutritional journal containing technical reports of research being conducted in nutrition. Publications such as *Food for Us All: The Yearbook of Agriculture* (1969) or *Nutritive Value of Foods* (Home and Garden Bulletin No. 72) are available from the U.S. Department of Agriculture. *Recommended Dietary Allowances* (latest edition) is available from the National Research Council. Material dealing with proper nutrition is also generally available from state departments of agriculture or state universities. Many libraries have books that contain good information on the subject.

IN REVIEW

1. Many Americans are seriously overweight, and many fail to maintain a proper weight.
   a. Appetite and hunger are controlled by the appestat, which in turn is controlled by emotions, body chemistry, and inheritance.
   b. There are a number of reasons for maintaining a desirable weight: look attractive, live longer, fewer diseases, fewer painful conditions.
2. A healthy diet must provide sufficient amounts of all nutrients. It should also provide a variety of textures and sufficient water.
   a. Recommended Daily Dietary Allowances (RDAs) are nutritional recommendations based on the needs of normal, healthy individuals.
   b. Tables of dietary needs show requirements for all nutrients. Pregnant and breastfeeding women have specific needs. Individuals need fewer nutrients as they grow older.
3. Desirable weight can be determined by a number of processes: averages for large numbers of people, individual calculations, or body measurements.
   a. Childhood and teen-age growth rates have been charted by the University of Iowa.
   b. Adult height-weight relationships are listed on tables or calculated by a series of body measurements.
4. A distinction should be made between obesity and overweight. Obesity involves excessive deposits of fat, while overweight is simply overheaviness.
   a. Obesity is due to an accumulation of fat caused by overconsumption of calories, and a deficiency of caloric expenditure because of lack of exercise. Overeating is complicated by a number of factors.
5. Weight is lost by reducing food intake and increasing physical activity levels.
   a. Each pound of human fat is equal to about 3,500 calories. A diet program must reduce food intake by 3,500 calories for each pound to be lost.
   b. Having reached an ideal weight, an individual needs a plan to maintain it throughout life. Weight monitoring plans include keeping steady track of weight on a graph, exercising regularly, and acquiring responsible eating habits.
   c. Reducing involves a number of specific "don'ts."
   d. Underweight can also be a problem for some individuals. There are certain things they can do to increase their weight.

CHAPTER 10

TOTAL FITNESS

According to Dr. Roger Bannister (the first person to run a mile in four minutes), *fitness* is one of the most misused words in the English language. It can mean everything from a general, overall joy in living to a specific suitability for particular kinds of mental and physical tasks.

THE MEASURE OF TOTAL FITNESS

Total fitness is the ability to function at an optimum level of efficiency in all aspects of daily living. A totally fit individual has the strength, speed, agility, endurance, and social and emotional adjustments appropriate to his or her age. Each of these characteristics involves the expenditure of energy.

All body activities require energy. Very simply, energy is produced by breaking down foods (carbohydrates, fats, and proteins) in the presence of oxygen. The body can store food, but it cannot store oxygen. If the body takes in more food than is needed, it uses what it needs and stores the rest for later. Not so with oxygen. We cannot store oxygen, so we breathe in and out every moment of our lives to keep the supply coming in. If the oxygen supply were suddenly cut off, the oxygen stored in the body would not last more than a few minutes. The brain, the heart, and all body tissues would cease to function, and we would die. The oxygen in the air is readily available; as we need it, we breathe it in. The problem is getting enough oxygen to all parts of the body where food is burned.

Most of us produce enough energy to perform ordinarily daily activities — that is, to walk, talk, think, or study. However, as activities become more vigorous, we sooner or later reach our maximum performance, or *maximum oxygen consumption*. A person's maximum oxygen consumption is also known as *aerobic capacity*. The range between our minimum oxygen requirements (amount of oxygen used at rest) and our maximum oxygen consumption is the major factor in the physiological measure of our total fitness. The most totally fit persons have the greatest range of aerobic capacity; the least fit, the narrowest range. In some persons, minimum energy requirements and maximum oxygen uptake are almost identical. A totally fit individual should have, among other qualities, adequate aerobic capacity and physical strength to engage in physical activities, such as tennis, swimming, bicycling, or handball, without producing undue fatigue.

Fitness Development

As early as the preschool years, the growth and maturation of the neuromuscular system and the establishment of locomotor movement patterns lay the foundations for the individual's future skill learning and development and attitudes toward physical fitness. Strong physical play forms the basis for a person's strength, agility, and coordination. To keep these skills, an individual must maintain a routine activity program throughout life. In school, activities for children can be planned with specific objectives to develop muscular strength, endurance, aerobic capacity, and skill. Actively

Physical play prepares children for physical fitness programs later in life.

maintaining this development enables a person to perform in sports and physical activities at a *social level* throughout life.

The development to the social level of physical fitness or the redevelopment of fitness (for those who have been inactive for a number of years) require regular periods of physical activity. Individuals must start with low-energy-use programs (such as walking). Then they should gradually increase the stress in terms of speed, workload, and duration of activity (termed *overload principle*) until the level of performance they are striving for is

reached. High-energy-use programs such as tennis, running, swimming, bicycling, handball, and rope skipping are reached through hard work and time

SLEEP

Sleep is crucial to good physical fitness. Lack of sleep (sleep deprivation) can make you groggy and irritable. Over longer periods of time, sleep deprivation can produce hallucinations and other psychotic symptoms and, ultimately, competely emotional collapse.

The Stages of Sleep

Most people think of sleep as just the opposite of being awake. Actually, there are different stages of sleep. As shown in the following figure, the stages of sleep are distinguishable by progressively slower brain-wave (EEG) patterns. As we pass from stage to stage, heart rate, breathing rate, body temperature, and muscle tone all decrease. We sleep in cycles, with deeper stages of sleep alternating with periods of light sleep and dreaming. Typically, adults progress through four to six cycles per night, each cycle averaging about 90 minutes. The first dreams usually occur about 90 minutes after falling asleep. Dreaming is marked by rapid electrical activity in the cerebral cortex, the part of the brain where consciousness lies. The eyeballs move dur-

Sports can help children organize their time and make physical fitness part of their lives.

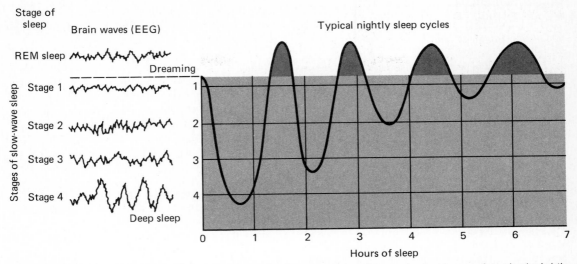

Sleep occurs in four stages, during which brain waves become increasingly slower. A typical night's sleep includes cycles of light sleep, deep sleep, and REM sleep, when dreams occur. A person reaches deep sleep soon after retiring and spends a greater amount of time dreaming as morning approaches.

ing dreams; thus, this phase of sleep is termed *rapid eye movement* (REM) sleep. During REM sleep, the skeletal muscles have almost no tone; and spinal reflexes, such as the knee jerk, are lost. REM sleep is also called *paradoxical sleep* because, even though the brain waves indicate light sleep, relaxation is complete and the individual is very difficult to awaken. This indicates that REM sleep is actually very deep. Males frequently experience penile erections during REM sleep, while females experience corresponding vaginal lubrication. This occurs even during nonsexual dreams (Schneider & Tarshis, 1980).

Why We Dream

Dreaming is apparently important to our emotional well-being. In dream-deprivation experiments, human subjects were awakened as soon as their brain-wave readings showed the beginning of REM sleep.* During the days after such dream-deprivation, the subjects became irritable and confused. They had difficulty concentrating, experienced memory lapses, and became extremely hungry.

*Dement, W., "The Effect of Dream Deprivation," *Science, 131* (1960), 1705–1707.

When they were then allowed to sleep without interruption, the subjects showed above-normal periods of dreaming for several nights, as if they were "catching up" on their dreams.

Dreams apparently help us integrate the events of our daily lives with the massive amount of information stored away in our unconscious minds. During our waking hours, it seems, our brains are fully occupied with processing the constant barrage of sensory input. Sometimes our dreams are directly related to recent experiences. Sometimes they seem to have little relationship to our "real" lives. But psychologists believe that all of our dreams are important and, with proper interpretation, may reveal some of our buried feelings.

● **ASK YOURSELF**

1. What is the major physiological measure of total fitness? How do you measure up to this criterion of total fitness?

2. What qualities does a totally fit individual have? How many of them do you have?

3. What is meant by "social fitness"? What type of fitness do you possess?

■ ASK YOURSELF

1. Outline the stages of sleep, and the time intervals of each stage.
2. Tape-record or write down some of your dreams immediately upon waking. How do you think they are related to your daily routine?
3. How much sleep do you need?
4. Explain what "insomnia" is. If you were to have insomnia, what actions could you take to resist it and obtain a good night's sleep. Outline these actions.

How Much Sleep?

Sleep requirements vary from person to person. The majority of people "need" about seven to eight hours of sleep per day for optimum physical and mental functioning. However, some people do quite well with less sleep while others require more.

Inherent physiological differences seem to account for some of the variation in sleep requirements. Another factor is a person's emotional state. Some people respond to emotional conflict by sleeping more, others by sleeping less. Finally, habit is a big factor in how much sleep we feel we need. Our "biological clocks" program us to fall asleep and wake up at certain times.

The "quality" of sleep also determines how much sleep we need. Sleep is apparently most effective when it alternates between deep sleep and REM sleep about every hour or so. Anything that interferes with these cycles, such as the neighbor's barking dog, reduces the effectiveness of our sleep. Alcohol and other drugs, including sleeping pills, may diminish the effectiveness of sleep. While drugs may cause one to fall asleep rapidly, they inhibit normal periodic REM sleep. In addition to the psychological benefits of dreaming, the body muscles relax most completely during REM sleep (Schneider & Tarshis, 1980). Thus, if a night's sleep does not include enough dream cycles, you wake up still feeling tired.

Sleep Disorders

Just about everybody has trouble sleeping on occasion. Worry, excitement, overeating, noise, pain, and unfamiliar surroundings are just a few common interferences. But at some time in life, about one out of every five to ten people suffers from severe

insomnia, the inability to sleep—not for just a night or two, but on most nights. Physical and emotional health suffer.

Insomnia has many different causes. Sometimes no cause is apparent. Some insomniacs have obvious underlying emotional disorders, but many do not. Some people merely overestimate the amount of time they spend lying awake. Some people mistake their periods of dreaming for being awake. They are so worried about not sleeping that they dream that they are awake. Some people simply go to bed too early for their own sleep needs. Some suffer from a condition called *sleep apnea* in which breathing stops when they fall asleep. They must wake up frequently (sometimes every few minutes) to breathe.

Insomniacs should avoid taking any drugs, if possible. Since most sleeping medications block REM sleep, even large doses do not produce normal, restful sleep. Further, regular drug use will establish a dependency cycle. If sleeping medications are suddenly discontinued, an extreme rebound of REM sleep occurs, including vivid nightmares. Withdrawal from sleeping pills must be very gradual, in some cases as long as 18 months (Adams, 1981).

Several more desirable alternatives to sleeping pills exist. Increased exercise is one. Most people sleep better when their day has included adequate physical activity. Relaxation training is another alternative. In addition to progressive muscular relaxation (*see* Chapter 1), an individual is taught to focus attention on pleasant, monotonous internal sensations. This automatically blocks stressful sleep-preventing thoughts. Sleep arrives quickly and its depth is enhanced.

Behavior therapists recommend that sufferers from insomnia reserve their sleeping quarters strictly for sleep. Studying, working, eating, watching television, and so forth should take place elsewhere. The insomniac should retire to the bedroom only when drowsy and leave the bedroom if unable to fall asleep within ten minutes after retiring. This program conditions one to associate the bedroom with sleep and usually assures a rapid onset of sleep. Of course, if insomnia is secondary to a more basic problem, such as severe depression, it is necessary to treat the basic problem.

Less common than insomnia, but more dangerous, is *narcolepsy*. This is the tendency to suddenly

The nonexerciser. The average American is a spectator and does not participate in a physical fitness program.

fall asleep at inappropriate times such as when driving or operating machinery. These "attacks" of sleep last up to 15 minutes. They occur abruptly, without any preliminary feeling of drowsiness. Individuals with narcolepsy enter immediately into REM sleep, while other people require about 90 minutes to do so (Adams, 1981). Thus, narcolepsy is viewed as a disorder of the REM sleep system. Some cases may be due to physical injury. There is currently no real cure, though some individuals are able to control the attacks through medications.

THE MEANING OF TOTAL FITNESS

Total fitness implies that a person is "in condition." Such a level of fitness is also called *overall fitness, endurance fitness,* or *working capacity* (the ability to do prolonged work without undue fatigue). If an individual is working toward a level of fitness needed for competitive sports, he or she should do so only under the direction of a professional physical educator. In this section we are restricting our definition of "in condition" to that of the American Medical Association's Committee on Exercise and Physical Fitness. It stated in 1967 that fitness is "the general capacity to adapt and respond favorably to physical effort. The degree of physical fitness depends upon the individual's state of health, constitution, and present and previous activity." This is also our definition of a person's *total fitness.*

Nonexercisers

Many individuals are *nonexercisers.* They possess only passive fitness and make no effort to keep their bodies fit. They do only what is necessary during their daily routines. There is nothing physically wrong with this kind of person—not yet—nor is there anything really right with them. If lucky, they may remain like this for years. But their bodies begin to deteriorate and will continue to do so unless they increase their physical activity.

Many nonexercisers are ill, but do not know it. It takes years of inactivity before recognizable diseases such as heart attacks or strokes appear. A physician, with the aid of tests, may be able to detect biological changes in time to prevent these extreme illnesses.

Biological Aging

The vitality and health of an individual is determined by many factors, including optimum diet and total fitness. An estimation of normal aging becomes an estimate of the decline of one's vitality. Under optimum conditions of energy supply, oxygen supply, waste removal, and periods of functioning and rest, the cells and the body tend to live longer. Body-age estimations based on such factors

The Meaning of Total Fitness

Nonexerciser   Balanced Exercise Activities
| |
Premature Biological Aging   Totally Fit Individual

1. Joint & muscle problems	Has adequate aerobic capacity
2. Decrease in muscle mass	& physical strength to engage in
3. Diminished oxygen intake	strenuous physical activities
4. Reduced heat production	without undue fatigue
5. Fall in total blood quantity	

1. Excess weight
2. Increased risk of heart attack in 30s.
3. Loss of body control
4. Inability to adjust to stress
5. Disease & premature death

determine *biological age*. This bears no relationship to one's *chronological age* (age in years).

The body shows biological age by changes in structure and functioning of the major body systems. Various physiological functions change within these systems at different rates. Some changes take place rapidly. Others appear to be age resistant. The biological aging process can be speeded up by a deficient food supply, inactivity, infection, traumatic injury, and physical irritation. Rapid biological aging leads to deterioration of the body, disease, and *premature death* (death prior to the estimations for the general population).

Premature death The struggle for a better life has brought about impressive improvements in our living conditions. The infant mortality rate has been lowered, and over the last century, the average life span in the United States has increased. This has been due to the control of *communicable diseases,* such as pneumonia and influenza, which caused the majority of deaths in the past. With control of the communicable diseases over the past 100 years, life expectancy in the United States has been extended from 40 to over 69 years for males and to nearly 77 for females. More people should be living into their 80s.

However, these longer-living individuals often are unable to function adequately in the later years

and die of diseases that formerly were uncommon. Today's deaths are more commonly caused by various *degenerative conditions* (the result of biological aging). These diseases cause (or are the result of) biological changes in the structure and functioning of the body: atherosclerosis, coronary thrombosis (cardiac infarction), cerebral thrombosis (stroke), as well as cancer of various organs.

The Control of Aging

Exercise physiologists show that physical exercise, if sufficiently intensive and regular, can override the various phenomenona of aging. Signs of aging are: decrease in muscle mass, diminished oxygen intake, reduced heat production, and fall in total blood quantity. Physical exercise also stimulates intense cerebral activity. It stimulates the neural controls of metabolism, respiration, blood circulation, digestion, and the activities of the glands of internal secretion. A correct combination of alert mental activity and physical exercise is at present the best method of preserving, for as long as possible, at a high level, the activity of the brain cells.

This control of the phenomenon of aging should begin as early as possible, before the completion of physical development (between 14 and 22 years of age). It is considerably more difficult to control premature biological aging when it has already set in. A regular exercise program contributes to vitality and healthful good looks through the middle years. A person who has maintained a successful fitness program can enjoy middle and later years to the fullest. This is why the habits of physical exercise and total fitness should be formed during earliest childhood.

Total fitness *Total fitness* is produced when someone engages in balanced activities that strengthen all body systems, particularly the cardiovascular, respiratory, nervous, and muscular. Total fitness, furthermore, is produced by *optimum* intensity and duration of physical activity. The amount and duration of physical activity required differs from person to person, female and male. The value of total fitness is the same for both.

Contrary to some common belief, strenuous physical exercise programs and sports do not develop bulky muscles. Programs that *overdevelop* specific muscles are unhealthy for everyone. Balanced exercise improves the figure by "normalizing" it and causing it to become better proportioned. If the arms or legs are too heavy, exercise works toward slimming; if too thin, exercise develops them.

ACTIVITY PROGRAMS

In order for people to maintain a physical fitness program over the years, it must maintain their interest. Most of the well-publicized physical fitness programs that stress exercises, machines, or rely on an athletic club membership require great motivation to be maintained throughout life. The best exercise programs are individual activities and dual sports. But you will not enjoy a sport that you are not in condition to perform. Consequently, your general conditioning program must be routine (usually three days a week) and must be easily maintained.

While in school or college, you should develop skills in several different activities to participate in throughout life. You should consider five major factors as you plan your physical fitness activities: You should develop (1) muscular strength, (2) muscular endurance, (3) circulatory endurance, (4) flexibility, and (5) skill (coordination). Physical education courses should inform you about activities and sports that develop and maintain these factors.

The concepts of exercise, but not the principles,

Exercise "normalizes" the figure and improves its proportions.

● ASK YOURSELF

1. How would you describe someone who is a nonexerciser?
2. Explain the difference between *biological age* and *chronological age.*
3. What is the ultimate outcome of the premature aging of a nonexerciser?

How Fit Are You?

Finding out how *unfit* you are is the first step in becoming fit. If you are between 15 and 60 years of age, this simple test will help determine your fitness level:

1. Stand nude in front of a mirror. Look for areas of loose or flabby skin. Are you completely satisfied with what you see?
2. Pinch your waist or the back of your upper arm. If this fold of skin is more than an inch thick, you are *obese* and need an exercise program.
3. Can you hold your breath for more than 45 seconds? This is a good test of your lung condition.
4. Stand up straight, with your eyes closed and your arms at your sides. Raise one knee while standing on one leg. Can you stand like this for 15 seconds without losing your balance?
5. Find your pulse rate at rest (*see* p. 213). Then run in place for three minutes. Your pulse rate should be under 120 beats per minute. It should also take less than one minute to return to your resting pulse rate.

If you fail this test and are under 30, begin a regular fitness program. If you are over 30, consult your physician before beginning an exercise program. After six months of regular exercise, you should pass this test with flying colors.

have changed drastically in recent years. Modern total fitness programs are the result of laboratory studies that have added greatly to our knowledge. Exercise is accepted today as an essential counterbalance to our overly sedentary life style. The question then becomes, "How much and what kind of exercise?" Dr. Kenneth H. Cooper (1970) points out that rhythmic physical activity supplies the body with needed oxygen. During exercise, the blood becomes richer in oxygen and nutrients and more effectively eliminates wastes from the muscles and other organs. Activities that promote such efficient body functioning include walking, running, swimming, cycling, dancing, and tennis. For exercise to be effective, it must also be routine and done at your capacity. But what is your capacity?

Before Starting an Activity Program

If your current total fitness status is adequate and you are satisfied with your ability to function, then simply maintain this level. But if you are dissatisfied with your physical condition and capacity and want to improve it, you should do several things before starting a total fitness program:

1. Have a complete medical examination, especially if you are over 30 years of age. (No one

over 30 should start on a total fitness program without first consulting a physician.)

2. A major part of your medical examination should be an *exercise capacity test* (bicycle ergometer or treadmill test). It may be necessary to ask your physician for this test.
3. If you have not exercised regularly for a number of years, you must "recondition" yourself. To prevent sore and injured muscles, you should participate in a warm-up exercise program for a minimum of six weeks for each year you have been out of condition.
4. Have an individual program worked out for you by a competent exercise physiologist or physical educator. It is important that you have a *balanced* activity program that improves the condition of your lungs, cardiovascular system, muscular system, and total fitness of your body.

Too often sedentary persons start an exercise program without considering these four points. Many individuals who die while jogging, running, or swimming have rushed into a daily exercise program without consulting a physician and a physical educator. Others become sore or injure themselves because their muscles and joints are not ready for strenuous activity or for carrying their body weight while exercising. A proper warm-up program helps to stretch and loosen ligaments, increase timing, improve muscle strength, and increase the cardiovascular activity in preparation for a more vigorous exercise program.

Activity Program Principles

Exercise tolerance *Exercise tolerance* is the level at which the body responds favorably to exercise. An individual's exercise tolerance is his or her ability to perform a series of exercises, participate in a sport, or enjoy a walk without undue fatigue. All exercises should be adapted to an individual's tolerance level. Activities that are too easy or are impossible should not be attempted.

Overloading The body has great ability to adapt to stress and increase its exercise tolerance. Therefore, if you wish to improve your performance and physical condition, you should continually increase the duration and intensity of the exercises. Extending yourself beyond your usual physical effort is called *overloading*. This involves increasing physical stress by:

1. Gradually and progressively increasing the *speed* of performing an exercise, or
2. Gradually increasing the total *resistance* (amount of weight to be lifted), or
3. Increasing the total *distance* being run or *time* spent exercising

Fatigue may be delayed by reducing the work load (resistance), by slowing the rhythm, and by breathing regularly and deeply. In using overloading, you may alternately run and walk to give yourself periods to recover from fatigue. The principles of overloading should help increase efficiency and performance. As you master an exercise program, progress to more strenuous exercises. All fitness programs should provide for progression.

Pulse Rates

The intensity of an exercise program should be as great as possible, based upon your current level of fitness and tolerance. The intensity of exercise should increase as the level of tolerance and fitness

It is easy to take your pulse from your carotid artery while you exercise. Place your fingers (not your thumb—it contains its own pulse) on one side of your Adam's apple, and move them around until you feel a pulse beat. Now, using the second hand of your wristwatch, count the number of beats in 10 seconds and multiply this number by 6 for the number of beats in 1 minute. Practice a few times while standing still before trying it while exercising.

rises. A good indication of your current level of total fitness is your *pulse rate* (a regular throbbing in the arteries caused by the contractions of the heart).

A good place for taking the pulse, especially while exercising, is at the carotid arteries on either side of the neck.

Resting pulse rate Your heart rate at rest is an indication of out basal level of total fitness. As your fitness level increases, you strengthen your heart and your resting pulse rate *decreases*.

A nonexerciser's heart works harder, faster, and less efficiently than a fit individual's. Someone in condition and regularly active may have a resting

TABLE 10.1 TARGET PULSE RATES (BASED ON PERCENTAGES OF PREDICTED MAXIMUM HEART RATES)

Age	Predicted maximal heart rate	85% of maximal heart rate	60% of maximal heart rate
15	212	180	127
25	200	170	120
30	194	165	116
35	188	160	113
40	182	155	109
45	176	150	106
50	171	145	103
55	165	140	99
60	159	135	95
65+	153	130	92

TABLE 10.2 DR. KENNETH COOPER'S TARGET PULSE RATES

Age ranges	Target pulse rates
15–30	180
31–40	170
41–50	160
51–60	150
61–70	140
71–over	130

Source: Dr. Kenneth Cooper, The New Aerobics (New York: Bantam Books, 1970), p. 150.

heart rate of between 55 and 60 beats per minute or lower. A nonexerciser may have a resting pulse rate of 70 or more.

Recovery pulse rate The *recovery pulse rate* is also important, because it indicates when an activity is too strenuous for you. After exercising, if your pulse rate is 5 beats per minute or more above your resting rate after 30 minutes of rest, the physical activity has been too strenuous. You should reduce the intensity to an acceptable level at the next session.

Target pulse rate For exercise to increase your fitness to an acceptable level, your pulse rate should reach a predictable *target pulse rate* and be maintained during the exercise session.

There are many methods of obtaining your target pulse rate. The following four are commonly used for pulse rates ranging from low to high:

1. Conservative exercise physiologists feel that a pulse rate of 151 beats per minute for a person under 30 years of age and 131 beats per minute for a person over 30 shows that the heart is working at 60 percent of capacity. After 30 years of age, a person should not let his or her pulse rate go too much over 131 beats per minute, because of the greater chance of heart attacks or strokes with increasing age.

2. To *maintain* your present level of fitness, raise your pulse rate to 60 percent of maximum. To *increase* it, raise your pulse rate to 85 percent of maximum (*see* Table 10.1).

3. Many exercise physiologists use a rule-of-thumb method for obtaining maximum pulse rate. They never let the pulse rate increase above one's age (after 30 years of age) subtracted from 200. For example, 200 minus 50 equals 150 beats per minute is the target pulse rate for a 50-year-old.

4. Dr. Kenneth Cooper uses the rates in Table 10.2 to determine what an individual's maximum pulse rate should be during an activity.

The following graph compares the four different ways of obtaining a target pulse rate. Remember, you should not try to reach a target pulse rate until you have gone through a warm-up period.

Kinds of Exercise

Before beginning any total fitness program, you must possess the muscular strength to support your body weight easily. You must also have the cardiac endurance needed to complete the activity. There are two types of muscular activities: *static exercises* and *dynamic exercises*. Static exercises are produced by *isometric* muscular contraction. Dynamic exercises are produced by *isotonic* muscular contractions. Most physical activities are a mixture of both static and dynamic exercises. Activities that use dynamic exercise maintain or increase cardiovascular fitness. Static exercises are actually dangerous to someone who may, unknowingly,

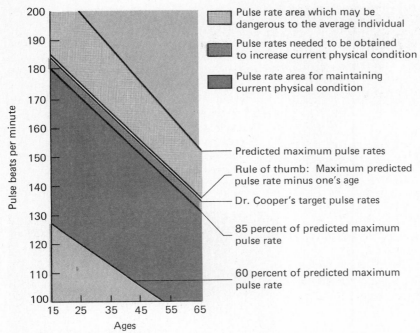

The four target pulse rates.

have a heart problem. Therefore, it is crucial to engage in the right kind of activities.

Static exercises Isometric muscle contractions increase tension within the muscles. Such tension is produced by pushing or pulling against an immovable object, such as a heavy desk or the steering wheel of a car. Static exercises limit the range of muscular motion; examples include weight lifting or skiing (both snow and water). These exercises may damage the heart because they markedly increase blood pressure but not the amount of blood pumped by the heart; such exercises actually decrease the amount of oxygen available in the body.

Static exercises tense muscles and squeeze blood vessels so that less blood can pass through. This produces an abrupt rise in blood pressure, which causes the heart to work much harder to overcome the resistance in the blood vessels. For a person who has no underlying heart condition, this extra demand does not present any hazard; for someone with a coronary condition, it may be deadly. Someone with a known heart disease, or

● **ASK YOURSELF**

1. What is overloading, and how do you accomplish it?
2. What produces your pulse?
3. What is the resting pulse rate?
4. What is the target pulse rate? How is it calculated?
5. What is the recovery pulse rate?

anyone who might have one should not participate in static exercise activities without a physical examination and a consultation with a physician.

While weight lifting and other forms of static exercise increase the strength, size, and tone of the muscles being used, little if any cardiovascular fitness results from such isometric activities.

Dynamic exercises Isotonic contractions are produced by raising, lowering, or moving the body or its parts. Dynamic exercises, such as walking, running, swimming, rowing, and bicycling, involve

Skiing is a static exercise. While skiing produces great tension in the leg muscles, it does not allow them a full range of movement. It is therefore not a good cardiovascular exercise.

muscles is much higher in an endurance-trained athlete.

3. Working muscles extract more oxygen from the blood. There are about 20 ml of oxygen in each 100 ml of blood. At rest the body uses only about 4.5 ml of oxygen per 100 ml of blood. During strenuous physical activity, the tissues of a fit person can consume as much as 15 ml, a 233 percent increase in oxygen consumption. This increase produces a tremendous "oxygen transport reserve," which saves the heart considerable work, even during mild physical activity. The better

large numbers of muscles. They generate a great demand for oxygen in the muscles needed for the physical activity. The oxygen is supplied by the heart and the circulatory system (cardiovascular system). Basically, three physiological changes take place during dynamic exercises.

1. The heart pumps more blood per minute. At rest, the heart pumps about 5 to 6 quarts (or liters) of blood each minute. This amount can increase over twenty times during strenuous exercise.

2. A greater portion of the blood is directed to the working muscles. With training the body redistributes the blood from the visceral organs and unused muscles to the muscles being used. The blood supply to the working

Running is an excellent dynamic exercise. The world-class runner is perhaps the ultimate dynamic exerciser.

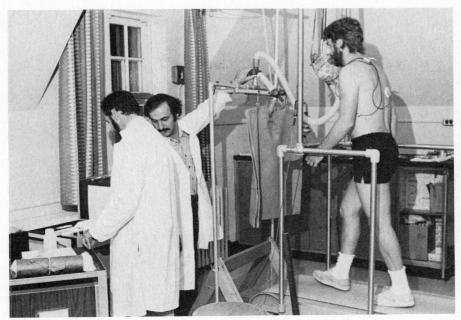

The motor-driven treadmill measures a person's maximum oxygen uptake.

trained an individual, the higher the oxygen transport reserve.

The maximum amount of oxygen that the blood can transport from the lungs to the muscles and other organs of the body is called the *maximal oxygen uptake,* a standard measure of cardiovascular fitness. This sets the upper limit of endurance. Maximal oxygen uptake is also known as *conditioning effect, training effect,* or *aerobic capacity.* A decrease in maximal oxygen uptake occurs in sedentary and passively fit individuals. It also occurs as a part of the biological aging process.

Maximal oxygen uptake is measured by performing work on a motor-driven treadmill or a stationary bicycle (bicycle ergometer). During these tests the heartbeat, electrocardiogram (ECG), blood pressure, and sometimes the amount of oxygen being consumed are measured.

Conditioning effects of dynamic exercise An increase in maximal oxygen uptake increases stamina and diminishes fatigue during prolonged work or exercise at any level. The heart muscle cells, like those of any muscle that is exercised, become stronger with conditioning. The conditioned heart is able to pump more blood with each beat, requiring fewer beats to deliver the same amount of blood and oxygen as before conditioning.

Many highly conditioned athletes have a resting pulse rate of approximately 45 beats a minute, compared to the average of 72 to 80 beats a minute for the passively fit individual. The heart of such an athlete may pump twice the volume of blood with each beat as that of an average person. If conditioning lowers the heart rate only 10 beats a minute, the heart will beat 14,000 fewer times a day or *5 million fewer beats a year.*

● **ASK YOURSELF**

1. What are isometric exercises? What are isotonic exercises?
2. What are the differences between static and dynamic exercises?
3. List some static exercise activities. How can these be incorporated into an exercise program?
4. List some dynamic exercise activities. How should these be used in an exercise program?

Even pregnant women can exercise. A simple walking program is best for the last trimester.

Women in Fitness Programs

Women can participate in any fitness program in which men participate. A general conditioning program is as necessary for women as for men. And, as discussed earlier, a reasonable exercise program does not overdevelop muscles, but instead firms them, improves muscle tone, and increases circulation. It also promotes total health. However, women should remember that they are generally smaller in bone and muscle structure than men. They should start at lower levels of weight and intensity to keep from damaging their muscles. A

pregnant woman should consult her obstetrician about any exercise program. If she has been participating in a regular daily exercise program, which includes activities to strengthen abdominal and back muscles, she should be able to carry her child easily, deliver easily and swiftly, and recuperate rapidly after delivery. Normally, a woman should be able to continue her regular exercise program, if it is not too strenuous, up to the sixth month of pregnancy. During the last three months, she should engage only in a simple walking program.

Blood and Circulatory Changes

The total blood volume (amount of blood in circulation) first falls and then rises above normal during a four-to-nine-week training period. This increased blood volume will persist for about four weeks after cessation of training. Average individuals can expect their blood volumes to rise 10 to 19 percent as a result of regular exercise. Some athletically trained people have 41 to 44 percent higher blood volumes than nonexercisers. This higher blood volume increases the number of red blood cells.

A greater blood volume brings more oxygen to the muscles and other organs of the body. Also, it helps the blood return to the heart more quickly and easily, which allows the heart to work more efficiently. Thus, the heart can beat more slowly at rest, producing the low resting pulse rates of totally fit individuals.

Physical Differences Between Males and Females

The greatest difference between males and females is the ratio of strength to weight. Adult males have about 30 percent more muscle than women of equal size. Well-designed physical exercises do not produce overdeveloped muscles in women; instead, they produce firmer, sleeker bodies that are more pleasing to women themselves and to men.

Women have larger stores of fatty tissues in their bodies. This, however, is not always a disadvantage; for example, women swimmers are more buoyant than men and lose less body heat in cold water.

There is also a sex difference in ratio of heart weight to body weight. From the age of ten to sixty, a woman's heart is 10 to 15 percent smaller than a man's for the same body weight. After sixty the woman's grows faster and becomes very similar in size to a man's. Female hearts pump about 18 liters of blood per minute as compared to 24 liters per minute in males—a 30 percent difference in cardiac output. From the late teens to about thirty years of age, men have about 15 percent more hemoglobin per ml of blood and about 6 percent more red blood cells per ml. Thus, men's blood has a greater oxygen-carrying capacity.

Women's capillary walls are more easily broken than men's, and women are therefore more susceptible to bruises.

FITNESS PROGRAMS

You should set certain goals before beginning any exercise program. You should be able to say what purposes you want the program to serve. A professional physical educator can design a *balanced* total fitness program that does not overemphasize any one aspect of physical development. Isometric exercises and weight training improve both muscular strength and endurance, but they have minimal value for circulatory fitness. Brisk walking, jogging, and running are excellent circulatory exercises, but they do little for the abdominal, back, shoulder, and arm muscles. Therefore, they should be accompanied by exercises that strengthen these regions. Few activities exercise all parts of the body equally well.

Although more heart attacks occur during rest periods than during exercise, certain precautions are necessary. Every physical conditioning program should involve three phases: *warm-up, workout,* and *cool-down.*

Warming-up and Cooling-down

Each time you exercise or participate in a sport, you should warm your body up for a minimum of 3 to 5 minutes; a warm-up can consist of walking or limbering-up exercises (such as those shown) of gradually increasing intensity. Individuals with cardiac problems or who are over forty years of age should warm up for 5 to 10 minutes.

A proper warm-up increases body temperature, stretches ligaments, and slightly increases cardiovascular activity in preparation for more strenuous exercise.

Just as the body needs warming up, it also needs cooling off after exercise. This helps return the blood to the heart and get the body temperature back to normal. One should keep moving for several minutes after vigorous activity until breathing has returned to normal, the stress of the activity has subsided, and the body has cooled.

Intensity of Exercise

Because of the relationship between heart rate (pulse) and the intensity of exercise, checking the

---

● ASK YOURSELF

1. What restrictions apply to women's fitness programs?
2. How do pregnant women benefit from fitness programs? What restrictions should they observe?

The Importance of Cooling-down

It is a good idea to keep your body moving for a short time after a work-out. This allows the body to return to a resting state. Cooling-down is essential for normal blood circulation. Without a proper cooling-down, the blood pools in the muscles of the arms and legs and reduces the blood volume throughout the body. This can cause the body to go into shock or to hyperventilate (short, fast, shallow breathing). Because it washes too much carbon dioxide (CO_2) out of the blood, hyperventilation can produce muscle cramps or cause you to pass out.

pulse during and immediately after a work-out is the best means of monitoring the intensity of exercise.

Research has shown that a work-out need not be exhausting to achieve the desired conditioning effects. Intense exercise should produce a heart rate in the range of 60 to 85 percent of predicted maximal heart rate (target pulse rate). In very sedentary nonexercisers, some benefits, even if minimal, are derived from prolonged activities that require as little as 25 percent of maximal heart rate.

Duration of Exercise

The time required to produce a desirable conditioning effect depends on the intensity of the exercise: the lower the intensity, the longer the duration. For example, an hour of walking at a 50 percent target pulse rate equals 20 minutes of jogging at a 75 percent target pulse rate. Any exercise should last long enough to produce sweating, mild fatigue, and breathlessness (usually from 10 15 minutes at the least). Obviously, the longer the work-out the better the conditioning effect.

Frequency of Exercise

Work-outs three to four times a week will rapidly raise your conditioning level; five or six times a week will raise it even more rapidly at the beginning. However, reaching a level of fitness is one thing; staying there is another matter. Once a desired level of fitness is reached, three work-outs a week are sufficient to maintain an adequate level of conditioning.

Cutting back on the number of work-outs each week reduces the level of fitness. Reducing the number of work-outs from three to one a week will reduce your fitness level by half within ten weeks.

Stopping your work-outs completely will reduce your fitness level to that of a nonexerciser within four to six weeks. If, for some reason such as illness you must reduce or eliminate your work-outs, re-establish them by very gradually raising their direction and intensity to give your body time to adjust.

Progression in Exercise

The body adapts to the overload of conditioning by increasing its maximal oxygen uptake gradually, over a period of several weeks or months. As conditioning takes place and maximal oxygen uptake increases, the intensity of the work load diminishes. Thus, to increase your level of conditioning, your work-out must become longer or more intense. This is not an open-ended progression, however, because once you reach an ideal level of fitness you can maintain it with slight variations in the time or intensity of your work-outs.

Varieties of Exercise Programs

Many different exercise and recreational activities can be used for physical conditioning. Walking and jogging provide the most direct conditioning program, but other kinds of exercise are also beneficial. A variety of exercise programs can help to maintain interest and allow for changes in a daily schedule.

Calisthenics Calisthenics exercise specific groups of muscles or the whole body. The following illustrations show a series of test exercises that can provide a foundation for a calisthenics program. You should not, however, go on to more vigorous calisthenics (progression) until you can complete these tests. Many basic calisthenic programs are designed for the whole body. In the early 1960s the

Royal Canadian Air Force originated two programs that have proved to be very successful. *The 5BX (Five Basic Exercises) Plan for Physical Fitness* for men and the *XBX Plan* (a series of ten exercises) for women are simple calisthenic exercises that can be completed in 11 minutes each day. You should either follow an organized program such as these or have a physical educator design a program especially for you.

Walking Walking is the most natural of all exercises. A person may walk at any time with almost no medical risk. Brisk walking for a period of time sufficient to accelerate the pulse strengthens heart, lungs, and leg muscles. Extremely inactive people may obtain endurance effects with a regular walking exercise program. However, they must increase their rate of speed to obtain further benefits. Inactive, sedentary men (more than women) over 50 years of age should walk for exercise. They should not try jogging or running because of the increased chance of heart attack after 50.

Jogging The next step up from walking is *jogging* —steady, slow running. It may be alternated with breath-catching periods of walking. Jogging is pleasant, free, easy, and relaxing. It can be done alone or in groups. Jogging will maintain a level of fitness and circulatory endurance, but to increase circulatory endurance, an individual must progress to running at some point in the schedule. College students should alternate jogging with intervals of hard running for a month or two before entering a regular running program.

A jogging program begins with a short period of trotting and walking. As jogging progresses, the jogger covers greater and greater distances.

Jogging may be satisfactory to an individual during the initial phase of conditioning. After the goal of a mile is achieved, however, running often replaces jogging. Running can provide a challenge as well as good circulatory endurance. The progression of walking and jogging outlined previously may be used to progress toward the goal of running a mile. However, you may want to increase the time period in the schedule to two months. Before you try to run a mile, be sure that you can jog comfortably for this distance. Don't worry about your time at first. When you can jog for a mile, start pacing

● **ASK YOURSELF**

1. Why do you warm up? How long should you warm up before working out? Design a good warming-up set of exercises for yourself.
2. How many work-out sessions a week are recommended for increasing your fitness level? How many sessions a week are needed to maintain an adequate level of fitness?
3. How, and for how long, should you cool down after working out?

yourself and reduce your time. You should pace yourself so that you are running at a constant rate. Avoid any bursts of speed; they greatly reduce your efficiency and cause fatigue. Reduce your time by 10 seconds a week until you can run a mile in 6 minutes.

Running in place When space or weather restricts your ability to run, *running in place* can be a very effective means of maintaining circulatory endurance. The important factors in running in place are the cadence of the step and the duration of the activity. The most comfortable length of time seems to be five minutes. To reach a five-minute goal, begin by running in place for one minute. Then increase your time by 30-second intervals. Run for five sessions at each time level before progressing to the next interval. When you can run in place for five minutes, you may begin to increase your step cadence.

TABLE 10.3 PROGRAM FOR RUNNING IN PLACE

Number of sessions	Running time (minutes)
1 to 5	1
6 to 10	1½
11 to 15	2
16 to 20	2½
21 to 25	3
26 to 30	3½
31 to 35	4
36 to 40	4½
41 to 45	5

A trial calisthenics program.

Deep breathing exercise. *Stand tall, rise on your tiptoes, and inhale deeply. As you inhale and rise, raise your arms until your hands come together over your head. Hold this extended position for 1 or 2 minutes. Then lower your arms and drop back to the standing position as you exhale.*

Knee-to-nose kick. *Get down on your hands and knees on a soft surface. Bend your head down and bring your left knee as close to your nose as possible. Hold for 2 seconds. Next, extend the left leg back and up, at the same time bringing your head up. Hold for 2 seconds. Repeat action with right leg.*

Bent-knee sit-ups. *Lie on your back with your hands clasped behind your head, your knees bent, and your feet held down. Keeping your chin on your chest and your back rounded, roll up into a sitting position. When you are in a sitting position, straighten your back, lift your head, and press your elbows back. Hold for 2 seconds, then drop your head, round your back, and roll slowly down to the starting position.*

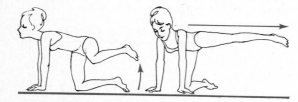

Leg extension. *Start on your hands and knees on a soft surface. Keeping your left knee bent, raise it to the side until it is level with your hips. Extend the leg straight out to the side, keeping it at hip level. Then bend it back and return it to the floor in the starting position. Repeat with right leg.*

Body rotation. *With your legs apart and hands on hips, lean forward and bend at the waist. Now rotate your body from the waist in great, slow circles. Lean far enough to the right, left and front so you feel the muscles stretch. Now rotate to the left, rear, right, and front.*

Arm rotation. *Extend your arms straight out from the shoulders, and rotate them so that your hands are tracing circles about 1 foot in diameter. Rotate arms forward, then backward.*

Standing body rotation. *Stand with your feet well apart, and extend your arms at shoulder level. Twist your upper body all the way around to the right, following it with your eyes; hold for 1 second. Twist back to the left as far as possible; hold for 1 second.*

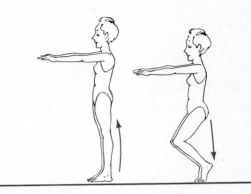

Half knee bends. *Stand with your legs together and feet parallel. Rise to the toes for a count of 1 and bring your arms forward for balance. Tighten the muscles of the seat, abdominal area, thighs, and knees. Keeping your back very straight, lower yourself until your knees are only half-way bent. Do not go into a deep knee bend. Rise again to your toes and then lower yourself to your heels.*

Windmill. *Stand with your feet well apart and arms extended at shoulder level. Bending at the waist, touch right hand to left foot, keeping your left arm extended. Return to an upright position and repeat action, this time touching left hand to right foot.*

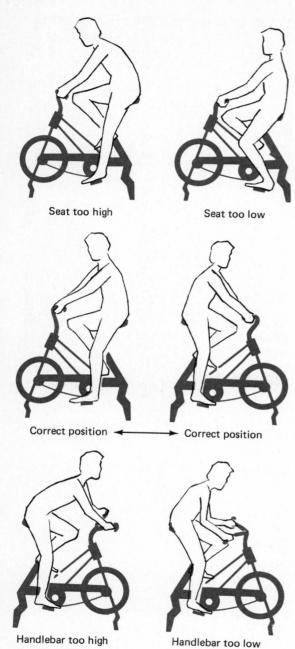

Seat too high

Seat too low

Correct position ⟷ Correct position

Handlebar too high

Handlebar too low

A stationary cycle is an excellent exercise machine. Arrows indicate correct seat height and handlebar position. The knee should be slightly bent when the toes are at the lowest reach of the pedal. The body should be relaxed and lean slightly forward.

Bicycling Both as a form of transportation and as a dynamic exercise, bicycling is very popular. Because of the great variations in gears, bicycle weights, and terrain, it is impossible to stipulate the amount of cycling necessary for a conditioning program. Stationary bicycles eliminate some variables and permit exercise indoors during bad weather or when city traffic and pollution create hazardous conditions. Reaching a desired target pulse rate and maintaining it for a sufficient time is a good way to determine the effectiveness of cycling.

Swimming Swimming is an excellent exercise for dynamic conditioning. Since swimming exercises the muscles of the arms, back, and shoulders, it works well in conjunction with jogging. Swimming is an excellent exercise for persons with problem knees, backs, or other orthopedic problems. Walking or jogging in chest-deep water is also a good way to exercise the lower part of the body, because the bouyancy of the water lessens the strain normally imposed on the joints.

The great variation in swimming ability makes it difficult to stipulate the amounts of swimming needed in a conditioning program. Here again, checking the pulse rate provides an accurate gauge of performance.

Rope jumping This is an excellent cardiovascular exercise. A ten-minute daily program of rope skipping improves and maintains cardiovascular endurance. Rope skipping may be either a program in itself or a bad-weather substitute for a jogging or running program.

Obtain a piece of rope anywhere from six to nine feet long. The correct rope length is twice the distance from your armpit to the ground. Tape the ends so they will not fray.

Variations within a rope-skipping program can add interest and incentive. They may also provide progression. (Progression may also be achieved by performing a specific number of jumps in a certain amount of time.) The normal skipping style may be modified by jumping on one foot, alternating feet, or jumping with both feet together. Running with skipping increases timing and coordination. Jumping backward is a simple maneuver in which various foot styles can be used. Make some forward-to-backward changes and then some backward-to-forward. This is difficult because you must jump an extra time as the turn is completed.

A double jump is challenging. This is done by spinning the rope faster and jumping a little higher. When you have increased your skill, shorten the rope slightly by winding it within the hand. Stay in the air longer by bending the knees and keeping them high. You may even achieve a triple jump. Double and triple jumps have an effect similar to that caused by sprints in running; they rapidly increase the heart rate.

A front cross may be achieved by crossing the arms when the rope starts downward. This makes a loop through which you may jump. On completion of the jump, the arms are uncrossed and the next jump is made in the normal manner. If you have difficulty performing a front cross, lengthen the rope slightly and either lower the hands as the arms cross or cross the arms far enough to bring the elbows together. These changes will give you a wider loop to jump through. A back cross may be done by performing a regular back jump with arms crossed in front of the body (cross your arms so your elbows touch each other).

Make your own modifications. Try double jumps with a front cross, or try to run while you change jumping forms. Again, make some forward-to-backward changes and then some backward-to-forward.

Aerobics *Aerobics* is a total fitness program for men first published in 1968. Its originator was Dr. Kenneth H. Cooper. The key concept in this program is oxygen consumption. Because oxygen cannot be stored in the body, it must be continually replenished. Consequently, the fatigue level is controlled by the ability of the respiratory and circulatory systems to supply oxygen to the muscles.

Dr. Cooper's aerobics system consists of a point count assigned to different physical activities that

Proper jump-rope length. The rope should reach to your armpits as you hold it down with your feet.

increase circulatory endurance. An individual progresses to the point where he or she can perform activities worth 30 points each week. The number of points represents the amount of physical activity necessary for maintenance of the cardiorespiratory system.

TABLE 10.4 DISTANCES COVERED IN 12 MINUTES OF RUNNING

Fitness categories	Distance (miles)			
	Under 30 years	30 to 39 years	40 to 49 years	50 to 60* years
Very poor	under 1.0	under 0.95	under 0.85	under 0.80
Poor	1.0 to 1.24	0.95 to 1.14	0.85 to 1.04	0.80 to 0.99
Fair	1.25 to 1.49	1.15 to 1.39	1.05 to 1.29	1.00 to 1.24
Good	1.50 to 1.74	1.40 to 1.64	1.30 to 1.54	1.25 to 1.49
Excellent	1.75 and over	1.65 and over	1.55 and over	1.50 and over

*It is not recommended that a man over 60 years of age begin an aerobics program.
Source: Dr. Kenneth A. Cooper, *The New Aerobics* (New York: Bantam Books, 1970), p. 30.

Aerobic dancing is an excellent exercise.

Points are obtained by performing specific circulatory endurance activities in a specific amount of time. Such activities include running, swimming, bicycling, walking, running in place, and participating in strenuous games of squash, handball, and basketball. The first test of aerobic fitness is the distance that can be covered by running for 12 minutes.

Women's aerobic capacity is smaller than most men's. This is because of women's smaller physical size and lung capacity; they have less blood circulating, less hemoglobin, and fewer red blood cells. According to Cooper and Cooper (1973), a female needs 24 points a week for a satisfactory total fitness level. But they do not discourage women from exceeding the 24 points, just as they do not discourage men from exceeding the 30 points per week if they are physically able to do so. Point evaluations for women can be found in *Aerobics for Women* (Cooper & Cooper, 1973).

Aerobic dancing (Eurhythmics) A series of dance movements can be used as an exercise program. Aerobic dancing can be either an individual or group exercise. By starting with slow music, increasing the tempo, and varying the type, you can make dance a complete work-out.

The warm-up period begins with movements to slow music. The tempo is gradually increased until the target pulse rate is reached; this is maintained for a time with music at a very fast tempo. Finally, music at a slow tempo is used for the cooling-down period.

Aerobic dancing is a very enjoyable form of dy-

● **ASK YOURSELF**

1. What does the word "aerobic" mean?
2. What types of exercises are considered "aerobic"?
3. Devise an aerobics program for yourself.

The Road to Physical Fitness

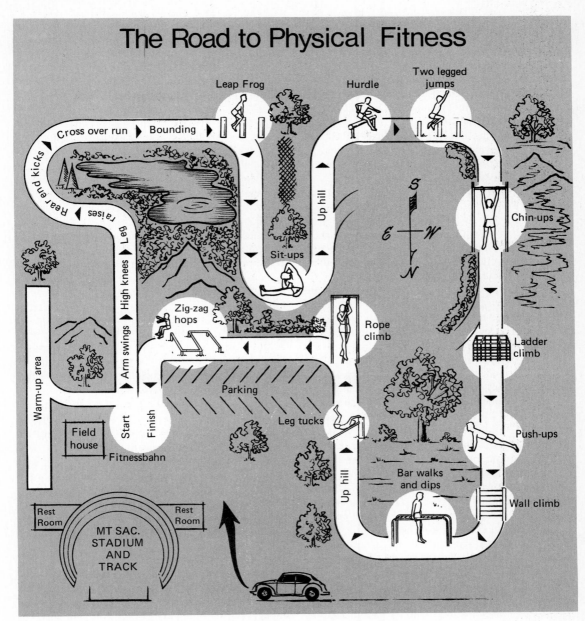

Exercise circuits are being laid out in many parks and recreational areas.

namic exercise. Also, varying the music keeps it from becoming boring. Dance is an excellent exercise to increase coordination and athletic timing. Many professional male athletes participate in dance programs to maintain their coordination and timing.

Circuit training A *circuit* refers to a number of carefully selected exercises that are arranged and numbered consecutively and located over a given area. Each numbered exercise within the circuit is called a *station*. An individual moves at his or her own speed from one station to another until the entire circuit is completed. In most cases the total circuit is repeated more than once (usually three times) and the total time of performance is recorded. This is a convenient and different approach to exercise — one that is physically, physiologically, and psychologically sound. Many cities are now establishing circuits in local parks. See if you can establish one in your area.

Circuit training involves two valuable activities: weight training and running. The best physical activity for developing muscular strength and muscular endurance is weight training. One of the most effective methods of developing and maintaining circulatory and respiratory endurance involves running (others are swimming, bicycling, and rope skipping). A well-planned circuit involves both weight training and running. Thus, circuit training increases muscular strength, muscular endurance, and cardiovascular endurance.

The value of circuit training lies in its extreme adaptability to a great variety of situations. A circuit can be designed to fit any individual, group, area, or condition (it is even adaptable to medical rehabilitation). Circuit training enables an individual or a group of people to progress through a series of exercises and to check their progress against a clock.

On a circuit, progression is produced by decreasing the time required to complete one circuit, increasing the work load (weight or sets), or a combination of both. An individual works at his or her present capacity and progresses as capacity increases. Circuit training provides a series of progressive time goals that are achieved step by step. This time factor provides built-in motivation; it encourages a person to improve. The circuit layout offers variety, and this is appealing to most people.

Calisthenic exercises may be used in a circuit. When this is done, the load is a person's own body weight. The load is increased by modifying the exercise. For example, the following changes will increase the work load in a push-up: standard push-up; push-up, pushing hands off the floor; push-up, pushing off the floor and clapping hands; pushing up, pushing off the floor and slapping the chest. Such modifications make calisthenic exercises very useful in circuit training. Load can also be increased by increasing the number of times the circuit is run. Laps may be added; in that case, a person runs the length of the circuit without performing at the stations. In maintaining good progression, the most important factor is to increase the work load gradually and at a rate that can be handled with ease and safety.

In planning a circuit, remember that your circuit must be based on your personal goals. Choose exercises that are strenuous. Each exercise should contribute toward progression by increasing both work load and work rate. This automatically excludes the very light warm-up exercises that should be performed before the circuit is run. Do not perform "duck waddle," deep knee bends, or any other exercise that can cause joint, ligament, or muscle damage if done too fast. Perform each exercise the same way every time so that you are able to observe and evaluate your improvement.

Select exercises that balance other exercises so that all groups of muscles receive proper exercise. Improper balance of strength between antagonistic muscle groups can produce permanent body damage. To avoid improper balance, group the muscles into three categories: the arm, neck, and shoulder group; the abdomen, back, and chest group; and the buttock, hip, and leg group. When arranging your stations, avoid consecutive placement of exercises that involve similar muscle groups. For example, both arm curls and chin-ups involve the arm, neck, and shoulder muscle group. If you include both of these exercises in a circuit, separate them. A circuit designed for general body conditioning should include exercises that involve all three of the muscle groups.

The amount of time you have available for performing a circuit is a factor in determining both the difficulty of exercises and the number of stations to be included. The spacing and the arrangement of the stations are determined by the amount of space you have available and by the kinds of exercises you want to emphasize. If your basic purpose is general body conditioning and cardiovascular endurance, you should choose a large area for your circuit, allowing for distances between stations for running and movement. There should always be enough room for exercise that is strenuous, yet free-flowing and uninterrupted. Given available space and facilities, several circuits (or variations of one circuit) may be organized, utilizing the same area or equipment.

Weight training Lifting weights does not develop circulatory endurance. It does increase muscular

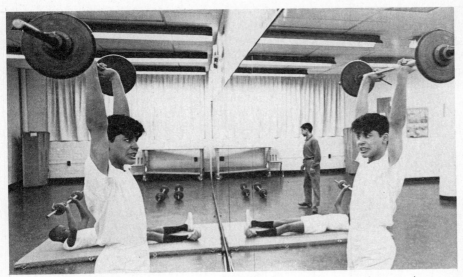

Weight training strengthens you for participating in more dynamic exercise programs and sports.

strength and endurance through application of the principle of *progressive resistance exercises.* In other words, as strength increases, resistance is increased by increasing weight lifted or repetitions. Progressive resistance occurs when a person moves a given resistance (barbells or weights) a definite number of times (repetitions). Resistance exercise programs may be as extensive or as simple as the individual desires (competitive weight lifting, body building, weight training). However, no program should ever deviate from the principle of *resistance progression.* A person who simple wants to see how much weight he or she can lift may injure muscles and joints. Weights should be increased only as strength increases.

Weight training, a combination of weight lifting and calisthenics, will help produce the strength needed for participation in medium- and high-energy sports. But weight training takes time. A person may not reach the strength needed by some sports for months or years of continuous progression.

If you are interested in weight training, you may want to follow a trial program. You should remain in it for three months to determine whether your strength has increased.

Remember that hard work is required to obtain results, and results occur very slowly. If you enjoy this program, ask a physical educator to design a weight training program for you. Weight programs are often part of a college or university curriculum, a men's club, the YMCA, or a gym. The YWCA and women's fitness and figure control studios are designing weight training programs for women. The same strength-building exercises performed by men to develop their muscles and reduce fat deposits can be performed by women. The only change is that women should start at a lighter weight and progress at five-pound increments.

SPORTS

Sports and other recreational activities can serve as conditioning programs. However, some individual and dual sports require so little physical activity that they are not adequate for a fitness program. Golf is such a sport. It has many psychological and social values but little physical value. The main energy expenditure in golf comes from walking. This walking is usually not vigorous enough to elevate the pulse significantly.

If a sport is to supply the requirements of a physical fitness program, it must be vigorous. Also, a person should participate in it for at least three sessions a week. Each session should last for a

● ASK YOURSELF

1. How do you use your leisure time? Does it include any exercise? How could you work a regular exercise program into your leisure time?
2. Work out a complete exercise program and make it part of your everyday activities. After one month, reflect on your level of fitness before and after the program.
3. Try to encourage others to work a complete exercise program into their daily routines.

minimum of 30 minutes. Ideally, a person who is using a sport as the basis of a fitness program should participate in the sport for 60 minutes every day.

When you participate in a sport, your expenditure of energy depends on several factors:

1. *The number of participants.* In calisthenics, weight training, and running, you control the expenditure of energy. In dual sports, however, you must consider the number of participants. Handball may be least demanding in a game of doubles, more demanding in a game of singles, and most demanding in a game involving three people ("cutthroat").
2. *The skill of the participants.* Generally, a high level of skill is reflected in greater efficiency. Thus a skilled person can participate at a lower energy cost. As your skill develops, you should either increase the amount of time you devote to a sport or pit yourself against people who are more skillful than you are.
3. *The duration.* The longer the duration of an activity, the greater the energy expenditure. Each participant should be vigorously active for 30 to 40 minutes. In a 60-minute game, then, you should be moving one-half to two-thirds of the time.
4. *The speed of the necessary physical movements.* Sports that require occasional bursts of speed are more demanding and require a greater expenditure of energy than do sports that require a steady pace. You should participate in such sports only after you feel your physical condition is adequate.

In choosing sports, remember the value of developing and maintaining circulatory endurance. Sports that improve circulatory endurance include individual activities such as swimming, scuba diving, snorkeling, hiking, running, and bicycling; dual activities such as wrestling and judo; and court games such as badminton, handball, squash, tennis, and volleyball.

Dual sports and court games require at least two participants. Court games further require a court. These requirements may limit your opportunities to engage in exercise. Individual activities usually offer more opportunities for participation.

USE OF LEISURE TIME

The automobile, television, and beer drinking are, unfortunately, the focal points of the leisure hours of many people; these are their means of "relaxing." If these things were abolished, however, many people would not automatically become involved in physical conditioning programs. People must want to change their life styles and exercise regularly. Motivation for these things depends on understanding how to condition yourself and knowing what activities produce physical fitness.

Psychological as well as physiological limits increase with exercise, making one feel better, be more decisive, and have a more positive and confident outlook on life. It enables one to participate in many activities that are part of a natural life style.

Leisure time is well spent in unmechanized, active pursuits that reacquaint one with the environment. One should seek out activities that provide healthful exercise, are pleasurable, and cost much less than a motorized recreational vehicle or a seat at a spectator sport.

Who can motivate you to exercise? No one can. You must motivate yourself. Also, are you likely to stay with an exercise program throughout life? Only you know. Anything that is enjoyable costs something. An active, enjoyable life costs energy. The energies of motivation and participation in a regular exercise program will pay off in a longer, more active, and more enjoyable life.

We can point out the importance of regular exercise in preventing or delaying chronic degenerative diseases (such as emphysema and heart disease). But only you will know what is a desirable level of exercise and fitness for you. Often, fitness pro-

Active use of leisure time.

grams are directed toward "big-time" athletics, not at an enjoyable level of "social athletics" for both men and women. It is very enjoyable to scuba dive, ski, hike, run on the beach, and grow *with* your friends and your children when you are in your forties and fifties. This, however, requires exercise on a regular basis so that you can perform when you want to.

IN REVIEW

1. Total fitness allows people to perform at an optimum level of efficiency in all daily activities. Totally fit people have the strength, speed, agility, endurance, and social and emotional adjustment appropriate to their age. The range between our minimum oxygen requirements and our maximum oxygen consumption is the major physiological measure of our total fitness.
    a. Fitness develops as early as the preschool years. To maintain a acceptable level of fitness, an individual must participate in an activity program throughout life. A physically fit person can perform in sports and other physical activities at a social level throughout life.
2. Sleep is crucial to proper fitness. Lack of sleep can make a person groggy and irritable.
3. The different stages of sleep are distinguishable by progressively slower brain-wave (EEG) patterns. Typically there are four to six cycles a night. These cycles alternate stages of deep sleep with periods of dreaming.
    a. Dreaming is important to our emotional well-being; they apparently help us integrate the events of our daily lives with our unconscious minds.
    b. Sleep requirements vary from person to person because of inherent physiological differences.
    c. Everybody has trouble sleeping on occasion. Many people suffer from in-

somnia—the inability to sleep. Drugs should be avoided in treating insomnia. Increased exercise and progressive muscular relaxation are good alternatives to sleeping pills.

4. Total fitness means that a person has the capacity for prolonged work without undue fatigue.

   a. Nonexercisers are only passively fit and make no effort to maintain their bodies. They are often ill, but do not know it because degenerative diseases caused by inactivity take years to appear.

   b. Normal aging is usually measured by the decline of vitality. Body-age estimations based on the physical condition of the body determine *biological age*. Chronological age is one's age in years. There is no relationship between these two types of aging. Rapid biological aging leads to deterioration of the body, disease, and premature death.

   c. The types of fatal diseases have changed since the early 1900s. Today's deaths are more commonly caused by various degenerative conditions caused by biological aging. Many people die prematurely—that is before their expected life span of between 75 and 80 years.

   d. Physical exercise that is sufficiently intensive and regular can override the various phenomena of aging. People who have maintained successful personal fitness programs throughout their lives can enjoy their middle and later years to the fullest. The habits of physical exercise and total fitness should be formed during early childhood.

5. While in school or college, individuals should develop skills in several different activities that they can participate in for the rest of their lives. One's persistence in a physical fitness program depends on how interesting it is. Physical exercise is an essential counterbalance to our overly sedentary life styles.

   a. Several steps should be taken before starting a total fitness program. A proper warm-up program must precede a more strenuous exercise program.

   b. All exercises should be adapted to individual tolerance levels. As people master one exercise program, they should progress to more strenuous programs. All fitness programs should include such a progression.

   c. Total fitness level is indicated by resting pulse rate. As the level of physical fitness increases, the resting pulse rate decreases. The pulse rate indicates the intensity of a physical activity. To increase fitness, you must reach a predictable target pulse rate during an exercise session. After exercising, your resting pulse rate should be reached within 30 minutes (recovery pulse rate).

   d. There are two types of muscular activities: static exercises and dynamic exercises. Static exercises are produced by isometric muscular contractions. Isometric muscle contractions increase tension within muscles. These activities are detrimental to the heart but strengthen the muscles. Isotonic contractions generate great demand for oxygen in muscles. These activities increase cardiovascular fitness.

   e. A rise in maximal oxygen uptake increases stamina and diminishes fatigue during prolonged work or exercise. The conditioned heart is able to pump more blood with each beat. Thus, conditioning strengthens the heart and reduces the resting pulse rate.

   f. Women can participate in any fitness program in which men participate. Women should start at lower levels of weight and intensity to keep from damaging their smaller muscles. A pregnant woman should consult her obstetrician about any exercise program.

6. Every physical conditioning program should involve three phases: warm-up, work-out, and cool-down.
   a. The body should be warmed up for a minimum of 3 to 5 minutes in exercises of gradually increasing intensity.
   b. For conditioning, an exercise should produce a heart rate of 60 to 85 percent of the predicted maximal heart rate (target pulse rate).
   c. The time required to produce a desirable conditioning effect depends on the intensity of the exercise. An exercise should last long enough to produce sweating, mild fatigue, and breathlessness.
   d. How soon a desired level of conditioning is reached depends upon the frequency of exercise. Once a desired level of fitness is achieved, three work-outs a week are probably sufficient to maintain an adequate level of conditioning.
   e. To increase the level of conditioning, work-outs must become longer or more intense.
   f. A variety of exercise programs can help maintain interest and allow for changes in a daily schedule.
7. Sports and other recreational activities can serve as conditioning programs. However it is better to exercise regularly in order to participate in the sports you enjoy.
8. Leisure time is well spent in unmechanized, active pursuits that put us in touch with the environment. We should seek out activities that provide healthful exercise, are pleasurable, and cost much less than mechanical or spectator sports.

CHAPTER II

CONSUMER AFFAIRS

People can do more for their health than can any physician. As enlightened health-care consumers, they can also save money and time. They can treat many health problems at home; and they can recognize when it is important to see a physician or go to the hospital. Such self-care should be universal and should be the foundation of any health-care system. But consumers need educating.

Everyone should be able to identify, for example, the stress-creating factors on the job, in family life, and in sexual relationships that lead to overeating, smoking, and hypertension and effectively reduce them. A person's own medical decisions, habits, and lifestyle have far more impact on his or her health than any actions or recommendations by a physician.

Pharmacies and the drug counters in supermarkets are filled with nonprescription, over-the-counter (OTC) products, many of them ineffectual. But when used excessively or under the wrong conditions, they can be dangerous. These are good reasons for observing some rules in the use of aspirin, antihistamines, and eye washes. All labels on nonprescription and prescription drugs should be written in language people can understand. Moreover, beyond deciding which products to select and how to use them, people should begin to question whether they need *any* product whatever.

Self-care has its limits, however. Obviously, people should not run to a physician for every scrape, bruise, ache, or pain. But some symptoms do require the attention of a physician. Here are five warning signs that indicate a physician should be consulted:

1. *Severe symptoms.* Any type of attack in which the symptoms are severe or alarming — such as severe abdominal or chest pain, or bleeding — should obviously receive prompt medical attention.

2. *Prolonged symptoms.* Any symptoms — such as cough, headache, constipation, or fatigue — that persist day after day should be checked by a physician, even though the symptoms are minor. Serious chronic disorders are often revealed through persistent minor symptoms.

3. *Repeated symptoms.* Symptoms, even though minor, that recur time after time should be reported to a physician because, like prolonged symptoms, they may indicate a serious problem.

4. *Unusual symptoms.* Any symptoms that seem to be unusual, such as unusual bleeding, mental changes, weight gains or losses, digestive changes, or fatigue, call for a visit to a physician.

5. *If in doubt.* If in doubt, the safest action is to see a physician. If there is a serious problem, it can be corrected in its early stages; if there is no problem, then you have paid a very small price for your peace of mind.

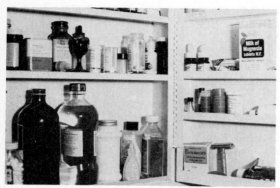

Potions, nostrums, elixers, "wonder drugs," tonics — people swallow all manner of things in their quest for good health.

DRUG PRODUCTS: GOOD AND BAD

Pharmacy counters are filled with a huge array of OTC drug products from which consumers must choose. In the following discussion, we shall describe various nonprescription *and* prescription drugs to help you make safe, intelligent decisions about purchasing and using these products. Our discussion is far from exhaustive. For purposes of identification, the federal Food, Drug, and Cosmetic Act defines drugs as *articles that are intended to be used for diagnosis, cure, lessening, treatment, or prevention of disease.*

Aspirin

Aspirin is probably the most effective medical substance that can be bought without a prescription. Aspirin, occasionally sold under its chemical name, *acetylsalicylic acid,* is the principal active ingredient of literally hundreds of nonprescriptive remedies. In many of these preparations, aspirin is the only effective ingredient, but the cost may be many times that of plain aspirin tablets. When advertisements refer to the "pain-reliever that doctors recommend most," they mean aspirin. Although certain brands of aspirin have been highly advertised as being more effective than other brands, there is really no significant difference between brands.

Aspirin has several beneficial properties. Its most common use is as a pain-reliever, or *analgesic,* especially for headaches and muscular pains. It also has the ability to reduce fever and inflamma-

tion. For some people, aspirin may act as a mild sedative.

"Glorified aspirin" products often contain aspirin, *phenacetin* (another analgesic), and caffeine (a mild stimulant), or just aspirin and caffeine, or aspirin with a buffering agent. These products have been shown to be no more effective for most persons than plain aspirin. Several products that originally contained phenacetin no longer do so. Although effective in relieving pain and reducing fever, phenacetin in larger doses over prolonged periods can damage the kidneys.

Patients receiving large amounts of aspirin over long periods may develop a condition known as *salicylism,* indicated by nausea, vomiting, ringing in the ears, trouble with hearing, and severe headache (which, of course, is not relieved by taking more aspirin).

Although the moderate use of aspirin is generally safe, even here there are certain precautions to follow. A few people suffer allergic reactions to aspirin (such as hives); others find that aspirin causes stomach irritation. This latter problem can often be prevented by eating before taking aspirin. The dosage recommended on the label should not be exceeded. Dosages above this level are *not* more effective and may be harmful.

Enteric-coated aspirin (which delay absorption of the aspirin until it reaches the intestine) is less likely to cause ulcers and erosions in the stomach than plain or buffered aspirin. Newer enteric-coated aspirin permits good absorption by the intestine.

Finally, aspirin should be used over long periods of time only on the recommendation of a physician, since it might otherwise relieve the symptoms of a serious disorder that needs medical attention.

Aspirin is among the most common causes of accidental poisoning of young children. Aspirin, like any other medication, should be kept where children cannot get to it, preferably in a locked cabinet. Specially flavored children's aspirin is a particular problem, since children may eat it as candy. Manufacturers have attempted to devise lids that prevent children from opening the bottles. But this product should still be carefully kept out of the reach of children. Some authorities even recommend splitting adult aspirin tablets for children rather than keeping flavored aspirin in the house.

A compound similar in its actions to aspirin is

acetaminophen. It is contained in over 60 OTC and prescription products, such as Excedrin and Tylenol. Acetaminophen is probably safer for people with aspirin allergy or bleeding disturbances. When acetaminophen is combined with aspirin, as in Excedrin, these advantages, of course, are lost.

Remedies for Coughs and Colds

Every year new "miracle" cold remedies are offered to the public in massive advertising campaigns, only to drop quietly out of the market a few years later when their manufacturers release newer "miracles." The fact remains that, despite the many advances in other fields of medicine, there is still no way to prevent or cure the common cold.

To put cough and cold remedies into their proper perspective, one needs to understand that a cold is caused by a virus and that, to date, only limited progress has been made toward producing drugs that will cure virus infections. In fact, no drug has been developed that will control cold-producing viruses.

A cough is a reflex action caused by the presence of foreign matter or other irritation within the respiratory system. The purpose of coughing is to remove the irritant. Cough and cold remedies may give symptomatic relief by deadening the cough reflex, opening a stuffed nose, or reducing fever, but they do not get at the basic cause of the trouble. By eliminating the symptoms, but not the source, cough and cold remedies may in the long run prove harmful. For example, by relieving the symptoms, persons with colds are tempted to go on with normal activities when they really should go to bed for 24 hours. This rest period, early in the course of a cold, can often hasten recovery and prevent secondary infections and other complications. The result of "fighting" a cold is often secondary bacterial infection of the middle ear, sinus cavities, or lungs. Coughs can be the result of many serious conditions that need medical treatment rather than just a deadening of the cough reflex. For example, a cough could indicate tuberculosis, pneumonia, other infections of the lungs, or lung cancer.

The reputations of many products for "curing" colds come from the fact that most colds are gone in less than a week. If a cold lasts beyond a week, it is probably caused by a secondary bacterial infection, and should be treated by a physician.

● **ASK YOURSELF**

1. When a member of your family is sick, what symptoms should convince you to call a physician?
2. For what symptoms do you use aspirin? How do these compare with the properties discussed?
3. How would you know when you are overusing aspirin? What symptoms would you look for?

Antibiotics have no effect on the virus phase of a cold, but may be useful in clearing up the secondary infections. The cold sufferer should not pressure his or her physician to prescribe antibiotics unless there is definite evidence of bacterial infection.

Most cold remedies contain an analgesic (aspirin, phenacetin, acetaminophen, or salicylamide) to relieve pain. Many contain an *antihistamine* to dry up the mucous membranes in the nasal passages, and a *sympathomimetic,* or nasal decongestant, which is most effective if applied directly to the mucous membranes by means of a spray or nose drops.

Antihistamines counteract the fluid release from cells that occurs during colds and allergy attacks. While this action may relieve the effects of a cold and certain allergies, antihistamines do not cure the cold. If they seem to cure it, there is a possibility that the actual problem was an allergy and not a cold at all. Antihistamines very commonly produce side effects such as dizziness and drowsiness, so they should not be taken when driving a car or operating machinery.

Sympathomimetics affect the blood supply to the swollen linings of the nose and throat. Diabetics should avoid sympathomimetics since they may raise blood sugar. Patients with rapid or irregular heart action are advised against using them because they may intensify high blood pressure and nervousness.

Pills for Relaxing, Sleeping, and Waking

The stress-loaded lives of twentieth-century Americans produce unwanted by-products of sleeplessness, nervous tension, and inability to relax. Remedies are offered for all of these conditions. Just what are these drugs?

Pills for sleeping Millions of people qualify as insomniacs, or people who can't sleep. To counter this, there is widespread use of nonprescription sleep-aids and prescription sleep medications.

Nonprescription sleep-aids may contain bromides, scopolamine, and various antihistamines. A common ingredient in remedies for colds, antihistamines have the side effect of producing drowsiness, as was discussed. Sleep-aids should be taken *exactly as directed on the label.* Higher dosages may produce dry mouth, dizziness, blurred vision, incoordination, loss of appetitie, and nervousness. As was pointed out earlier, individuals should never drive a car or operate machinery after taking any antihistamine drug.

The Institute of Medicine (IOM) of the National Research Council now recommends that physicians restrict the use of prescription sedative-hypnotic drugs to short-term treatment of insomnia. It found little evidence these drugs continue to be effective when used nightly over periods longer than 3 to 14 days. As we have seen, people sleep in stages, and some of these pills interfere with or suppress some important phases of sleep.

What, then, should insomniacs do? A physical exam should identify or rule out various physical ailments. A person's habits of eating, drinking, exercising, or relaxing might prevent a good night's sleep. Others need to deal with their fears of not being able to sleep, or of not sleeping long enough.

Pills for relaxing One class of drugs is called daytime sedatives. Sold without prescription, they are used to combat symptoms such as "simple nervous tension." Bromides, scopolamine, and antihistamines are common ingredients (as in sleep-aids). As such, their daytime use may do more than help a person cope with the day's problems. They will also make a person sleepy. There is a definite danger in the use of daytime sedatives when driving or operating machinery.

To avoid the risk of accident when taking drugs for sleeping or relaxing, the FDA suggests:

1. Read the dosage instructions carefully.
2. Ask your pharmacist or physician about side effects.

3. Ask your phamacist about the antihistamine content of nonprescription drugs.
4. Avoid alcohol when taking medications. A leading cause of highway accidents, alcohol is even more dangerous when mixed with most drugs.

Pills for staying awake Nonprescription pills that are advertised as helping people stay awake usually contain caffeine. Caffeine, of course, is contained in coffee, tea, and cola drinks. Though not everyone reacts in the same way, many people do find that caffeine acts as a stimulant that can prevent sleep. Caffeine also makes some people feel nervous ("coffee nerves"). So the sleep-preventing pills will do for a person just what coffee does. The effect of one tablet is usually about equal to one cup of strong coffee.

The situations that justify taking these pills are rather limited. Their use has been compared to "whipping a tired horse." They may help students study an extra hour or two, but beyond that the ability to learn drops considerably. These pills should not be used to extend driving for more than an hour or two because driving becomes very dangerous beyond that point.

Amphetamines are prescription drugs that are used (and abused) by drivers, among others, to stay awake. While they may increase alertness for a short time, excessive use interferes with the body's normal protective responses to fatigue. The feeling of exhaustion is short-circuited, causing drivers to use up reserves of body energy until they experience a sudden and total collapse.

Medications for the Muscles and Skin

At some time, almost everyone has sought relief from aching muscles or the painful skin that comes from sunburn, insect bites, or burns. There are now nonprescription drugs that can be applied directly onto the skin (*topical*) to bring relief. Some substances alleviate surface aches and itches by *depressing* skin receptors (nerve endings) that perceive pain, itching, cold, warmth, touch, and pressure. Other medications *stimulate* skin receptors.

Three types of drugs are used to depress pain receptors. *Topical analgesics* such as benzocaine, lidocaine, and dibucaine block pain without causing numbness. *Topical anesthetics* such as camphor, menthol, and phenol completely block pain but cause numbness. *Topical antipruritics* relieve itching and inflammation; they usually contain the chemical agent *hydrocortisone.*

Substances that stimulate skin receptors produce sensations such as burning, warmth, or coolness, all of which distract from deep-seated pain in muscles, joints, and tendons. These are called *topical counterirritants.* The most widely used of these is *methyl salicylate,* a common ingredient in substances used for athletic massages.

All of these products are for external use only. They should not be used around the eyes, and should be discontinued if the condition being treated gets worse or persists for more than a week.

Products for the Eyes

Eye discomfort is one of the more bothersome problems a person can experience. The effects of air pollutants, wind, smoke, and sunlight are the target of a group of health products. The value of the eyes as essential sensory organs makes discretion in the selection of products for eye care most important.

Eyewashes Many physicians warn against the self-treatment of the eyes with any kind of commercial eyewash or eyedrops. The natural flow of tears around the eye is the best means of cleaning the eye of dust, dirt, and other irritating materials. As with many kinds of self-treatment, eyewashes may be used to relieve the symptoms of serious eye disorders that really need prompt treatment by a physician.

Eyeglasses Eyeglasses should not be purchased from variety stores. Anyone having difficulty with vision should have a thorough eye examination by an eye specialist. An *ophthalmologist,* or *oculist,* is a physician who specializes in treating eye conditions. He or she tests the eyes, prescribes lenses, administers medications, and performs surgery. An *optometrist* (O.D.) is a nonmedical eye specialist who can test the eyes and prescribe lenses but may not administer medications or perform surgery. An *optician* is a technician who prepares lenses for eyeglasses.

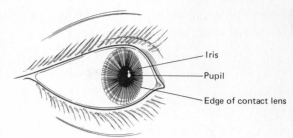

Plastic contact lenses float over the cornea on a layer of tears.

Contact lenses Contact lenses are small plastic lenses, most of which ride on the thin layer of tears directly over the cornea (the outer surface of the eyeball) and under the eyelids. Contact lenses must be properly fitted to the eye.

There are two common types. *Hard contact lenses* are made of hard plastic and may be worn each day for years. They correct most visual problems and are the least expensive of the contacts. They require a "break-in" period when placed in the eye. *Soft contact lenses* are made of soft plastic. They are popular because they are far more comfortable and require no break-in period; most users are unaware of their presence in the eye. Costlier than the hard lenses, they are also more easily damaged and must be replaced every year or two. They must be kept wet in storage and also must be disinfected daily. A third type, *intraocular lenses,* are implanted in the iris of the eye following removal of the natural lenses in cataract surgery.

● ASK YOURSELF

1. What have been your experiences in using over-the-counter cold remedies to treat colds?
2. In buying a cold remedy, what ingredients will you be on the alert for from here on?
3. Which common cold or cough products contain antihistamine? What side effects do they have?
4. Under what conditions would you *not* take pills for sleeping or relaxing?
5. How do you feel about using sleep medications now?

Tips for Buying Sunglasses

1. If you can see your eyes through the lenses in a mirror in the store's artificial light, the lenses aren't dark enough (unless they are photochromic).
2. Check for lens distortion by turning the glasses up to catch the reflection of an overhead light fixture. If the reflection is wavy, the lenses may distort your vision and are probably of poor quality.
3. Check to see that the frames are sturdy, just as you would any eyeglass frames. Pads, hinges, and screws should be secure, and the bows should not obstruct side vision.
4. Frames should fit snugly but gently at the nose bridge and over the ears. Eyelashes should not touch the lenses. The glasses should not slip off when you bend over.

Sunglasses Bright sunlight or glare can cause squinting, eyestrain, and headache. Ultraviolet rays in sunlight may damage the lens and retina of the eye, and infrared rays can damage the cornea. Sunglasses reduce the discomfort of bright sun and screen out much of the harmful radiation.

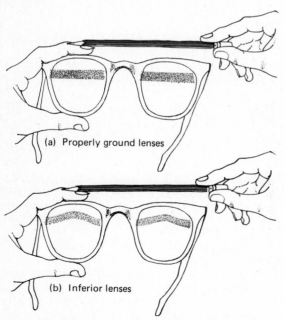

(a) Properly ground lenses

(b) Inferior lenses

A simple optical quality test for sunglasses. (a) Properly ground lenses reflect without distorting. (b) Poorly ground lenses transmit a distorted image to the eye.

According to the American Optometric Association, effective sunglasses should: (1) screen out from 75 to 90 percent of the sunlight present, (2) be large enough to prevent light from reaching the eyes from the side, and (3) have precision-ground optical-quality lenses. The best color for sunglasses is medium or dark gray; the next best is medium or dark green. Sunglasses in other colors, such as blue, rose, and other pastel shades, are not true sunglasses.

By law, sunglasses must have tempered glass or plastic lenses that meet impact-resistance standards. Plastic lenses are lighter in weight and, if treated properly, last as long and perform as well as glass lenses. However, they are not as scratch-resistant or infrared-resistant as glass lenses.

There are three types of sunglass lenses. *Tinted* or *coated* lenses are designed for high-glare situations. Some have a metallic finish that gives them a dark, reflecting (mirrored) surface useful for water and snow sports. *Polarizing* lenses have a special filter that screens out glare from smooth surfaces such as water, wet sand, or sunlit roads. *Photochromic* lenses adjust from light to dark and back according to changing light conditions. These lenses can adapt within one minute to provide maximum screening to bright light. Different photochromic lenses have varying screening capacities.

If you already wear glasses or contact lenses, a prescription can be duplicated in a pair of sunglasses. Prescription sunglasses are preferable to clip-on lenses, which can damage eyeglasses.

Cosmetic ads are a rich source of revenue for many magazines.

COSMETICS

The parade of cosmetic advertising in almost any magazine reveals their importance in most peoples' lives. But what exactly are cosmetics? Are they in any way the same thing as, or similar to, drugs? According to the federal Food, Drug, and Cosmetic Act, cosmetics are defined as "(1) articles intended to be rubbed, poured, sprinkled or sprayed on, introduced into, or otherwise applied to the human body or any part thereof for cleansing, beautifying, promoting attractiveness, or altering the appearance, and (2) articles intended for use as a component of any such articles, except soaps." At present, there are more than eighty types of products considered cosmetics by the Food and Drug Administration.

Cosmetics may also be considered drugs when they make claims to alter a body function. For example, a deodorant is regulated as a cosmetic, because it is intended only to prevent odor. But an antiperspirant is regulated as a drug because it is intended to reduce perspiration, which is a normal body function. If a cosmetic is actually classified as a drug, its active ingredients must be listed ahead of all other ingredients. This is done, for instance, on the labels of dandruff shampoos, hormone creams, antiperspirants, sunscreen products, and all medicated cosmetics.

Cosmetics for the Skin

Our personal appeal to others is of more than passing concern. Part of feeling good is feeling physically attractive. A favorable self-image, beyond being a message to others, is a way of saying that we know how to accept responsibility for ourselves. While caring for our skin, teeth, and hair may not take years off of our life, not giving them proper care may lead to skin and tooth damage that is both painful and unsightly. The multitude of skin products include cleansers, conditioners, and glamorizers. Several such products may be mentioned.

● **ASK YOURSELF**

---

1. For what conditions might you use topical skin and muscle drugs?
2. Have you or any of your friends had experience with contact lenses—either soft or hard? What advice would you offer someone considering getting them?
3. What are the basic types of sunglass lenses?

Skin conditioners Skin is nourished by the blood, and the way to make sure the necessary nutrients are available to the skin is to eat properly. Nutritional requirements for the skin are the same as for the rest of the body—a balanced diet. Dry skin will benefit from a simple oil, cream, or lotion. Since hormones can be absorbed through the skin in quantities large enough to have side effects, hormone skin creams should be used only on the recommendation of a physician.

Persons bothered by skin irritations may reduce or eliminate the irritation by using some common sense. Underarm irritation may come from rubbing too hard when drying, as well as from a cosmetic or drug. Shaving too closely may damage the skin. Clothing worn too tightly can chafe. Stiff, new clothing can also be a cause of skin irritation—especially some of the new permanent-press clothes. When a product containing alcohol, such as most deodorants, is applied to skin already irritated, a temporary burning sensation occurs.

Eye cosmetics Small amounts of carefully applied cosmetics can highlight a person's appearance and emphasize certain features, but the heavy application of cosmetics does not improve the appearance and may damage the complexion.

Cosmetics should be used around the eyes with extreme care. Some cosmetics cause rashes or other skin reactions for some people. (It is wise to test a new cosmetic on your arm before putting it on your face.)

More serious are possible eye infections brought on by the use and misuse of eye cosmetics. Preservatives in eye products retard the growth of hazardous bacteria. But during months of storage on the shelf of a pharmacy, the preservatives may lose their potency. Once the eye product is opened, microorganisms gain ready access to the cosmetic.

The human eye is bathed with secretions that keep in balance the normal skin microorganisms that migrate into the eye mucosa (underside of the eyelid) and onto the eyeball surface. However, the introduction of large numbers of these same microorganisms that have grown in a contaminated eye product severely challenges these protective mechanisms. If the cornea is scratched with a mascara brush, or irritated by wearing contact lenses or through the use of an eye-irritating shampoo, bacteris such as *Pseudomonas* (a type that can cause

eye infections) infect the eye and pose a serious hazard to a person's vision. A scratch on the cornea allows the bacteria to invade and infect the cornea. Only products of good quality should be used near the eyes, and use should be directed on the product.

Deodorants and antiperspirants Socially aware people not only want to be visually appealing, they also want to be fragrantly appealing. Conspiring against this are two problems: wetness and odor, each a product of a specific kind of sweat gland. Wetness comes from the *eccrine sweat glands,* and odor from the *apocrine sweat glands.*

About 3 million tiny eccrine glands prevent the body from overheating. They secrete a clear, odorless liquid of 99 percent water and 1 percent sodium chloride. When the water evaporates from the skin, it removes heat from the body. The eccrine gland responds to two kinds of stimuli: thermal (heat) and emotional (fear, pain, tension, and sexual excitement). Eccrine glands on the palms, soles, and underarms respond to both stimuli. Those on the rest of the body respond only to thermal stimulation, except in extreme emotional stimulation, which brings on a "cold sweat" over the whole body. Wetness alone does *not* cause body odor.

Apocrine glands serve no known useful purpose. Far fewer, they function primarily in hairy underarm regions. Present at birth, they begin to function at puberty and respond to emotional stimulation. Their activity reaches a peak with sexual maturity and diminishes with old age. When stimulated, they secrete a liquid composed of complex organic materials, which are decomposed by bacteria on the skin to form "body odor" or "underarm odor."

Since sweating causes two problems—wetness and odor—two types of products are used. The *deodorant* works against body odor. The *antiperspirant* works against both wetness and body odor.

There are several ways to control body odor. You can wash organic material away before the bacteria can decompose it. You can try to mask the odor with another odor (perfumes are used for this). You can inhibit the growth of the bacteria by removing the moisture necessary for their growth (except that no products completely stop eccrine sweating). Or you can kill the bacteria or inhibit their growth with antibacterial agents. Such agents stick to the skin even after washing, so that deodorant effectiveness builds up over a period of days (or tapers off if use is discontinued).

The best way to control body odor is with a combination of methods. Regular use of a scented deodorant or antibacterial soap will keep you free of natural body odor for up to twenty-four hours. However, deodorants only kill or inhibit the bacteria, so a deodorant is not a substitute for washing. One warning: Users of deodorant soaps may be more vulnerable to the effects of the sun and blister more easily.

An antiperspirant reduces the flow of perspiration from the eccrine gland through the use of metal salts such as aluminum. It also reduces body odor through the use of antibacterial ingredients. To be effective, an antiperspirant ingredient must penetrate the sweat duct. Antiperspirants work best when used on a daily basis. For the greatest antiperspirant action, apply the product when you are not already sweating. Instead of applying it immediately after a bath, when the body is warm, apply it when you are at rest and cool. Then lie down for a few minutes—you sweat least then and consequently the antiperspirant penetrates better. The ingredients are not harmful to the sweat glands. Normal sweating resumes usually within a week after the antiperspirant use is discontinued.

Deodorants and antiperspirant deodorants are available in many forms: liquid, pads, creams, roll-ons, sticks, powders, and aerosols. Read the label to make sure the product is actually an antiperspirant. One warning on aerosol products: While the propellants used in aerosol antiperspirants pose no direct hazard in normal use, some of the active ingredients may cause lung damage from long-term use.

Shampoos

The hair and scalp need to be washed frequently to clean away the normal buildup of cast-off cells, oil, bacteria, and dirt. The oil glands of the scalps of most people are very active. The oil these glands produce accumulates on the scalp and hair, becomes rancid and bad-smelling, and affects hair appearance.

While any ordinary bath soap will serve the purpose, the cleansing agent in most shampoos today is a synthetic detergent. Shampoos are classified as

● ASK YOURSELF

1. What distinguishes a drug from a cosmetic?
2. In view of the two types of sweat glands, why might you want to use an underarm product other than a deodorant? How is the product you use for underarms described?
3. What precautions should you take in order to keep your eye cosmetics free of microorganisms? Have you ever picked up any eye infections this way?
4. List the properties of the shampoo you commonly use. Does it contain a soap or a synthetic detergent?

cosmetics unless they contain a drug for correcting or preventing a bodily condition such as dandruff.

The major advantage of synthetic detergents is their efficient functioning in hard water. In hard water with much calcium in solution, soap reacts with the calcium to form deposits of a gummy material called "soap scum" (the familiar bathtub ring), which dulls hair luster. Synthetic detergents neither form scum deposits nor require an acid rinse to remove them; they do not require the extensive lather normally expected from soap to perform an adequate cleansing job.

The big problem with a synthetic detergent is that the more thoroughly it removes dirt and other unwanted material, the more likely it is to irritate the scalp, strip off hair dyes and tints, and remove the natural oil left on the hair from natural gland secretions.

Some synthetic detergents contain additives called *conditioners*, which take the place of the "hundred brush strokes" of the past to produce luster and sheen. Conditioners also give hair the appearance or feel of softness, impart smoothness and lubricity to the touch, make combing or brushing easier, give the hair "body" or bulk, add texture, and retain "set." No longer are shampoos confined to cleansing and scenting the hair. These additives perform no therapeutic function. There is no scientific evidence that shampoos which contain eggs or other forms of protein are medically helpful.

Suntans and Sunburns
Many lightly pigmented people believe that a suntan enhances their physical appearance. Almost

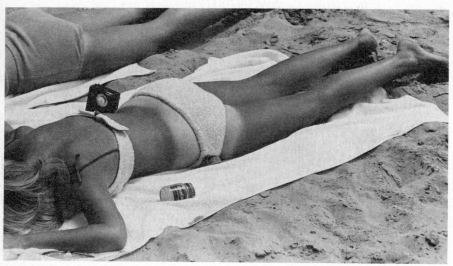

Many of the young and beautiful consider bronzed skin essential to their self-respect.

everyone experiences a feeling of well-being when warmed by the sun. Beyond this, there are few beneficial effects of sunbathing, and there are several adverse effects.

Sunbathing When you tan, ultraviolet radiation increases the production of *melanin* (the dark pigment beneath the surface of the skin). Melanin comes to the surface of the skin, darkens it, and shields the cells beneath it.

If you insist on getting a tan, you should take several safety precautions. Dark-skinned individuals have more immunity to sunburn than the fair-skinned. Yet even the person who tans well must be careful to avoid sunburn at the beginning of the summer. The key is to start with short periods of exposure and work up to longer periods. If you

have light skin, the first exposure early in the season should not exceed 15 to 20 minutes for each side, front and back. If you begin tanning later in the season, the length of first exposure should be even shorter. You can increase the exposure by about one-third each day. After a few days, exposures of several hours may be possible. Rays are reflected from sand and water, which means that you can burn even while sitting under an umbrella. You can burn on a hazy day, as well. When a person gets "sunburned," the body increases the flow of blood to the skin, giving a redness to the burn. A very severe burn can permanently destroy melanin-producing cells, making a person very sensitive to ultraviolet radiation the rest of his or her life.

Suntan preparations and sunscreens Suntan preparations are of varying value. Many lotions contain chemicals called *sunscreens,* which absorb and block ultraviolet rays to some degree. The better lotions allow you to stay in the sun longer. They do not shut out all the radiation; otherwise you would never tan. Despite the claims of advertisements, there is no way to screen out the "burning rays" of the sun while admitting the "tanning rays" since they are one and the same. Read the label before purchasing any preparation. Suntan lotions must be reapplied at least every two hours or after

● ASK YOURSELF

1. What routine have you developed for getting a suntan without getting badly sunburned?
2. List some brand names that have a good reputation for reducing sunburn damage. Which of these have been the most successful?
3. What has been your experience using a sunlamp for tanning? What precautions do you plan to take to adequately protect yourself?

Choosing Suntan Products

A huge, confusing array of suntan and sunscreen preparations is on the market now, and they all compete for our attention. The advertising budgets for some of these products must resemble the national debt. How do we choose?

A welcome addition to the labels of many products is the *sunscreen effectiveness rating*. This rating is expressed as a number, usually between 2 and 15. The higher the number, the greater the ability to screen out ultraviolet rays, the burning rays in sunlight. A rating of 2 indicates that half the ultraviolet is screened. With proper application, that product doubles the time you can stay in the sun. Products rated 6 allow you to stay out 6 times as long, and so forth.

Lower-rated products are useful for people with plenty of natural pigmentation or who have already developed a deep tan. Don't make the mistake of trying to tan faster by using a lower-rated product—you will just get sunburned.

Higher-rated products are helpful early in the season when you are starting to develop your tan. Even with a deep tan, they can be useful for long stays in the sun. For skiing or similar sports, you should use a maximum screen on your face to help prevent the premature aging (wrinkling) and skin cancers that plague outdoors persons.

each swim. A word of caution: Commercial sunburn preparations contain ingredients that may cause allergic skin reactions.

Those who like to tan very deeply every year should be aware of several possible effects of long-term or repeated exposure (even gradual tanning). Their skin may become prematurely wrinkled, leathery, mottled, or discolored. The skin may eventually look many years older than its true age. These changes, when caused by ultraviolet radiation, may lead to skin cancer. Eyes are very vulnerable to ultraviolet radiation. A severe eye burn can scar the cornea and permanently impair vision.

Sunlamps Exposure to sunlamps has become a popular means of staying bronzed year-round and of combating skin problems by creating a smoother skin. Sunlamps contain mercury gases and give off ultraviolet radiation. Similar to the sun, the radiation is more intense and takes less time to affect the skin.

There are three types of sunlamps: The *reflector*

Use Sunlamps with Caution!

- Don't use sunlamps consisting of a bare mercury tube.
- Don't use any lamp without a timer.
- Use only sunlamps that come with clear instructions for distance and exposure times.
- Get no closer to the lamp than the recommended distance.
- Don't exceed the recommended exposure time.
- Protect your eyes with goggles or some other eye device.
- Don't use sunlamps if you are especially sensitive to the sun.
- Be careful after replacing an older bulb with a newer more powerful one.
- Avoid taking a hot shower or sauna before using a sunlamp.
- Stop exposure immediately if you see any tanning or reddening while you are under the lamp.
- If the lamp requires a warm-up, don't begin to use it until it is warmed up.
- Use same precautions at health clubs as at home.

bulb screws into a socket with a metal reflector. Another type, the *fluorescent sunlamp* (often used in health clubs) resembles a conventional fluorescent light. Both types have a glass enclosure that partially filters out the most hazardous radiation. Sunlamps consisting of a *bare mercury tube* in a metal reflector may not filter out hazardous radiation.

Products for the Teeth

Few products have been advertised more heavily than certain *dentifrices* (toothpastes and toothpowders). Almost every brand is claimed to be the very best for the prevention of bad breath and tooth decay.

Tooth decay is one of the most common health problems in the United States today. An estimated 95 percent of the population has cavities. Basically, tooth decay comes from the destruction of the surface enamel of the tooth by acids. The acid that most violently attacks tooth enamel is *lactic acid.* Other acids may cause etchings on the enamel but lack the destructive action of lactic acid. Lactic acid is formed by the metabolic action of several kinds of tooth bacteria upon sugars from foods and drinks.

Dental research has shown that approximately 96 percent of all the harmful bacteria within the mouth live in the tiny crevice near the gum line between the tooth and gum. The damage done is not related to the *number* of bacteria present but rather to their *state of organization.* It has been found that bacteria must be organized or clumped together in tiny colonies or clusters called *bacterial plaques* before they are capable of producing harmful effects such as decay, tender and bleeding gums, and foul breath. But the presence of bacteria alone in a disorganized or unclumped state produces no harmful effects. Once the bacteria have been disorganized, it takes 24 to 30 hours for them to reorganize.

The bacterial action suggests how a person might prevent tooth decay. If once each day a person disorganizes the bacteria in the mouth, he or she should be free from decay and gum problems such as periodontitis. This should be done by means of a thorough brushing of the teeth. Flossing should be done daily around each tooth and in the crevices around each tooth.

Toothpastes do not kill the acid-forming bacteria, nor do they neutralize the decay-producing acids. The basic ingredients of toothpastes and powders are abrasives, detergents, flavoring, and in some, fluorides. Cavities (*caries*) can be prevented by brushing teeth regularly with dentifrices containing sodium fluoride, sodium monofluorophosphate, or stannous fluoride. (Read labels to discover which brands contain these substances.) Fluoride rinses (for swishing around the teeth) and gels (for applying to the teeth) are also effective. Baking soda (sodium bicarbonate) contains abrasives but no fluoride. As such, it has no effect against cavities.

Fluorides make tooth enamel more resistant to decay. Fluorides are available to the teeth in drinking water, by topical application to the surface of the affected area, or in dentifrices. The water supplies of some localities naturally contain adequate amounts of fluorides; many municipalities add it where the water supplies are deficient in it. Sodium fluoride, a form contained in drinking water, is especially effective with newly erupted teeth. In areas where the water supply contains adequate amounts of fluoride, the fluoride contained in a toothpaste is of little additional value. Fluoride concentrations above one mg per liter, although not harmful, tend to discolor the tooth enamel.

Professional cleaning by a dentist or dental hygienist once or twice a year, depending on one's particular mouth chemistry, is essential. This should guarantee that tooth loss does not become a problem later in life.

The Safe Use of Cosmetics

According to the FDA, today's cosmetics are among the safest products available to consumers. Yet no rules, regulations, or precautions by government or industry can protect the person who does not follow the label directions and warnings.

Some basic unwritten rules for good judgment that can prevent adverse reactions to cosmetics are:

1. Read and follow all directions and warnings on the product. If patch testing (trying a portion of the product first on a very small area of your skin) is suggested, do not skip this wise step to determine your sensitivity to the product. And sensitivity can change: hair color, skin cream, or another type of product that should be patch-tested may irritate you the next time you use it, even though it has not in the past. Your body chemistry is always changing.

2. Maintain basic cleanliness—in other words, wash your hands before applying a cosmetic. Also remember to close containers after each use; dust and germs can easily settle into any product left uncovered.

3. Never borrow another person's cosmetics. You may be borrowing trouble.

4. When water must be added to a cosmetic before it can be used—such as cake eyeliner—it is dangerous to substitute saliva. This practice can transfer the bacteria from the mouth to the eye and cause an eye infection.

5. If you do develop an adverse reaction, do not try to "wait it out." See your physician immediately. And to speed diagnosis, take the suspected cosmetic with you.

These actions are the consumer's own "voluntary program" for safety. They can be enforced only by you.

QUACKERY

Today's quacks are much more sophisticated than was the snake-oil salesman of the past, but their goal is still the same—to separate the suckers from their money. Many quacks seem to be sincerely interested in a person's health, but their real interest is in making money—lots of money. The cost of medical quackery in the United States today is in the billions of dollars each year. Present-day quackery takes several common forms. It may involve a direct "doctor-patient" relationship, the mail-order or house-to-house sale of worthless products, or the sale in drug or "health food" stores of products that cannot do what is claimed for them.

Who Are Quacks?

A *quack* may be defined as a boastful pretender to medical skill. A quack is anyone who promises medical benefits that cannot be delivered. Quacks may attempt to go beyond the limits of medical science or the limits of their own training.

The quack usually inspires confidence. Some are well intentioned, others are sinister. Yet in either case their attempts to diagnose physical and nutritional problems can be dangerous to the patient.

Quacks have various kinds of training. Once in a while, a licensed medical physician enters into an area that may be considered quackery. A few best-selling books have been written by such doctors.

● ASK YOURSELF

1. What steps does your dentist recommend in keeping your teeth clean?
2. If you have children, what rules of dental hygiene are you going to insist they follow to have fewer cavities than you have had?
3. What is your understanding of the role of fluorides in reducing the incidence of tooth decay?
4. Have you ever experienced any adverse reactions in the use of cosmetics? How might one best avoid such bad experiences?

Much more common is the quack who has had more limited training, perhaps in methods of chiropractic or naturopathy (a mode of therapy using air, water, light, heat, massage, and so on). Many convicted quacks show no record of any formal higher education. College degrees and transcripts are obtainable from print shops and "diploma mill" colleges. The word *Doctor* in front of a name, or the letters of a degree after it, mean nothing in this situation. Sometimes the "quack" is not an individual at all, but a corporation that makes false claims for its products. And of course, self-treatment is a form of quackery because most individuals are simply not qualified to diagnose and treat their own illnesses.

Who Turns to Quacks?

All kinds of people become the victims of quacks—the old and the young, the rich and the poor, and everyone in between. But quacks do seem to prey particularly on the extremes—the young, the elderly, the very rich, and the very poor.

The young person often becomes the victim of mail-order quackery. There are many deceptive advertisements in the magazines that appeal to young people. The products offered promise good looks, popularity, and sex appeal. Some products claim to help gain weight, lose weight, build muscles, enlarge the breasts, and cure acne. Seldom are these products of any real value.

Elderly people find appeal in products that promise to renew their lost youth and vigor. They often waste their limited money on useless treatments and products that claim to relieve arthritis, impotence, prostate conditions, gray hair, baldness, and "tired blood." The nutrition quack caters to el-

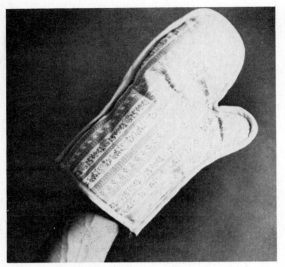

The Uranium Wonderglove was sold to many sufferers of painful arthritis in the hand. The promotion of this gadget was based on the alleged therapeutic value of radiation from uranium ore.

derly people who are led to believe that all their aches and pains can disappear through the use of certain food products or food supplements. In addition, the elderly are attracted to so-called clinics and health ranches that claim cures for various chronic diseases through chiropractic, fad diets, and other limited methods.

Quacks are interested in the rich simply because they have money and will pay for "miraculous" cures. The poor are receptive to quackery because good medical services are also scarce in low-income areas, and the cost of ethical care may seem prohibitive. The urban minority resident, who may not have a private physician and for whom an emergency room visit may take upwards of three hours, must often rely on self-prescription and past experience in treating common illnesses.

Much quackery preys on *fear*. People who have been told by a physician that they have an incurable disease live in fear—fear of death, pain, surgery, or of the unknown. They may "grasp at any straw" of hope offered by the quack, no matter how unscientific or expensive the treatment may be. Sometimes fear keeps people from seeking an ethical physician in the first place. Afraid of surgery, they may turn instead to a quack who promises a cure without surgery. Some are fearful of rejection if they don't

measure up to some mythical image of beauty or sex appeal.

Ignorance and *gullibility* are strong allies of the quack. Millions of people unquestioningly accept almost anything they hear as reliable information and anything they read as absolute gospel. A sharp quack can make his or her sales talk and literature seem entirely believable to such people. Even well-educated people who should know better often fall for quack schemes because no one can be completely knowledgeable in all areas.

Major Types of Quackery Today

While quackery knows no seasons, the popular targets of quackery shift with the major causes of disability. Capitalizing on public awareness of major health concerns, the quacks center their pitches of "sure cure" and "money-saving self-treatment" on cancer, arthritis, food selection and

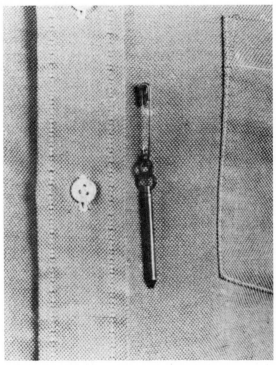

The Vrillium Tube, containing a fraction of a cent's worth of barium chloride (a chemical used in photography), came with a safety pin so that it could be attached to the user's clothing. The gadget was advertised as a therapeutic source of radiation. Thousands were sold at $306 each.

TABLE 11.1   PRODUCTS SEIZED BY THE FOOD AND DRUG ADMINISTRATION

Product	Charge
Hair implant assemblies and implant syringes (10/19/79)	False and misleading claim as to FDA registration number; labeling fails to reveal that implantation of these fibers into scalp may result in infection, rejection, and permanent scarring.
Ovultron for "L-Field" measurement (10/23/79/	Labeling contains false and misleading claims for detecting preparations for and act of ovulation, for being designed to indicate changes in the female "electrocynamic field," or "L-field," and other familiar false and misleading claims.
Acut-dot adhesive magnet patches (10/30/79/	False and misleading claims for the temporary relief of occasional minor aches and pains in muscles and joints.
Intraocular lenses (6/14/79)	A medical device without an effective approved application; and persons granted exemption failed to comply with regulations.
Aspirin tablets (9/26/79)	Circumstances of manufacture, processing, packing, and holding not in conformity with current good manufacturing practice; label fails to bear expiration date.
Phenobarbital tablets (12/11/79)	Strength differs from, and quality falls below, U.S.P. standards.
APC aspirin compound tablets (11/20/79)	Circumstances used for article's processing and packing not in conformity with current good manufacturing practice.
Ovultron electrometer (10/13/79)	False and misleading claims for detecting the time of ovulation, as well as other changes in the body; fails to bear adequate directions for safe use for such purposes.
Candy novelties of marzipan (9/10/79)	Contain the nonconforming color additives FD&C Red No. 2 and External D&C Orange No. 3.
X-ray unit (10/9/79)	Article's quality falls below the component certification since the article fails to comply with the standards. Dangerous to health when used as directed, because it emits radiation beyond preset exposure time.
Prophylactics, rubber (5/14/79)	Quality falls below claims; it contained holes.
Catheter tray kits, disposable (4/27/79)	Prepared, packed, and held under unsanitary conditions.

preparation, weight-reduction, and sexual dysfunction. Examples of these and other forms of quackery are given in the chapters to which they relate.

Specific Cases of Quackery

The cases of quackery in the features have been selected from recent issues of *FDA Consumer,* an official publication of the Food and Drug Administration. They should serve to illustrate the types of fraud and deception to which the public is being subjected. (See also Table 11.1 listing products seized by FDA.)

Many people naively believe that they do not have to worry about quackery because the government does now allow it. In reality, worthless treat-

ments and products are always being offered to the public. Although government agencies are active and successful in combating quackery, a quack may escape government detection for a period of time. Then the government must gather evidence for a case. (The burden of proof of the fraud is on the government's shoulders.) The case may be tied up in the courts for many years until every possible appeal is exhausted. In the meantime, the fraud often persists.

Public Protection

At every level of government, efforts are being made to control fraudulent health practices. The

Signs of Quackery

Public efforts at combating the more flagrant cases of quackery are largely successful. The government must investigate, file charges, and seek prosecution. Conviction, however, brings appeals, which may run for many months or years.

It is important for the individual to be able to recognize quackery and to know how to avoid it. Some of the signs apply mainly to the sale of nonprescription remedies through health stores, drugstores, and mail order. Other signs apply to clinical quackery, where a "doctor" examines the patients and administers drugs.

Diagnosis by mail   This is *pure* quackery. Not even the most skilled physician could accurately diagnose all disease by a mailed-in description of symptoms or sample of blood or urine.

Free trial package   Many mail-order health or beauty aids offer a free or low-cost trial package. Other send a free book that promotes the product. The general provider of the "free 30-day trial supply" knows that because people are creatures of habit, after 30 days the users will probably continue the product, which they will then purchase at a higher price.

"Limited supply—act at once"   "Don't miss this once-in-a-lifetime opportunity." This sales approach, so commonly used in all lines of business, is intended to stampede the customer into acting without taking time to think about the offer and to check its validity.

"Recommended by doctors and nurses"   The advertisement seldom states *which* doctors and nurses or even what type of doctor.

"Approved by independent research laboratories"   It is not difficult for the patent-medicine producer to find some chemist who is willing to set up an "independent research laboratory," perhaps in a garage, and, for a fee, to approve almost any product.

Testimonials   No importance can be placed on letters of testimony. At worst they can be outright lies. At best they are probably written by or purchased from naive people who never had the disease or whose recovery had nothing to do with the product. Remember that 50 to 75 percent of all physicians' office visits are the result of psychosomatic complaints that often disappear through suggestion, even with no useful treatment.

Location   Ethical practitioners usually prefer a professional environment for their offices. Medical offices, for example, are often near a hospital. The practitioner who rents space in a department store or discount store is not necessarily unethical, but one should be alert for further signs of quackery.

"New scientific breakthrough"   These claims often refer to scientific studies or reports in leading medical journals but never mention the journal, authors, or institutions by name. Such claims in tabloids and newspapers should not be accepted without further proof.

Guarantee of cure or satisfaction   Ethical physicians never guarantee a cure. They do the best they can, but medical science has not progressed to the point where results are that certain. Even with a "guarantee," the quack is seldom known to refund any money.

A "secret cure" Since quacks cannot legally use surgery or drugs, they often claim superiority over these methods of treatment. Some become very defensive and claim they are being persecuted by medical associations and the government. Look carefully at any physicians who resort to such a line. It probably indicates that they are in legal or professional trouble.

Effective for a wide variety of ailments There are no cure-alls! Quacks often claim ability to cure conditions ethical physicians cannot always cure or remedy.

Federal Trade Commission is active in cases involving fraudulent or deceptive advertising. The U.S. Postal Service may move rapidly in cases of mail-order fraud. The Food and Drug Adminsitration regulates the purity, safety, and proper labeling of drugs and food products moved across state lines. Certain state, county, and city governments are also active in suppressing quackery by enacting laws that make fraudulent practices a felony.

Several privately financed groups actively participate in the restraint of health frauds. Among these are the Bureau of Investigation of the American Medical Association, the Better Business Bureau, and the Chamber of Commerce. Although these organizations have no legal regulatory powers, they can bring cases of fraud to the attention of the public and the proper legal regulatory authorities.

When in doubt about the merits of a particular product or treatment, it is often worthwhile to check with a local chamber of commerce, Better Business Bureau, local medical society, or licensed and reputable physician.

Guidelines for Filing Consumer Complaints

When the top of a table is ruined by a furniture polish, when a can of tuna smells tainted, when a garage does not make promised repairs, or when a skin conditioner causes a rash, there is more to do than throw the product away and vow to buy a different brand the next time or shop elsewhere. You can, and should, complain to the proper government agency. The problem is in knowing how to register a complaint and to whom.

Here are some steps to follow *before* you report, in *how* you report, and *where* you report.

Before you report Before you report violations or hazards, you should ask yourself these questions:

1. Have I used the product as directed?
2. Did I follow the instructions carefully?
3. Did an allergy contribute toward the bad effect?
4. Was the product old when I opened it?

Take all these factors into consideration first to confirm that the problem is due to the product, not to your misuse of it.

How to report Report the complaint as soon as possible. Give your name, address, telephone number, and directions on how to get to your home or place of business. Clearly state the complaint. Describe in as much detail as possible the label of the product. Give any date or code marks that appear on the container (on canned goods these are usually stamped or embossed on the end of the can). Give the name and address of the store where the article was bought and the date of purchase. Save whatever remains of the product or the empty container for your physician's guidance or possible examina-

---

● ASK YOURSELF

1. Have you ever been "suckered in" by a quack? What was the product and what was your experience?
2. What simple rules might you have ignored in falling for this quack's "line"?
3. What actions does the Food and Drug Administration take to save you from the products of quacks?

tion by the FDA. Retain any unopened containers of the product you bought at the same time. If an injury is involved, see your physician at once. Report the suspect product to the manufacturer, packer, or distributor shown on the label and to the store where you bought it.

Where to report Ten government agencies have jurisdiction over products. The three most common are:

1. *The Food and Drug Administration.* The FDA deals with cases of mislabeled or harmful drugs, food, cosmetics, and health devices. See your local phone directory for the FDA office nearest you, or write: Food and Drug Administration, Rockville, Md. 20857.

2. *The U.S. Postal Service.* The Postal Service handles complaints of fraudulent or misleading promotions involving use of the mails (either to advertise or to receive orders for products). Contact your local postmaster, postal inspector, or write: Inspector in Charge, Special Investigation Division, U.S. Postal Service, Washington, D.C. 20260.

3. *The Federal Trade Commission.* The FTC is concerned with cases of suspected false advertising. Write: Federal Trade Commission, Sixth Street and Pennsylvania Avenue, NW., Washington, D.C. 20580.

Other agencies that may be notified are:

1. *Consumer Product Safety Commission:* toys; laundry, cleaning, and polishing products; home repair and paint products; hobby items; and automotive fluids

2. *U.S. Department of Agriculture:* meat and poultry products.

3. *U.S. Department of Justice, Bureau of Narcotics and Dangerous Drugs:* illegal sales of narcotics or dangerous drugs (such as stimulants, depressants, and hallucinogens)

4. *Environmental Protection Agency:* pesticides; air and water pollution

5. *State Health Department:* products made and sold exclusively within your state

6. *State Board of Pharmacy:* dispensing practices of pharmacies and drug prices

7. *Poison Control Centers:* accidental poisonings

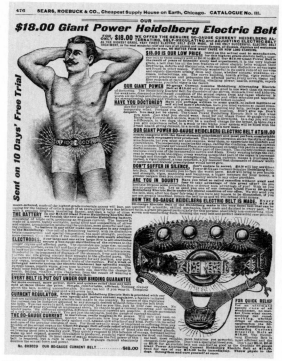

This cure-all advertisement appeared in a 1902 Sears catalogue.

The Consumers Union in Mt. Vernon, New York, which publishes a monthly periodical called *Consumer Reports,* also takes complaints and often relays them back to the manufacturer, packer, or distributor on items that have been tested.

The Persistence of Quackery

How does quackery persist in spite of intensive efforts by government and individuals to eliminate it? The answer is that, although quacks might not be very skilled in treating disease, they are very adept in other areas. They often operate at the very borderline of legality, perhaps obeying the letter but not the spirit of the law. When they are convicted, they usually serve a short jail sentence and pay a stiff fine (which they can well afford). Then, immediately, they change location and perhaps their names, and are back in business again.

Often, getting a conviction for a quack proves to be very difficult. Juries may be swayed by the emotional testimonies of former "cured" patients of the quack. Large corporations engaged in the sales of proprietary compounds retain excellent lawyers to fight their battles with the authorities, and the corporations often win. For example, it took the federal government sixteen years to get the word *liver* removed from the name of Carter's Little (Liver) Pills on the basis that the pills had nothing to do with the liver.

The private citizen can aid the campaign against quackery by reporting incidents of suspected quackery to the local district attorney's office or the local medical society. It is often only through such complaints that authorities are alerted to a fraudulent operation. Today, as always, it is the responsibility of the individual to be alert to both health fraud and quackery and to avoid them.

● ASK YOURSELF

1. What questions should you ask yourself before reporting consumer complaints?
2. What information should you include in any consumer complaint?
3. If you felt you were the victim of consumer fraud, what public agencies might you report it to? What sorts of complaints would you give to which agency?
4. Has someone you know done anything to expose quackery? If so, what?
5. Can you spot examples of apparent quackery in newspapers or magazines you usually read? Cite several examples.

IN REVIEW

1. Everyone has the responsibility for becoming aware of and adopting habits and life styles that promote good health. This is especially true of self-care. Everyone should learn which symptoms require self-care and which need the attention of a physician.
2. Drug counters are filled with a vast array of drug products, both useful and useless. Some are nonprescription (over-the-counter) drugs, and others are prescription drugs.
3. Aspirin has several useful properties. It relives pain, reduces inflammation, and reduces fever. Its active agent is acetylsalicylic acid, which can cause stomach irritation. Consult a physician before using large amounts of aspirin or using it over a long period of time. People allergic to aspirin may use a substitute, acetaminophen.
4. Cough and cold remedies are a big business. Colds are caused by viruses, for which there is no antibiotic cure. Symptoms of discomfort, however, can be reduced through the use of pain-relievers (analgesics), mucus-drying agents (antihistamines) and nasal decongestants (sympathomimetics).
5. Stress-reducing pills include:
   a. Pills for inducing sleep: These produce drowsiness and should be used in moderation. The underlying reasons for continued sleeplessness need to be identified and dealt with.
   b. Pills for relaxing: These are daytime sedatives; they must be used with care to prevent oversedation, which can cause accidents.
   c. Pills for staying awake: These pills combat fatigue; they usually contain caffeine. They are no substitute for sleep, and prolonged fatigue can lead to total collapse.
6. Medications for the muscles and skin include topical (skin surface) pain-relievers (analgesics), pain-blockers (anesthetics), itch-relievers (antipruritics), and producers of warmth or coolness (counterirritants).
7. Eye products include:
   a. Eye washes (eye drops): These are of questionable value.

b. Eyeglasses: These should be prescribed by an eye specialist, who usually examines the eyes first.

c. Contact lenses (hard or soft).

d. Sunglasses: These screen out sunlight and reduce eye discomfort.

8. Cosmetics are for cleaning, beautifying, or altering the appearance of the exterior of the body. They are considered to be drugs if they alter body function.

9. Skin products include:

a. Skin conditioners: These may oil or cleanse the skin, but do not nourish it.

b. Eye cosmetics: These highlight one's appearance, but should be used with care.

c. Deodorants and antiperspirants: Deodorants eliminate perspiration odors; antiperspirants reduce wetness. Both may be contained in the same product.

10. Shampoos: These soaps cleanse the hair and scalp; in addition, some condition the hair.

11. Sunbathing and suntans may enhance feelings of well-being, but some precautions are in order:

a. Sunbathing can be dangerous to light-skinned persons; overexposure to the sun can cause severe burns.

b. Suntan preparations and sunscreens block ultraviolet radiation to varying degrees.

c. Sunlamps help people maintain year-round tans; there are various types, and they must all be used with great care.

12. Products for the teeth.

a. Dentifrices (tooth pastes and tooth powders) are used to clean the teeth.

b. Tooth decay is the result of destruction of surface enamel by lactic acid (formed by the action of bacteria upon sugars from foods).

c. Bacteria organize into colonies called bacterial plaques. These can be disorganized by use of dentifrices and flossing.

d. Teeth should be thoroughly brushed and flossed every day, and cleaned by a dentist or dental hygienist periodically.

e. The use of fluorides in drinking water, topical applications, and dentifrices makes tooth enamel more resistant to decay.

13. Cosmetics should be tested periodically on a small area of skin. They should also be kept free of contamination.

14. Quackery is a multibillion dollar business in the U.S.:

a. A quack knowingly sells fraudulent products or methods to unsuspecting customers.

b. The young, the elderly, the very rich, and the very poor often turn to quackery.

c. Quacks take advantage of people's fears, ignorance, and gullibility.

d. Quackery is rampant in treatments for incurable diseases.

e. Sure-fire signs of quackery include: diagnoses by mail, free trial packages, limited supplies, proven cures, testimonials, physicians' and nurses' recommendations, and independent laboratory approvals.

f. Both state and federal agencies exist to control and eliminate quackery and to protect the public.

g. Consumer complaints: Make sure the problem is due to the product and not to your misuse of it. Then, follow established guidelines in filing complaints.

h. Quackery persists because it is highly lucrative, and quacks are difficult to prosecute. Many fraudulent operations are exposed only through consumer actions and complaints.

HEALTH SERVICES

Americans place high priority on freedom from disease. They spend more per person for health care than anyone else in the world. Yet many are not getting their money's worth. The worry over medical bills hangs like a cloud over the daily lives of many people. Many are bewildered about what kind of health-care professional to seek out for health counsel; or having located a professional, they find the fees asked are more than they are able to pay.

Fees charged by both physicians and hospitals are out of control. Too many health-care professionals view medicine and dentistry more as a means of attaining an elegant life style than as a compassionate ministry to human suffering.

The health-care professionals are not fully to blame. Some people make no attempt to understand health insurance. Many have not the foggiest notion that the more insurance coverage they use, the higher the cost of the insurance goes.

Private insurance, whether paid directly or as a wage-fringe benefit, encourages expensive treatment in hospitals and discourages preventive care in physician's offices. A national health-care program in the United States still is little more than a hope, while the one in Canada has been benefitting Canadians for many years. A rational means of financing adequate health services for all is vital to an improved general level of health in this country.

In this chapter some fundamentals of providing and paying for health care will be looked into.

THE HEALTH OF AMERICANS

A greater proportion of personal income is spent for health care in the United States than is spent in any other country in the world. Yet this country does not rank first in some of the key criteria of national health. In infant mortality, we presently rank behind sixteen other countries. In the number of years adults live, we rank behind seven other countries.

Half of the civilian population in the United States has one or more chronic conditions. Every month about three out of four American adults experience at least one episode of ill health or injury. But only one out of every three or four of these episodes is ever discussed with a physician. This is an enormous burden of illness.

Loss of Working Time

While many Americans receive excellent health care, others do not. Maladies that affect ability to earn a living are particularly important. In 1977, according to the National Health Survey, wage earners lost 315 million workdays as a result of some acute condition. This amounted to an average of 3.5 days off the job per worker.

But further, the survey revealed a distinct difference in days of disability between the low- and middle-income brackets. The average American had 17.8 days of restricted activity as a result of an acute and/or chronic condition. Persons from fami-

TABLE 12.1 INFANT MORTALITY BY COUNTRY

Rank	Country	Infant mortality per 1000 births
1	Sweden	7
2	Finland	8
3	Japan	8
4	Netherlands	8
5	Denmark	9
6	Norway	9
7	Switzerland	9
8	France	10
9	Iceland	11
10	Australia	12
11	Belgium	12
12	Canada	12
13	East Germany	13
14	Hong Kong	13
15	Luxembourg	13
16	Singapore	13
17	Spain	13
18	United Kingdom	13
19	*United States*	*13*
20	New Zealand	14

World	97
More-developed countries	20
Less-developed countries	109

Source: Population Reference Bureau, "1981 World Population Data Sheet" (Washington, D.C., 1981).

TABLE 12.2 LIFE EXPECTANCY BY COUNTRY

Rank	Country	Life expectancy at birth (years)
1	Iceland	76
2	Japan	76
3	Netherlands	75
4	Norway	75
5	Sweden	75
6	Switzerland	75
7	Canada	74
8	Denmark	74
9	Israel	74
10	Puerto Rico	74
11	*United States*	*74*
12	Australia	73
13	Belgium	73
14	Cyprus	73
15	France	73
16	Greece	73
17	Hong Kong	73
18	Ireland	73
19	Italy	73
20	New Zealand	73
21	Spain	73
22	United Kingdom	73

World	62
More-developed countries	72
Less-developed countries	58

Source: Population Reference Bureau, "1981 World Population Data Sheet" (Washington, D.C., 1981).

lies with an annual income of less than $3,000 averaged 34.6 restricted activity days, while those from families with incomes of $25,000 and over had an average of 12 days.

The average number of bed-disability days was 6.9. Those with incomes under $3,000 averaged 13.7 bed-disability days. Those with incomes of $25,000 and over averaged 4.4 bed-disability days.

A *bed-disability day* is one on which a person stays in bed for all or most of the day because of a specific illness or injury. A *restricted-activity day* is one on which persons cut down on their usual activities for the whole of that day because of an illness or injury. *Family income* includes the total income of all members of the family.

Causes of Death in the United States

The causes of death among different age groups sheds further light on this health-care scenario. The five leading causes of death, arranged by age group and ranked from one (most frequent) through five, are:

1. *Infancy (under 1 year of age):* (a) conditions of early infancy, (b) congenital defects, (c) respiratory distress, (d) sudden infant death syndrome, and (e) low birthweight

2. *Preschool children (ages 1–4):* (a) accidents, (b) congenital anomalies, (c) cancers, (d) influenza and pneumonia, and (e) homicide

3. *Elementary and junior high school students (ages 5–14):* (a) accidents, (b) cancers, (c) congenital malformations, (d) homicide, and (e) influenza and pneumonia

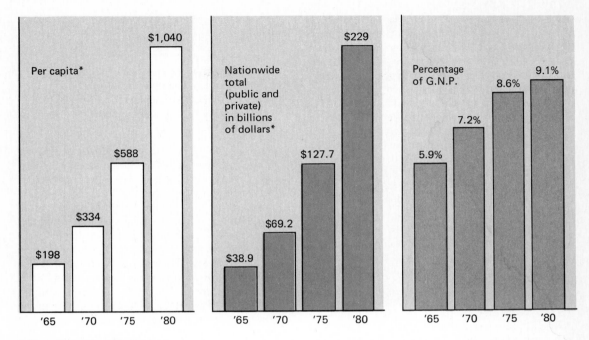

*Not adjusted for inflation.
Health care expenditures for fiscal years 1965-1980 (1980 figures are estimates).

4. *Senior high school and college-age persons (ages 15–24):* (a) accidents, (b) homicide, (c) suicide, (d) cancers, and (e) heart diseases
5. *Young adults and parents (ages 25–44):* (a) accidents, (b) cancers, (c) heart diseases, (d) suicide, and (e) homicide
6. *Middle-aged persons (ages 45–64):* (a) heart diseases, (b) cancers, (c) stroke, (d) accidents, and (e) cirrhosis of the liver
7. *The elderly (ages 65 and over):* (a) heart diseases, (b) cancers, (c) stroke, (d) influenza and pneumonia, and (e) hardening of the arteries

Notice that accidents top the list from childhood through age 44. The high incidence of heart disease from the teen years through old age tells its own story of shortcomings in preventive medicine. The ranking of homicide and suicide among adolescents and young adults is a sign of our failures in living with ourselves and each other. Obviously, many health needs are decidedly unmet in a country that spends hundreds of billions on health care.

THE HIGH COST OF HEALTH CARE

Health care is an increasingly costly item for the American public. The total national health bill now stands at an estimated $229 billion a year, or more

● **ASK YOURSELF**

1. Considering the loss of working time, what conclusion might you draw when comparing income brackets with the average number of bed-disability days? How many days was someone from your family bed-disabled last year?
2. What causes have any of your friends died from during the past several years? Within their particular age group, where do these causes rank in the five leading causes of death?
3. Do you have any ideas why the United States ranks where it does worldwide in infant mortality (rather than being number 1)?
4. If the United States spends the most per person in health care, why don't Americans have the greatest longevity worldwide?

than $1,040 per person per year. Beyond the effects of inflation, this is four times what it was 30 years ago. About $1 of every $11 spent in the U.S. today goes toward health care (compared with $1 of every $22 in 1950). At this pace, health costs will double every five years.

Consider some typical examples: In 1969, Massachusetts General Hospital charged $80 a day for a semiprivate room; now it charges $189 per day. Ten years ago a baby could be delivered at Manhattan's New York Hospital for $350; today, without medical complications, the cost is $2,800 ($1,300 going to the hospital). In 1965 the nation spent $38.9 billion for health care (5.9 percent of total spending for all goods and services). By 1980 it spent an estimated $229 billion (9.1 percent of total spending for goods and services). At this pace, health costs will also double every five years.

Each of the following are partly responsible for the escalation in health costs:

Hospitals

The excessive number of hospital beds (up to 10 percent are empty) and the unnecessary duplication of the latest costly equipment and personnel have been major factors. Hospitals have rejected mergers and efforts to reduce beds and duplicated services. Hospitals have exploited both public and private health plans that reimburse hospitals according to their charges rather than according to fixed schedules.

Physicians

Many physicians don't worry about the cost of services they perform or order performed in hospitals. Nor do some of them even know what these hospital costs are—particularly when treating patients who have insurance coverage. Physicians in general continue the fee-for-service system of charges, rather than cut costs through the prepaid group practice method. Furthermore, organized medicine has done little to correct the gross imbalance between excess numbers of physicians in the high-paying specialties and too few in the lower-paying primary-care field.

Insurance companies

As long as policyholders have tolerated rising premiums, private insurance companies have done little, if anything, to challenge cost increases by physicians, hospitals, and other health-care providers.

Labor/Management

Providing health insurance as a fringe benefit in salary contracts has removed workers' anxieties about medical bills and lessened their sense of responsibility for their own health.

The public

People in the United States have been spoiled into expecting the biggest, best, and most. They abuse their bodies and then want miracle-making medics to cure their ailments. If not cured, they sue, driving up the cost of physicians' malpractice insurance and, indirectly, the physicians' fees. People demand superequipped and superstaffed hospitals nearby that are open 24 hours a day. They haven't learned that it's impossible to have all these things and still control health costs. The cost of medical care is borne by everyone. It should be used discreetly and only at those times it is required.

PAYING MEDICAL BILLS

The traditional method of financing health costs was for each family to pay off its medical bills as they arose. This "pay-as-you-go" method forced some families to worry if the week's money would hold out. Others had to budget and set money aside. If all medical expenses were "average," it would be a bit easier to budget money each month for such a purpose. But medical expenses are often unpredictable, so budgeting becomes almost impossible. Some families are hit repeatedly by heavy medical expenses that go way beyond the average. As a result, there has been a greatly increasing trend toward group financing of medical expenses.

The health-insurance companies boast that over 63 percent of our civilian, noninstitutional population holds some form of health insurance. Studies show that medical care is somewhat cheaper in the form of a prepaid plan. In this manner people are protected against sudden large medical expenses by being forced to lay away funds systematically.

The current methods of collective financing of medical costs fall into two basic categories: public (tax-supported) and private (voluntary) health insurance.

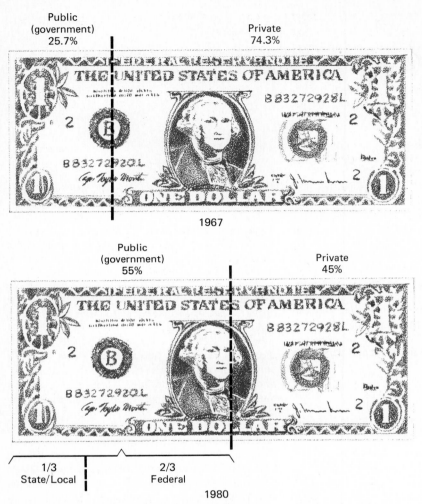

National health expenditure, 1967 and 1980.

PUBLIC HEALTH CARE

Public health care is provided through various government programs—local, state, and federal. The federal government spends about two-thirds of the health-care dollars, and state and local programs about one-third.

Federal Programs

Started in 1798, tax-supported public health-care programs have grown until today almost half of all health care in the United States is paid for by the government. Today seven major groups benefit from these programs:

Low-income persons (Medicaid) Under the Social Security Act, states receive matching federal funds to assist persons with low incomes. In 1978, the poverty level was estimated to be $3,900 for a four-person household. According to the Bureau of the Census, over 11 percent of all Americans live at the poverty level. A disproportionate number of these poor are nonwhite. While 8 percent of all whites are poor, 33 percent of all nonwhites live at or below the poverty level.

● ASK YOURSELF

1. Are there any ways in which you or your family have unnecessarily aided in the escalation of health costs during the past year?
2. How do your (or your family's) out-of-pocket health costs compare to your total income for the past year? How did health costs compare to total income years ago?
3. If there are any elderly members in your family, what benefits have they received from Medicaid or Medicare during the past year?
4. Have you availed yourself of any of the services provided by your local or country health department during the past year? If so, which services? Are there any other services offered you could make use of?

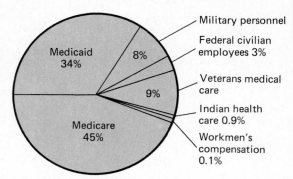

Percentage distribution of federal health care dollars, 1977.

Two-thirds of all Medicaid payments go to those 65 years of age and older and to families with dependent children. One-third goes to the blind, the permanently disabled, and to those whose incomes are below the poverty level.

Medicare *Medicare* is a government health insurance program under the Social Security Administration that helps the elderly (those 65 years of age and older) pay for medical care. It has two parts:

1. *Hospital insurance.* The hospital insurance portion of Medicare helps pay for medical care received by hospital inpatients and for certain follow-up services. It does not pay physicians' bills.
2. *Medical insurance.* The medical insurance part of Medicare helps pay physicians' bills, as well as a number of other medical items and services not covered under hospital insurance. This plan is voluntary and is financed by monthly *premiums* (the cost of the insurance) paid by those who enroll and by the federal government.

Military personnel and dependents All active and retired military personnel and their dependents are eligible for medical treatment at any Department of Defense medical facility.

Federal civilian employees Civilian employees of the federal government may participate in a variety of insurance programs.

Veterans medical care The Veterans Administration (VA) operates 172 hospitals, and many other clinics for the care of veterans. Priority is given to service-related ailments.

Indian health services The federal government provides health-care services to American Indians and Alaskan natives.

Worker's compensation medical care Worker's compensation insurance pays for the medical care of on-the-job accidents. Most of the cost is paid by the employer, but a small amount is paid by the federal government.

State and Local Programs

State health agencies, either directly or through county or local health departments, maintain health programs. These services may include: maternal and child health, communicable disease, dental health, mental health, maintenance of vital records, air pollution control, and public health education.

PRIVATE HEALTH INSURANCE

Private health-insurance programs exist because individuals cannot successfully budget against the potential costs of all illnesses or accidents. This is especially true of illnesses, such as cancer, that may require long hospitalization and that could easily bankrupt the average family. One hope for protection against these larger medical expenses, then, lies in large numbers of persons "pooling" the risks through health plans. This way insurance companies can spread both the risks and the costs.

Unlike public insurance plans, the costs of private health coverage are borne solely by the beneficiaries or their employers. An advantage of private health coverage is that it permits the insurer and the insured to agree upon specific provisions and types of coverage. Most private health-care policies now terminate at age 65, or when the insured person becomes eligible for Medicare.

Types of Policies

Several types of health insurance coverage are provided by insurance companies.

Hospital expense insurance This provides specific *benefits* (the services for which the insurance plan makes payments) for daily hospital room, board, and usual hospital services and supplies during the hospital stay. Room-and-board benefits may be provided in two ways: (1) reimbursement for actual room-and-board charges up to a stated maximum amount, or (2) a service-type benefit that equals the semiprivate room-and-board charge of the hospital.

Surgical expense insurance This provides benefits for the cost of surgical procedures performed as a result of an accident or sickness. Benefits might be paid according to: (1) a preset schedule of surgical procedures, or (2) by paying the physician's fee up to a "reasonable and customary" charge for the procedure performed.

Physician's expense insurance This provides benefits that help pay physician's fees for nonsurgical care in a hospital, home, or physician's office.

Major medical expense insurance This insurance provides large amounts of coverage for major unpredictable expenses. They are not designed or intended to pay for smaller medical expenses that can be easily paid out of pocket, or that can be covered by a regular health insurance plan. Major medical plans usually contain a deductible clause that excludes payments on medical expenses below a given amount (often $100 to $500).

Disability income insurance Disability income benefits in accident and health policies provide for the partial replacement of income lost by employed

● ASK YOURSELF

1. What type of health insurance are you covered by?
2. Who pays the premiums on your health insurance: school, employer, union, parents, yourself?
3. For what types of health care would major medical insurance be of the greatest benefit?
4. Do you feel your health insurance is adequate? If not, how would you provide yourself or your family with better coverage?
5. If you or your partner become physically disabled, are you covered by any sort of disability insurance?

persons as the result of accident or illness. The amount of the benefit varies depending on: (1) whether the disability is total or partial, and (2) how long the benefits are payable.

Dental expense insurance Dental expense insurance reimburses expenses for dental service and encourages preventive care.

Types of Benefits

Insurance companies pay benefits according to two general plans:

Service plans These are generally in the form of contracts between the insurance company and the hospital or physician. The hospital agrees to provide certain services to any policyholder, who must present a policy identification card.

Cash indemnity plans Cash indemnity plans pay benefits in the form of cash directly to the policyholder. The insured person is paid a specified amount toward the covered expenses, unless he or she assigns the benefit to be paid directly to the hospital or physician. No contracts exist between the insurance organization and the hospital or physician.

Types of Subscriptions

Persons wishing to buy a health insurance policy may do so as individuals or as members of a group.

UNITED STATES BUSINESS (in billions of dollars).

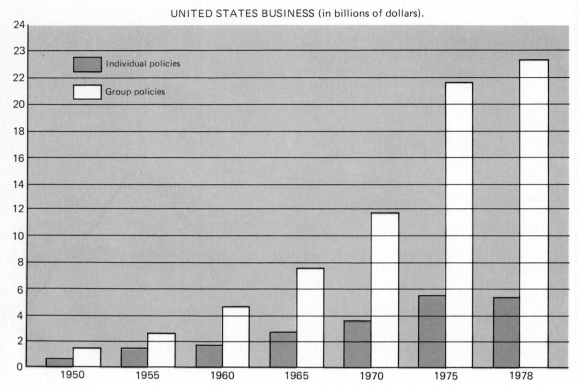

Ratio of health insurance premiums of insurance companies by type of policy. (Adapted from the Health Insurance Institute, Source Book of Health Insurance Data, 1980–1981 [Washington, D.C., 1981].)

Individual subscriptions Individuals may buy health insurance and choose the type of "tailor-made" coverage they want. This approach generally costs more than when similar coverage is obtained through a group.

Group subscriptions The most favorable premium rates are given to groups of subscribers. Formed by a number of individuals or families, group plans are generally available through an employer, a union, or professional organization.

Types of Insurers

There are over 700 different private health insurance companies in the country today. Their plans fall into several main categories:

Blue Cross and Blue Shield These are nonprofit health-care membership groups. Organized on a geographic basis, each plan offers individual and group coverage. Set up on a service-type basis, they contract to reimburse hospitals and physicians for covered services.

Blue Cross, through seventy-four plans in the U.S. and Canada, insures members against costs of hospital care. Even though all these plans bear the name Blue Cross, each is sold exclusively within a given geographical area and is governed independently by an area board. The coverage ranges from several weeks to a full year of hospital service. The most widely sold policies cover the partial or complete cost of thirty days of hospital care. Some plans also cover services of physicians, drugs, and laboratory tests. Most plans now include some major medical coverage.

Blue Shield, through sixty-nine different geographical plans, pays for surgical and medical services performed by a physician.

Points to Look for When Buying Health Insurance

Insurance companies can reduce their obligations to their customers in a number of ways. Here are some crucial provisions to look for when buying health insurance:

1. *Insuring clause.*   Be sure the coverage includes the types of benefits you desire.
2. *Persons covered.*   Usually policies cover spouses and unmarried dependent children. These names must be listed on the policy.
3. *Age limits.*   Some policies specify maximum and minimum age limits. Most policies will cover a dependent child only up to a stated age. Coverage may end when you reach age 65.
4. *Waiting periods.*   There may be a time interval between the beginning of the policy and the date selected benefits are payable. Examples might be maternity and preexisting illnesses.
5. *Maximum benefits.*   Maximum benefits should be no less than $10,000.
6. *Deductibles.*   These should not exceed $100 on hospital insurance and 5 percent on major medical. The fewer deductibles the better.
7. *Copayments.*   A copayment is where the policyholder pays a percentage of the cost. These should not exceed 20 percent of any claim.
8. *Exclusions.*   The fewer the better. Many policies exclude cosmetic or elective surgery and preexisting illnesses.
9. *Conversions.*   Children or spouses covered under a family plan may wish to convert to an individual policy due to moving away from home, death, or separation.
10. *Premiums.*   Policy rates should not go up solely with your age.

Health Maintenance Organization (HMO) The HMO is a group of physicians, surgeons, dentists, or optometrists who provide care on a contract basis with subscribers. In return, the subscribers make fixed periodic premium payments. The HMO provides complete medical care, including medical, dental, and hospital services; outpatient surgical care; acute mental illness care; and ambulance service. HMOs may be sponsored by hospitals, county medical societies, corporate industries, commercial insurance companies, Blue Cross/Blue Shield, unions, consumer groups, and government bodies.

Commercial Most commercial health insurance plans are written by independent insurance companies and one usually of the cash-indemnity type. These companies do not set up contracts with hospitals or physicians. Commercial insurance plans are expected to make a profit for the company.

PURCHASING HEALTH INSURANCE

The novice can easily become confused when first buying health insurance. However, such a transaction should not become a battle of wits between the buyer and seller. In purchasing health insurance, the buyer must be careful not to end up with a policy that does not give needed coverage or requires payment for unneeded provisions. (Many employers automatically include health insurance as a fringe benefit, and an employee does not have to decide what provisions to choose.)

Generally, purchasers of health insurance should first determine: (1) the type of health-care expenses they want to be protected against, and (2) the amount of insurance they want on those health-care expenses. They should consider the other regular payments they are already making, such as auto loans, rent and loan payments on a home, education bills, furniture payments, and so on. Their

● ASK YOURSELF

1. Are you certain you are covered by health insurance? If you are still covered by your family's insurance, at what point do you lose this coverage?
2. Examine your, or your parent's, policy for the provisions you should have. Could you or your parents use better insurance?
3. In what way might conversion privileges in a policy be important to you?
4. In what way is it important to you to have a policy that is guaranteed renewable?

monthly budget should be able to cover the expected health insurance premiums. The family with many financial obligations needs health insurance as much as, or more than, the family with few financial obligations. Financially obligated persons who become burdened with huge unexpected medical bills that prevent them from keeping up car, refrigerator, or house payments stand to lose everything through repossession or foreclosure. Therefore, they particularly need adequate health insurance. Health insurance must not be something they plan to buy after they have paid all their other bills.

Some states, such as New York, now prescribe minimum benefits for certain types of policies. Policies failing to meet these standards must be so labeled. Check with a reputable insurance broker to see what the recommendations are in your state.

Premium and Renewal Provisions

Once the decision to buy insurance has been made, look into the premium and renewal provisions. Individual policies are categorized according to the following points:

1. *Noncancelable and guaranteed-renewable policies.* These policies are rarely sold today. They can be renewed to a certain age or for life, and the premium cannot be increased.
2. *Guaranteed-renewable policies.* These policies cannot be cancelled by the insurance company, but the premium can be increased on a statewide basis.

3. *Selectively renewable policies.* These can be dropped by the company for specified reasons other than changes in your health. Premiums are also subject to change.

The three most common health-care expenses are hospital, surgical, and diagnostic (X-ray and laboratory) expenses. Find out what the prevailing costs of these services are in your area and what benefits are offered by specific policies.

Your own physician's advice in the selection of a plan can be very useful. He or she can advise you of the record of a specific company in honoring its commitments.

It is wise periodically to review your policy. Changes in income, marital status, obligation to dependent children, and employment might substantially affect your policy and your needs for certain types of coverage. It is particularly important to be aware of employee plans that are carried by your employer. A change of job might terminate your coverage or might provide coverage that would make the insurance that you are presently paying for unnecessary.

Regardless of how a family provides for its medical care, it is obvious that some form of planning is essential. Most people are directly dependent on a continuous income, and financial protection against physical disability is a very important asset. Not only must a family protect its ability to earn money, it must also protect its savings and investments against large, unanticipated medical and hospital costs. A well-planned, well-balanced health-insurance program can do just this.

THE PEOPLE IN MEDICINE

The American physician belongs to a profession that is facing a deepening dilemma. While medical science is more prepared than ever to deal with disease, many people question its integrity and motives.

Physicians are better trained today than ever before. Acceptance into medical school is more competitive and medical training is more exacting. State licensing-board examinations are more difficult, and more states require in-service training to keep physicians updated.

N represents employed nurses
P represents all active physicians
in the United States and its territories
and possessions
D represents dentists in private practice

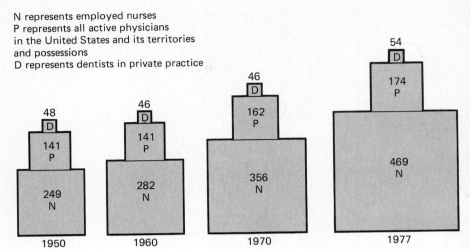

Estimated number of health professionals in the U.S. (per 100.000 population)

According to a recent Harris poll, almost 90 percent of the American public is pleased with the general quality of health care. Yet many polled feel that physicians are too well paid. More than half think the government should control physicians' fees. Over 80 percent think physicians should be required to disclose their fees and charges. The average physician now earns more than $75,000 a year.

Overall, the supply of physicians is up, and the government predicts a surplus by 1990. Today in the United States there is one physician for every 490 people, the highest ratio ever (in Zaire there is one physician for every 28,000 people). About 10 percent of all physicians in the United States are women.

Only one in six physicians (16 percent) in the United States is in general practice. Twenty-five percent of them practice in groups of two or more. Almost 40 percent provide primary (basic) care, while 60 percent specialize (restrict) their practice in some way. Specialists tend to cluster in well-to-do urban areas.

This situation is about to change, however. The U.S. Department of Health and Human Services (HHS) is pushing for more preventive medicine. New self-care programs are helping individuals assume greater responsibility for their own health. Holistic medicine is stressing the treatment of the whole person instead of isolated parts (as is done by many specialists). Many lay people want more objective information on what health services are available and who is best qualified to handle their care.

Requirements for the M.D.

Before being licensed to practice medicine, aspiring physicians must meet certain professional and ethical requirements. Although training standards vary somewhat from state to state, they must take three to four years of premedical college work, and then complete a three- or four-year training program in a medical school approved by the Association of the American Medical Colleges and the Council on Medical Education of the American Medical Association. In addition, most states require them to serve a one-year internship to gain hospital experience. To practice in a given state, physicians must be licensed by a board of medical examiners. This license is granted only after they pass either a state or a national board examination. If they desire to take a residency in a hospital to receive advanced training, or to meet the requirements of a given specialty in medicine, they must spend two to five additional years in training.

Standards for the training of physicians, as well as for the ethics of medical practice, are set by the medical profession itself. Local or county medical

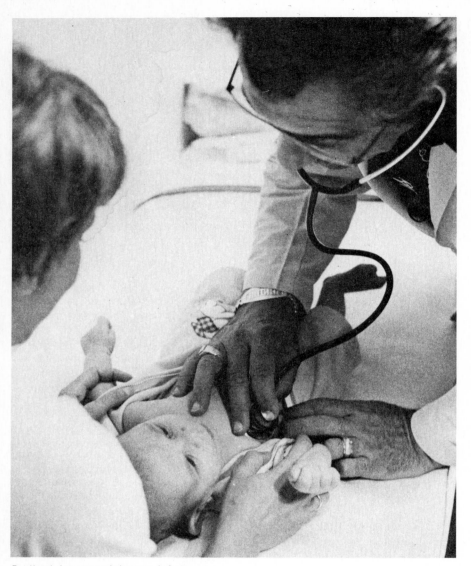

Pediatrician examining an infant.

societies assume the responsibility for regulating the professional conduct of their members. To do this, societies adopt ethical and professional standards and expect their members to adhere to them.

The Family Physician

Everyone should have a physician who can provide him or her with *primary care*. Such a physician may be known as a personal, or family, physician. The care family physicians provide is comprehensive and covers most medical needs. Such primary care may be provided by one of several types of physicians:

General Practitioners (GPs) General practitioners are licensed physicians who have no advanced medical training leading to a particular specialty. While some GPs may restrict their practice to certain types of care, they should not be confused with specialists who hold a *diploma* in a given specialty.

Keeping well-informed in many areas of medicine has been a problem for GPs. A high-quality practice demands keeping up with new medical information and techniques. This problem has led to the medical specialty of *family practice.*

Family Practitioners Family practitioners have completed a hospital residency in family practice and are diplomates of the American Academy of Family Practice. Training programs vary. Some residencies emphasized internal medicine, psychiatry, and surgery, whereas others stress obstetrics and pediatrics. Regardless, their specialty prepares them to provide primary care for individuals and families.

Pediatricians Pediatrics is a specialty that provides primary care for children from birth to adolescence. Pediatricians advise parents, oversee general physical and mental growth, treat diseases, and give immunizations.

General internists Specialists in *internal medicine,* general internists may provide primary care to adults. Some general internists choose not to provide primary care. Since internal medicine is directed primarily toward diagnosis, they are also equipped to coordinate the work of other specialists needed to treat specific problems people may have.

Other Specialties

Some physicians choose to be proficient in one body system or one area of medicine to the exclusion of general medicine. Such specialization is necessary for a number of areas such as neurosurgery, radiology, ophthalmology, orthopedic surgery, internal medicine, and psychiatry. Bona fide specialists must have completed regular medical school, served hospital internships, completed hospital residencies in the specialty of their choice, and passed the written examination of that particular specialty. They then receive certificates designating them as diplomates of the board of that specialty.

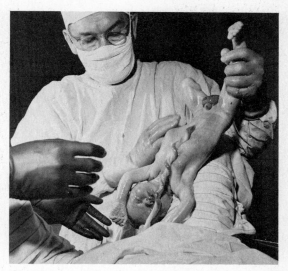

Obstetrician delivering a baby.

Brief descriptions are given here of several specialties:

Obstetrics and gynecology *Obstetrics* is the care of women during pregnancy and in child birth. It is frequently combined with *gynecology,* the care of women's genital disorders.

Surgery This specialty involves surgically operating on a patient to correct some physical condition. It is subdivided into specific areas, such as neurosurgery, thoracic surgery, orthopedic surgery, and abdominal surgery.

Psychiatry *Psychiatrists* are M.D.s who deal with emotional illnesses and disturbances and mental retardation by using verbal contact as well as drugs and other therapies. Some are also specialists in neurology. By distinction, *psychologists* generally hold a graduate Ph.D. or M.A. degrees (nonmedical degrees) and may engage in experimental, teaching, or clinical work. *Clinical psychologists* have had further training in a medical setting and diagnose and treat emotional and neurological disorders by the use of verbal/nonverbal methods rather than by medical measures. (By contrast, a *counselor* is anyone who gives advice to people with normal problems.)

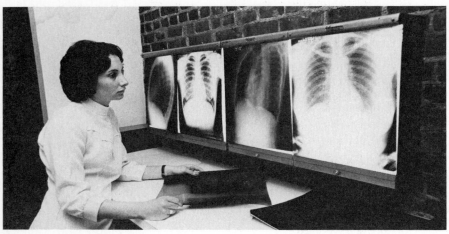

Radiologist with chest X-rays.

Some other medical specialties are:

1. *Anesthesiology,* the science of administering general and local anesthetics
2. *Dermatology,* the science of treating diseases of the skin
3. *Neurology,* the science dealing with physical diseases of the brain and nervous system
4. *Ophthalmology,* the medical branch treating the eye and its diseases
5. *Otorhinolaryngology,* the medical branch treating diseases of the ear, nose, and throat
6. *Pathology,* the study of the disease process, including the examination of functional changes in tissues and organs of the body, and identification of the disease causing the change, such as cancer; the pathologist may serve as a coroner and perform autopsies
7. *Proctology,* the medical branch treating diseases of the rectum and anus
8. *Radiology,* the science of using X-rays, radium, and other radioactive materials for the diagnosis and treatment of disease
9. *Urology,* the science of treating diseases and abnormalities of the urinary tract in the female and the urogenital tract in the male

10. *Emergency Medicine,* the administering of emergency medical procedures of all kinds; practice usually confined to the emergency room of a hospital

Overall, there are over thirty recognized fields of medical specialization today. Each is governed by its respective board for purposes of examination and certification.

Osteopathy The doctor of osteopathy (DO) is a physician who receives medical and specialty training similar to that of the M.D. except that the training is in an approved osteopathic college, and internship is in a hospital approved by the American Osteopathic Association. State licensing examinations are like or similar to those M.D.s take. Osteopaths use musculoskeletal manipulation in combination with all other accepted diagnostic and therapeutic practices. Only about 25 percent of DOs complete the requirements for an osteopathic residency. Osteopaths constitute about 4 percent of all active physicians in the country.

Selecting the Right Physician

In selecting medical care, individuals must determine their personal or family needs and then must find out what is available. Individuals belonging to independent health groups may have to choose a physician from that group.

Some people prefer a general practitioner, while others select several specialists in selected branches of medicine. Since some general practitioners restrict themselves to a narrow branch of practice, it is necessary to know the nature of a physician's practice. If your family includes young children, the physician you choose should enjoy working with children. The same point applies to elderly people. When you cannot find one ideal physician for the entire family, you should choose several specialists, such as a pediatrician for children and an internist for adults.

In selecting a private physician, either for general practice or to deal with a specific problem, certain procedures are worth following:

1. *Consider the reputations of various physicians in your community.* Contact a local *accredited* hospital for the names and addresses of the physicians who practice through that hospital. There is usually a relationship between the quality of a hospital and the quality of the physicians who practice there.

   You may wish to look into the educational credentials of a particular physician. These are known to the local or county medical society, or the local hospital in which he or she practices, or are listed in the *AMA Directory* or in *The Directory of Medical Specialists.* These sources will tell you whether the physician is licensed to practice in your state, where he or she has taken basic medical training, and where and when he or she has taken postgraduate work. These books may be found in some public libraries or in a hospital library.

   If you are considering physicians who have been recommended by friends and relatives, you should be prepared to do some independent investigating. Other people's attitudes toward their physicians and their illnesses might not be useful to you. It is wise, however, to stay away from physicians who consistently cause dissatisfaction among people whose judgment you respect.

2. *Visit the office of the physician you are considering.* The office should be within easy

● **ASK YOURSELF**

1. Do you have a family physician?
2. Does your library have *The Directory of Medical Specialists?* If so, see if you can find a listing for a specialist you or your family may have used.
3. What type of physician would you select to provide you or your family with primary care?
4. Since almost any physician could claim to be a specialist, what proof might you ask from a so-called specialist to confirm that he or she is a bona fide specialist and has the necessary additional training?

traveling distance. Find out if the physician is accepting new patients. See if the office is neat, clean, and orderly. Discuss with the physician such general questions as whether he or she can furnish you or your family general medical care: whether he or she makes house calls; if he or she is usually available for emergencies; what other physician provides these services in the event the physician is out of town; the fee schedule (you can later compare it to that of other physicians in town).

The physician should strike you as a person you can confide in and who appears interested in your family's health and well-being. If you are satisfied regarding these points, you have found your physician.

The Patient-Physician Relationship

Patients are entitled to receive careful, professional service from their physicians. It may not always be possible for the physician to cure, but the patient should feel confident the physician is using the best skills.

There are some steps you as the patient should take if you wish to actively participate with your physician in your health care.

1. *Come prepared.* The physician will want to learn your medical history. The history is usually organized into these categories:

   The chief complaint and the present illness

1. Assume that you have moved to a new town and need to find a physician. How would you locate one you would be happy with?
2. In what ways have you not been fully satisfied with the care you have received from your physician? What could you do to improve this relationship?
3. Are there any ways in which you might have been a better patient for your physician in the past? List some.
4. Have you been to a dental specialist? If so, what was the specialty and why did you go to that specialist?
5. If you feel that either your physician or dentist is charging you unfairly, what approach might you use to discuss the matter?

Your past medical history
The family medical history
Your personal and social history
A review of your major body systems

2. *Ask about any diagnostic tests the physician may order.* You have the right to understand tests being called for as well as:
   The purpose of the tests
   What risks are involved
   Their cost
   How long you ought to wait for results
   After the results are in, what are the meanings of the exact values, and what the normal range of values should be
   The physician's view of things

3. *Ask about all available forms of treatment.*

4. *Feel free to seek a second opinion from some other physician* before agreeing to any costly or potentially dangerous diagnostic procedure, or any surgery other than very minor surgery.

5. *If any drugs are presribed, ask about:*
   The drug
   Whether asking for the drug by its generic instead of its trade name will lower the cost of the prescription
   The drug dosage

6. *Be sure you understand what you are to do before leaving the office.* If not, ask the physician to write out the instructions.

7. *Your medical record should be available to you.* Ask to go over it with your physician if you have questions.

8. *Feel free to ask about fees, preferably in advance.* You are entitled to a full explanation of any charges. Ask for an itemized bill. If the cost of the care is imposing a genuine hardship on a patient or the family, the physician will want to know about it and may be willing to adjust fees. The patient should feel no embarrassment about raising such a discussion.

In return, the physician may expect certain courtesies and cooperation from patients. In non-emergencies, patients should make appointments and then keep them punctually. They should be prepared to pay medical bills promptly or make arrangements to pay them as soon as possible. Patients should follow the physician's instructions exactly. If medication is prescribed it should be obtained immediately and taken as directed. The physician has a right to expect the cooperation of patients.

Physicians are not necessarily under obligation to answer emergency calls from unknown individuals late at night. The physicians will have no medical history of the patient, may be subjecting themselves to physical hazard, or may be greatly fatigued. Some individuals moving into a new community may fail to contact a local physician until they need one in emergencies or late at night. Then, if they have difficulty in obtaining care, they make complaints against the medical profession. In such emergencies, the nearest emergency hospital should have physicians on duty who can provide care.

A physician can give the best service when confident that patients appreciate the service. As much as they would enjoy being able to cure every ailment, medical research has not provided all the answers to accomplish that aim. But a good physician will go just as far in diagnosis and treatment as ability, training, available facilities, and patients allow.

Dentistry
Dentistry treats ailments or abnormalities of the gums and teeth and attempts to prevent their recurrence. As in preventive medicine, there is increas-

ing emphasis on the prevention of tooth diseases by routine cleaning of teeth, fluoridation of drinking water, and adequate brushing. Most dentists are general practitioners who provide basic dental care. The several specialties in dentistry are:

Orthodontia: Straightening of teeth
Oral surgery: Extraction of teeth and surgical procedures
Pedodontics: Dentistry for children
Prosthodontics: Providing artificial replacements for missing teeth
Endodontics: Root-canal therapy
Oral pathology: Treating diseases of the mouth
Public health dentistry: Promotion of oral health through public efforts

Nursing

The nursing team is led by the professional, or *registered, nurse* (RN); it also includes practical (vocational) nurses, nursing aides, medical technologists, orderlies, and attendants.

There are three types of training programs for registered nurses. *Diploma* programs are conducted by hospitals and independent schools and usually require three years of training. *Associate degree* programs are usually located in community colleges and require approximately two years of nursing education. *Bachelor degree* programs usually require four years of study in a college or university, although some require five years. Additional training can be taken to qualify for a nursing specialty in obstetrics, pediatrics, psychiatry, or surgical nursing.

Registered nurses have traditionally handled the largest share of nursing services. They have administered medications and treatments prescribed by physicians; observed, evaluated, and recorded symptoms and the progress of patients; assisted in patient education and rehabilitation; improved the surroundings of the patients, and instructed other medical personnel and students. Some registered nurses after advanced training, become Nurse Practitioners and perform physical examinations and certain other services usually performed by physicians.

Many registered nurse duties are being delegated to the Licensed Practical Nurse (LPN),

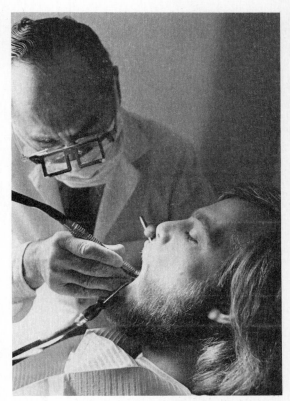

Dentists are emphasizing preventive dentistry.

sometimes known as the Licensed Vocational Nurse (LVN). Most LPNs work under the supervision of an RN in hospitals. Others work in private duty, offices, public health, industrial plants, and nursing schools.

Ancillary Professional People

Various other professional people serve in an auxiliary, or *ancillary,* manner on the health team. *Midwives* are professional people who help care for women during pregnancy, labor, delivery, and the postpartum period. The *nurse-midwife* is a registered nurse who has completed advanced training in midwifery. Successful completion of the American College of Nurse-Midwives certification examination entitles the nurse-midwife to use the title certified nurse-midwife (CNM). Nurse-midwives

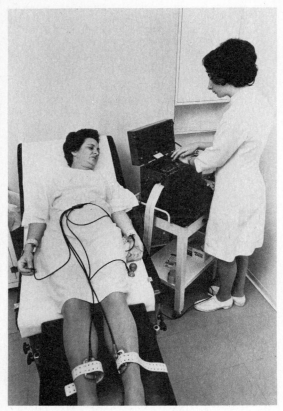

As centers for the healing arts, hospitals are one of the nation's largest industries.

FACILITIES FOR PATIENT CARE

As medical techniques have improved, so have facilities for patient care. Today's physician makes few home visits, particularly in urban areas.

Hospitals, Emergency Rooms, and Other Facilities

There are more than 7,300 hospitals of all kinds in the United States. They fall into several basic categories.

Voluntary hospitals These are private community hospitals set up on a nonprofit basis. The most common type of hospital in the country, they usually contain from 50 to 400 beds. Established by local communities, charitable organizations, and churches, they are often built with government funds. Run by a governing board of community leaders, they operate on money from patients, insurance claims, and public programs such as Medicare and Medicaid. Many of them are very adequate for the care required; most are accredited.

Proprietary hospitals These are private hospitals set up to make a profit. Sometimes called *community hospitals,* they are owned and operated by individuals or corporations. Some will admit only the more profitable types of hospital cases, taking few, if any, charity cases. The majority of these hospitals are small, and only about one-third of them are accredited.

Public hospitals This category includes city, county, public health service, military and Veterans Administration hospitals. They are generally large and contain from 500 to 1,000 beds. Some provide general care, others are restricted to certain persons or kinds of conditions. Many have the unfair reputation of providing service to poorer economic classes. They have permanent full-time staffs and physicians present at all times. Being larger, they offer more services. Many have physician interns and residents, which often indicates the presence of teaching physicians, and thus better care.

Teaching hospitals These hospitals are associated with medical schools. Large, they often range from 300 to 2,000 beds. As teaching facilities, they

function as a part of the obstetrical team of medical centers. In contrast, the *lay,* or *granny, midwife* provides assistance to women during childbirth in the absence of a medical practitioner. Trained largely through apprenticeship, lay midwives generally serve in low-income and rural areas, where the delivery of the baby usually occurs in the home.

The *physician's assistant* is qualified by academic and practical training to provide patient services under the supervision and direction of a licensed physician who is responsible for the performance of that assistant. Following a period of training, physicians' assistants work in physicians' offices, hospitals, or clinics. Graduates are also known as *physician's associates, clinical associates, MEDEXs, child health associates,* and *community health medics.*

always have interns and residents on the wards. Because of the access to superb technical resources, some of the leading events in surgery and therapy often take place in these hospitals. Patients involved in teaching exercises must sign a consent form which states that they understand the implications.

Hospitals are expensive. They are more than motels or spas. Acute general hospitals provide excellent, well-trained staffs with access to the best diagnostic and surgical facilities. Except for emergency cases, most of them accept patients only on the recommendation of the admitting physician, who must be a staff member of the hospital.

Emergency Rooms Emergency rooms are "physicians" for many patients. Not knowing where else to go, people use them for both emergency and nonemergency cases. They are accessible twenty-four hours a day and always have staff on duty or on call.

However, emergency rooms provide little or no continued care. A patient may not see the same physician twice in a row. The physician may not take adequate time to complete a full examination. Patients may wait for hours. Fees usually run higher than for a similar visit to a standard office. Health insurance may not cover the cost.

Short-term Surgery Centers These are facilities designed for surgery requiring a short stay (often overnight). Because the overhead is less than in a general hospital, the charges are often less.

Convalescent Facilities Nursing homes and various rehabilitation facilities provide less expensive care than that found in a general hospital, but care that is difficult to manage at home. Some provide excellent care, others are disreputable. The best are comfortable and homelike with fine nursing staffs and regular physician visits. The worst are understaffed, smelly, and pay little attention to critical medical problems. Before arranging for the admission of a relative to a nursing home facility, visit it yourself or have someone who is knowledgeable visit it. Check into fee schedules, menu planning, quality of food, and activities programs.

Free Clinics Free clinics have appeared in many parts of the country to deal with undesired pregnan-

● **ASK YOURSELF**

1. A child can be delivered by a family physician, an obstetrician, or a midwife. Which one would you prefer for your next delivery, and why?
2. Suppose that this evening you were seriously injured in an auto accident near where you live. What hospital would you be taken to? How good is it? Is it accredited? How would your care be paid for?
3. Does your hospital display a JCAH Accreditation certificate in its lobby or business offices? What might such a certificate tell you about your hospital?
4. If any elderly members of your family are residents of a nursing home or some other nonhospital facility, what is their care costing? How does that compare with semiprivate rates in your local hospital?

cies, drug use, and sexually transmitted diseases. Some have appeared because general medical care is poor or inaccessible; others have come into existence out of a basic sympathy for the needs of the patients. Some have provided a very important service for many people. Some operate with limited facilities and volunteer help, and are financially unstable. Many deserve stronger community support than they have so far received.

Accreditation Hospitals and hospitallike facilities are accredited by the Joint Commission on Accreditation of Hospitals (JCAH). The commission sets national standards for hospital care, accredits hospitals meeting these standards, and periodically reviews accredited hospitals. Upon application, a hospital is examined for cleanliness, laboratory operations, food handling, and records, as well as for the practice of its staff physicians. The hospital may be accredited for one or two years depending on how well it qualifies.

It is increasingly important for a hospital to have JCAH accreditation. A nonaccredited hospital may not train interns, residents, or nurses. Some medical insurance companies have refused to make payments to nonaccredited hospitals. Today, hospitals representing over 89 percent of the country's total hospital bed space are accredited. Most of these hospitals are voluntary ones.

IN REVIEW

1. Americans spend more on health care than anyone else in the world, yet America does not rank first in many key areas of health and health care.
   a. America ranks nineteenth in infant mortality and eleventh in life expectancy.
   b. Illness among Americans accounts for a great loss of working time.
   c. The major cause of death from childhood to age 44 is accidents.
   d. Suicide and homicide are major causes of death among young adults.
   e. The major causes of death for all age groups are heart disease and cancer.
2. The estimated cost for health care in the nation is $229 billion a year. Health costs are expected to double in five years. These high costs are caused in part by
   a. *Hospitals.* Hospitals resist efforts to regulate the costs for services, and often exploit public and private health plans.
   b. *Physicians.* Like hospitals, physicians often exploit existing health plans. There are gross excesses between the number of highly paid specialists and lower paid general practitioners.
   c. *Insurance companies.* Because there is little resistance to rising premiums among policy holders, insurance companies do little to reduce the rising costs of health care.
   d. *The public.* Americans do not do enough to insure their personal health through prevention and healthful lifestyles.
3. The traditional "pay-as-you-go" method of paying for health care has been replaced by group financing of medical expenses. It is estimated that 63 percent of our civilian population is covered by some form of medical insurance. There are two basic types of health plans—public health programs and private health insurance.
4. Public health care is provided through local, state, and federal programs.
   a. *Medicaid* is a federal program that provides matching funds to states to assist persons with low incomes. Approximately 11 percent of Americans live at or below the poverty line.
   b. *Medicare* is a federal program under the Social Security Administration that provides health care funds for the elderly. *Medicare* provides both hospital and medical insurance.
   c. Other federally sponsored health programs include medical care for military personnel and their families, veterans, health care for Native Americans, and workman's compensation for work related illnesses and injuries.
5. Private health insurance programs are funded by beneficiaries and employers and are provided by insurance companies.
   a. Various types of policies cover: hospital, surgical, physicians', and major medical expenses, as well as dental costs and disability income insurance.
   b. Insurance companies pay expenses according to two plans—service plans and cash indemnity plans.
   c. Individuals may subscribe to health plans as individuals or as members of a group.
   d. Over 700 companies provide health insurance plans. The largest of these are Blue Cross and Blue Shield and Health Maintenance Organization (HMO).
6. When buying health insurance, purchasers should consider the type of health expenses they want to be protected against and how much coverage they want.

Purchasers should consider premium and renewal provisions. Purchasers should review their health plans periodically.

7. America has an average of one doctor for every 490 people. Most physicians are located in well-to-do urban areas. There is a shortage of medical personnel in rural areas.
   a. Physicians are licensed to practice in the states in which they work.
   b. Physicans undergo four years of training in medical school and some period of internship to gain hospital experience.
   c. Primary care is provided by general practitioners, family practioners, pediatricians, and general internists.
   d. Some medical specialties are surgery, obstetrics and gynecology, psychiatry, and osteopathy.
   e. Dentistry is the treatment of problems of the teeth and gums. Some dental specialties are orthodontia, oral surgery, pedodontics, and prosthodontics.
   f. The field of nursing has many specialties ranging from nursing aides to Nurse Practitioners.
8. As medical technology has developed, so have the ways in which medical services are delivered. Home visits by physicians are no longer common.
   a. Services are provided at many kinds of hospitals, emergency rooms, short-term surgery centers, convalescent facilities, free clinics, and in physicians' offices.
   b. Hospitals are accredited by the Joint Commission on Accreditation of Hospitals (JCAH).

SECTION II

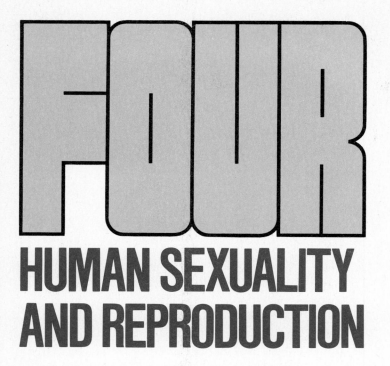

FOUR
HUMAN SEXUALITY AND REPRODUCTION

We are sexual beings from birth to death. Human sexuality and reproduction contain more myths than any other part of human existence. Some major aspects of our sexual lives are sexual behavior, partner selection, and the decision to reproduce.

Our behaviors develop through emulation. But most often the sexual behavior of parents is not openly available to their children. Therefore, unlike most behaviors, sexual behavior is taught, not emulated. Sexual education is needed so that all may learn to separate myths from facts. Sexual behavior can be extremely pleasurable or it can be extremely painful. Exposure to all aspects of sexuality would allow each of us to freely select those aspects of sexual behavior that are relevant to ourselves.

Sexual partnerships can last for the entire adult lifespan. Thus, great care should be used in the selection of the partner, and the type of partnership, that each of us chooses. Sexual partnerships can be the major joy of life.

Biologically, the only reason we are alive is to reproduce. This should be a most natural process in life. Yet, because of social, cultural, and economic reasons, the decision to reproduce has become a major life decision. Some people want children, but cannot reproduce because of medical problems. Others can produce children, but do not desire them for various cultural and social reasons. Both should have their desires fulfilled. Reproductive education can help both groups to fulfill these desires.

CHAPTER 13
HUMAN SEXUAL BEHAVIOR

Only in recent years has human sexuality become a "respectable" field for physiological and behavioral research.

Objective sex research is of value to society for several reasons. It places sexuality in its proper perspective as a normal, healthy part of human physiology and behavior. It removes the cloak of secrecy from sexuality. It reassures people that their own sexual feelings, responses, practices, and problems are typical of millions of other people, or if not, how they differ. It helps to dispel anxiety-producing misconceptions and inhibitions. And it provides a sound basis for overcoming sexual problems.

Much of the information in this and the following two chapters has been established relatively recently. The Kinsey group did most of its research in the late 1940s and 1950s; Masters and Johnson in the 1960s and 1970s; and current research is being carried out by many others. We have tried to avoid imposing our own behavioral standards on the conclusions of these researchers. Rather, we have attempted to provide the reader with the factual basis on which he or she may make sound personal decisions about sexual behavior.

THE CONCEPT OF SEXUALITY

Human sexual instincts operate from earliest infancy. Continuing throughout the entire life cycle, these sexual instincts and feelings slowly change as a person moves from one life period into the next. From these sexual feelings arise our sexual attitudes and conduct—our sexual behavior. Such behavior varies according to what we perceive as being approved, tolerated, or disapproved.

Our *sexuality* is nothing more than the sum of these sexual feelings and behavior. Such a concept of sexuality differs from the usual meaning of the word *sex* as something we do with our genitals. Our sexuality is a fundamental part of our life-long personality; it is not something that emerges at puberty and tapers off during late middle age.

Gender Identity and Gender Role

How we see ourselves as male or female is known as *gender identity*. In early childhood we acquire a relatively fixed internal sense that we are either boys or girls. These self-labels determine our future preferences and how we respond to circumstances.

The activities that are linked to gender identity are called *gender roles*. Our culture defines these roles as appropriate to a particular gender identity. Gender roles become the basic framework for our masculinity and femininity. As such they shape our future sexual adaptation.

A child's awareness of being male or female occurs between the ages of eighteen months and three years. Several factors determine how an individual develops gender identity.

Chromosomal factors Chromosomal sex is determined at the moment of conception. If a fetus carries an XX chromosomal complement, it generally produces female hormones and develops into a female. A fetus with an XY chromosomal comple-

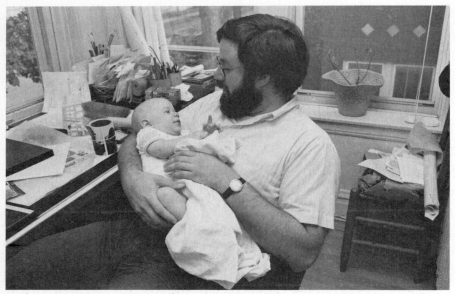

From positive parental reactions to their bodies, children develop good feeling about their own sexuality.

ment usually develops into a male. Genital differences make their appearance early in fetal development. Originally, however, the genital organs are the same in both males and females. After the seventh week the fetus begins to develop either male or female characteristics. Later, gender differences appear in rate of growth, rate of food use, physical energy, and heart action.

Hormonal factors In the genetically male fetus, masculine sex tissue develops only under the stimulus of the *androgens* (masculinizing hormones). Deprived of androgen, the sex organs of a genetic male differentiate into those of a female. In the female fetus, on the contrary, the development of the female sexual structures occurs whether or not the female sex hormones are produced. But if exposed to an excessive dose of androgen, the female embryo may become masculinized. Thus, gender identity is influenced by the sex hormones the fetus produces.

Social factors One's gender identity is not fully determined at the time of conception. Some authorities, however, believe that life experiences have more influence on a child's gender identity than do biological factors. In this sense, gender identity is *learned*. Our culture, for example, expects males to be more sexually active, dominant, aggressive, and less emotional than females. It expects females to be more domestic, loving, and less assertive than males. Parents are important in this process. In covert and overt ways they can convey to the child what they consider appropriate behavior. Parents reinforce gender roles in their children by rewarding the behavior that they wish their children to display.

SEXUALITY IN THE LIFE CYCLE

Many modern societies are age stratified—that is, they divide human life into the age periods of childhood, adolescence, adulthood, and old age. They then further subdivide these divisions—for example, childhood consists of the newborn period, infancy, toddlerhood, and so on. During each of these life periods, sexual behavior follows more or less regular patterns. Some of these patterns arise from biological factors, others are culturally imposed. For example, between the age of twelve and eighty a male is capable of impregnating a female. But our society limits the impregnating period of

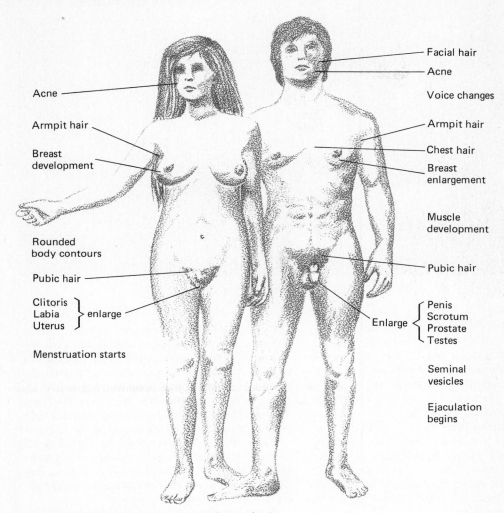

Acne

Armpit hair

Breast
development

Rounded
body contours

Pubic hair

Clitoris
Labia   } enlarge
Uterus

Menstruation starts

Facial hair

Acne

Voice changes

Armpit hair

Chest hair

Breast
enlargement

Muscle
development

Pubic hair

Enlarge { Penis
Scrotum
Prostate
Testes

Seminal
vesicles

Ejaculation
begins

Female and male body changes during puberty.

most males to the years between the late teens and early forties.

Considerable variation exists within each life stage. For instance, a person may be chronologically middle aged, but be psychologically youthful. Nor are there distinct breaks between the life stages. One stage shades into the next.

Sexuality in Childhood

Childhood in our society begins with birth and ends at about the onset of puberty. Early childhood is an active phase of sexual development. Young chil-

dren clearly respond to stimuli that produce sexual reactions, such as having their genitals or other erogenous zones caressed.

Children learn about their sexuality in various ways. They explore their bodies out of curiosity and discover pleasurable sensations. From parental reactions they learn which activities are approved and which are disapproved. In such ways they begin to develop their first deep-seated feelings toward sexuality, and whether such feelings are to be cherished or regarded with shame and guilt. Children ask many questions about life process.

Most young adults find partners and develop sexual relationships.

How parents handle a child's sexual curiosity may determine his or her sexual attitudes in adulthood. For example, relaxed or punitive parental attitudes may cause the child's later sexual interests to be either open or inhibited.

Adolescent Sexuality

Adolescence is a period of physical and sexual changes. Puberty arrives at different ages for different persons, but with each sex it follows a certain sequence. Structural, hormonal, and psychological changes occur. Both males and females become biologically mature.

Adolescence is also a period of psychosocial stresses. Sexual stress is brought on because while most adolescents are physically ready for heterosexual sex, they are not socially mature. Teen-

age males, and to a lesser degree, females, become obsessed with masturbation. Adolescents learn that sexual excitement and erection of the penis or clitoris can be initiated at will, and that climax and the release of sexual tension can be quickly brought about by self-stimulation. Masturbation thus helps them develop a control over their sexual impulses and prepares them for heterosexual intercourse.

Many adolescents pass through a period of homosexual experience as part of their development. In their desire to test out their own sexuality, they may turn to others of the same sex, since heterosexual experimentation is discouraged.

There are many ways to be heterosexual between puberty and marriage. Some people experiment with many of them, some with only a few. Many young people are left to their own sexual devices. Our society neither clearly endorses nor condemns sexual experimentation. Since our social attitudes are so fragmented, young people receive ambivalent signals on the question of premarital heterosexual experimentation.

Sexuality in Adulthood

In young adulthood we expect to settle down. After a period of experimentation or trial marriage, we may look for and select a future sexual and marital partner. In marriage we expect to behave het-

---

● **ASK YOURSELF**

1. In what ways does the concept of sexuality differ from the usual concept of sex?
2. How might one's gender identify differ from one's gender role?
3. What major factors are responsible for gender identity?

erosexually with our marriage partner, both for pleasure and to have children.

Because of a widespread tendency toward sexual privacy, the sex lives of some become isolated, even in marriage. All they know is their own experience. Many compensate with fantasies. Young adults are led to believe that sexual intercourse is so much more remarkable and romantic than it actually is. They are surprised to learn that the mechanics of sexual intercourse must be practiced and mastered. Yet, after all the learning of techniques, they must still own up to the question of caring—for the other person and for themselves. The more difficult questions have to do with relating to someone else, not performing sex with them.

Sexual concerns may shift somewhat toward the middle years. The early romantic days of marriage may be gone, and the partners may be unsure about their feelings in the years ahead. Some reaffirm their vows of fidelity to each other; others drift toward separation, divorce, or extramarital sex. Those who have never married may begin to view their unmarried status as a permanent life style.

Older adults may develop other concerns. Those who have based their identity on their masculine prowess or their feminine sex appeal may be distressed at the prospect of these qualities fading. Others derive satisfaction from their social and professional achievements; they enjoy the sense of esteem they find within themselves and from their associates. The quality of the marriage relationship provides them with more satisfaction than the frequency of coitus.

Sexuality in Old Age

Age takes its toll on all parts of the body. Hormone production drops, affecting the appearance of sex organs and the way in which sexual response occurs. All females go through a "change of life." Males, at different ages, experience a diminishing sexual capacity.

But physical changes by themselves do not fully determine the sexual behavior of older adults. They must work through questions of sexual attractiveness, sexual self-confidence, fears of rejection, and the availability of sexual partners.

Sexual relations do not stop automatically at a certain age. Some older adults continue their sex lives into their 80s and 90s. The cessation of sexual functioning in old age is unnecessary. Sexual rela-

● **ASK YOURSELF**

1. Children learn from parents. How might parents help their children develop good attitudes towards sexuality?
2. How is masturbation a positive force during adolescence?
3. In what ways could parents and society help adolescents develop positive self-images?
4. Many changes in sexual interests occur during adulthood. Trace these changes from early adulthood to old age.
5. What fears may adversely affect the sexual behavior of older adults?

tions in later years can provide a much-needed resource when other sources of fulfillment and self-confidence begin to wane.

NONMARITAL SEXUAL ADJUSTMENT

It is necessary to make some distinctions between various types of sexual relations. *Nonmarital sexual relations* occur between persons who are not married. *Premarital relationships* occur before an intended marriage. *Marital relations* occur between spouses during marriage. *Extramarital relations* are carried on by either spouse with someone outside the marriage while the couple is married.

The sexual behavior of unmarried people has been a source of interest throughout history. The traditional belief was that sexual tensions, if ignored, would somehow just go away. This attitude is obviously unrealistic. The biological sexual drive of unmarried persons is no different from that of married persons of the same age. But traditional attitudes have condemned any form of sexual satisfaction outside of marriage.

Today in American society there is no agreement about what is right or wrong in matters of sexual conduct. During the past fifty or sixty years, attitudes toward sex have become very liberal. But individual codes of behavior still range from total sexual freedom to strict prohibition of nonmarital sexual contact. Today, as always, unmarried individuals and couples must decide on their own course of sexual conduct. Various forms of sexual activity are available to unmarried persons.

Petting

Petting is defined as all relations more intimate than kissing but short of actual sexual intercourse. Petting is manual (hands) or oral (mouth) caressing of any part of the body. It typically includes the breasts and/or genitals. It may or may not lead to orgasm in either the male or female.

Any value judgment regarding petting should be based on the age and emotional maturity of the individuals and their attitudes and backgrounds. Petting is a normal step in the development of psychosexual maturity. It enables people to learn about their own sexual responses and about those of the opposite sex.

Petting generally stimulates sexual desire more than it relieves sexual tensions. Sexually arousing, petting naturally and gradually becomes more and more intimate until it culminates in sexual intercourse. Consequently, the sexually aroused couple finds it very difficult to stop short of intercourse. It is such unplanned intercourse that most often results in nonmarital pregnancy, since the couple did not take adequate steps for contraception. Many couples develop techniques of petting to mutual orgasm as an alternative to intercourse. If this is done, it is important that no semen be allowed to fall near the vaginal opening. Pregnancy can occur even without vaginal penetration by the penis.

Masturbation

Stimulating one's own genitals is *masturbation*. It may lead to orgasm. Despite the elaborate mythology about the supposedly harmful effects of masturbation, it is perfectly harmless. As described earlier in the chapter, it is a part of normal sexuality in both the male and the female. Over 90 percent of all males masturbate at some time in their lives, as do over 60 percent of all females. There may be various reasons for this difference. The female genitals, being less externally apparent than the male genitals, may receive less self-stimulation. Males are often more intensely sexual in their teens than females. They often find the need for the release of sexual tensions more compelling because the buildup of seminal fluid is going on continually. If not released through masturbation or sexual intercourse, the only other outlet is nocturnal emissions ("wet dreams").

Most individuals masturbate as a substitute for sexual intercourse when the latter is unavailable. As a result, masturbation is more common among unmarried individuals. Married and otherwise sexually active men and women may masturbate when their partners are pregnant, menstruating, or unavailable for intercourse. When sexual partners have died, as with widows and widowers, or when partners are unable to respond sexually, masturbation may provide a necessary release of sexual tension without the need to seek out new sexual partners.

Masturbation may begin at any age. The frequency of masturbation among normal unmarried males varies considerably, from about once a month to several times daily. Females typically masturbate less frequently than males.

Prostitution

Prostitution is the exchange of sexual favors for a fee—usually promiscuously and without affection. The most common form of prostitution is female and heterosexual. Male heterosexual prostitution as well as male and female homosexual prostitution also exist. The distinction between prostitution and transient sex is sometimes vague. Examples would be a person exchanging sexual favors for grades or job advancement, or accepting expensive gifts from a wealthy lover.

Prostitution is illegal throughout most of the United States. It is often singled out as a prime example of victimless crime since it involves a mutually agreed-on exchange of fee for service rendered between two consenting persons. Both enter into the relationship voluntarily and, theoretically, both benefit. Those who favor the prohibition of prostitution usually cite grounds such as religious beliefs, fear of organized crime, suppression of sexually transmitted diseases, and degradation of both women and men.

Sexual Intercourse

Nonmarital sexual intercourse is a subject of considerable interest. In the last few years there has been some increase in its prevalence and a great amount of discussion about it. Nonmarital sexual relationships are now entered into more openly than at any time in the recent past. Many young people become sexually active in their mid-teens without

Sidewalk hookers.

● ASK YOURSELF

1. What is meant by the terms nonmarital, premarital, marital, and extramarital?
2. In developing a sexual philosophy that includes heterosexual behavior, what activities enlarge one's knowledge of oneself and of the opposite sex?
3. In what ways might involvement in prostitution, either as the prostitute or the client, hinder or help the development of a positive or sexual self-image?

The Responsibilities of Nonmarital Sexual Activity

Sexual activity may provide unmarried partners with great pleasure. It also immediately imposes some profound considerations on them.

Nonmarital pregnancy There is a basic motivation behind many of the legal, religious, and social regulations of sexual behavior and marriage. It is to provide a stable family environment for the child and to fix the responsibility for its support. Although attitudes toward sex have changed considerably, pregnancy outside of marriage is regarded as a serious problem by many people.

Contraceptive methods can reduce the chance of pregnancy to a very low level if they are used properly and consistently. Anyone engaging in nonmarital sexual relations should choose a highly effective contraceptive method and be certain that it is used properly. Sexually active partners should agree about what to do in case an unplanned pregnancy occurs. If a couple is not mature enough to discuss this possibility realistically, then it is questionable whether they are mature enough to have sexual relations. Some of the paths available to unmarried parents include single parenthood, marriage, adoption, and abortion.

Single parenthood The prospect of keeping and raising a child out of wedlock holds little appeal to many women. Yet in recent years, increasing numbers of unmarried women have been taking this option. The single mother must be prepared to provide the economic and human resources parenthood requires.

taking any precautions to prevent pregnancy. This occurs even though improved contraceptive methods are readily available. Many teen-age girls thus become pregnant. Some single women intentionally become pregnant and bear a child which they plan to raise by themselves.

As with petting, there is no universal answer to the question of whether or not to engage in intercourse outside of marriage. Many individual factors must be considered. For some, religious beliefs forbid nonmarital sex. For those below the age of consent, state laws proscribe such behavior. But, for the young adult, it becomes a highly individual question to be decided on the basis of personal values and philosophy.

Adoptive parents must be able to provide security and love to a child awarded them by a court.

Marriage Though pregnancy is one of the more common reasons for getting married, it may be one of the poorest. A high percentage of forced marriages turn into disasters, leading to divorce or, perhaps even worse, to meaningless, bitter relationships. Unless both parties truly want to marry, it is far better to take one of the other options.

Adoption In many cases, adoption is the best course of action. Often it assures the child a loving home where it is welcome rather than resented. There is currently a strong demand for newborn infants for adoption.

Abortion According to the U.S. Supreme Court, every woman in the United States has the same right to an abortion during the first six months of pregnancy as she has to any other minor surgery. In other words, during this period of time *any* pregnant woman is entitled to request an abortion.

Some precautions are in order if an abortion is being considered. An induced abortion may temporarily interfere with the body's hormonal system. Some women may need postabortion therapy. Repeated induced abortions may result in an inability to carry a child to full pregnancy. Abortion should not be viewed as just another method of birth control. At best it is to be used with caution (*see* Chapter 16).

A woman should be certain that abortion is philosophically and ethically acceptable to her. Because of religious or personal philosophical viewpoints, some women find abortion an unacceptable option. Abortion should be avoided if it is going to cause feelings of guilt or emotional recrimination. Regardless of her decision, a woman may still have questions. This is natural. She must make the best personal decision as she sees it and then let the matter rest.

Sexually Transmitted Diseases

The risk of sexually transmitted diseases depends on the pattern of sexual relationships. If the only relationship involves a mutually faithful couple, there is no risk of infection (assuming neither is infected to start with). If a person has casual sexual contacts or has intercourse with anyone who does have such contacts, the risk of sexually transmitted diseases is greatly increased. (For a further discussion of sexually transmitted diseases, refer to Chapter 19.)

Behavioral Needs

Sexual activity is often engaged in for reasons other than strong sexual desire. Some nonsexual motives may be quite desirable. One would be to express affection to a partner and to tell the partner he or she is wanted. Another would be the need to feel sexually attractive and wanted as a sexual partner. We all need to feel successful as lovers. The ego reinforcement obtained through successful sexual relations is very important to the total emotional adjustment of most adults.

Some reasons for sexual activity may be questioned. One is the use of sex as a bargaining tool, where sexual privileges are traded for various emotional or material rewards. A person may withhold sex in order to punish a partner for failing to provide such rewards. Sex is also misused when it is used to force a commitment from a partner who does not really feel any great degree of love in return.

It is important for each sexual partner to understand the motives of the other. If it is going to be "pure" sex—sex for physical satisfaction alone—then both partners should share this feeling. No pretense of love or other commitment should be made. Deception in this matter amounts to exploitation. It can only lead to someone feeling used and hurt.

The Effect on Future Marriages

One of the traditional concerns about nonmarital sex has been whether it might affect the success of any future marriage. This is really a very difficult question to answer, for several reasons. First, so many factors are involved in marital happiness that it is impossible to determine cause and effect. In addition, it is fruitless to make statistical comparisons of the divorce rate, orgasm rate, or any other "indicator" of marital happiness between persons who have and have not had nonmarital sexual experience.

The Double Standard

As noted earlier, different societies and different groups within a particular society hold varying attitudes toward nonmarital sexual intercourse, These range from total permissiveness to total prohibition. In the United States, we find an ambiguous situation, with many people disapproving the sexual activities of others but excusing their own. The "official" attitude here prohibits sexual activity before marriage. The "unofficial" attitude is often the highly discredited double standard, which is much

● ASK YOURSELF

1. You are nineteen, single, and pregnant. What would be some of the pros and cons of each of the following four options: (1) an abortion; (2) give birth and give the child up for adoption; (3) give birth and raise the child yourself; (4) marry your sexual partner and keep the child.
2. Abortion is a hotly debated subject. What are the reasons behind the prochoice and prolife positions?
3. Nonmarital sexual activity is based on a number of behavioral needs. What might be some nonsexual motives for it?
4. What legal concerns should a person be aware of before having sexual intercourse with another unmarried person?

more tolerant of the sexual activities of men than of women. This is an unfortunate carry-over from the era when women were not expected to seek and enjoy sexual intercourse but to be subservient to the desires of men.

The recent rapid changes in social attitudes toward sex have included a dramatic shift away from the double standard. As women expand their roles beyond those of wife, mother, and homemaker, the restrictive implications of the double standard are giving way. Replacing it is a more realistic view of the female sex drive and sexual behavior.

Legal Considerations

In a few states, any sexual intercourse between unmarried persons is illegal, but such laws are seldom, if ever, enforced. Of more importance are laws that prohibit intercourse with a woman below the age of consent, which varies among the states from 14 to 21, but is most often 18 years of age. A man having intercourse with a young woman below this age can be prosecuted for statutory rape (unlawful sex). This is a felony offense. Even if she appears to be older and lies about her age, the man can still be convicted.

Still another problem can arise when an unmarried woman becomes pregnant and sues for child support. Any man who has had intercourse with her during a given period of time may be named in the suit.

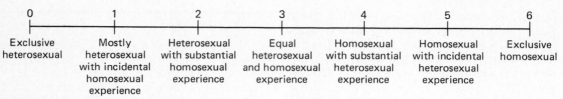

Kinsey's sexual preference continuum. Using this scale, individuals can determine their relative degrees of heterosexuality and homosexuality.

SEXUAL ORIENTATION

In Alfred Kinsey's studies of the human male and female (1948, 1953), it became apparent that the extent to which people participated in homosexual activities varied greatly. This led Kinsey to employ the concept of a heterosexual-homosexual behavior rating scale. It was apparent that there are not just two populations, heterosexual and homosexual. There is a continuum or scale of orientations ranging from the exclusive heterosexual person at one extreme to the exclusively homosexual at the other.

HOMOSEXUAL ORIENTATION

Homosexuality is sexual attraction to members of one's own sex. Significant numbers of both men and women experience it to at least some degree during their lifetime.

Different authorites give somewhat varying figures on the incidence of homosexuality. A conservative estimate based on a consensus of opinions seems to be that about 4 percent of American men and women are exclusively homosexual during adult life. A considerably larger group of both males and females (variously estimated at 25 to over 50 percent) passes through a transient period of homosexual feeling and/or activity during the preadult years. Most male homosexuals have a normal gender identity. They feel themselves to be true males, not females caught in a male body.

Many homosexuals give no overt indication of their sexual orientation. Most male and female homosexuals run the gamut from masculine to feminine in appearance. The majority are not visibly identifiable as homosexual. However, since social and professional difficulties may arise when a homosexual orientation becomes known, some homosexual people are guarded in their behavior. Generally, homosexuals meet each other through friends

and at bars and parties where they can be fairly certain of each other's sexual orientation. This is similar to the way heterosexuals meet, except that homosexuals often cannot meet in work situations or in other places where it might be hazardous to be open about their sexual orientation.

Homosexual Behavior Patterns

The typical pattern of homosexual relationships differs considerably between male and female homosexuals. The female tendency is to establish long-term homosexual relationships lasting for months, years, or even a lifetime. This is in accord with the socialization of women in our society toward love and long-term relationships. A study (Bell & Weinberg, 1978) found that females are more gentle, more sensitive, more thoughtful, and take more time in lovemaking than did men of either sexual orientation.

The typical pattern of male homosexuality is quite different. Male homosexuals, like their heterosexual counterparts, demand variety in sexual partners. Because men in our society are trained to be assertive and powerful, it is sometimes more difficult for two men to live together than it is for a dominant man and a submissive woman in a traditional heterosexual relationship. However, many homosexual males form close, long-term relationships with other men. When two men or two women live together in a love relationship, they often see the inadequacy and futility of traditional sex roles where one partner is dominant and the other submissive. In many of these relationships, there may be as much or more give-and-take and sharing as in many modern heterosexual marriages.

Among the practices used by homosexual males to achieve orgasm are mutual hand manipulation of the penis, oral-genital contacts, and anal-genital contacts. Techniques used in female homosexual

Homosexual rights demonstration.

contacts include kissing, manual and oral stimulation of the breasts, and manual or oral stimulation of the genitals.

Some homosexuals enter into heterosexual marriages, for a variety of reasons. Sometimes the individual is truly bisexual and enjoys relationships with both sexes. Marriage may be an attempt to live a "straight" life, in the belief that it may counteract homosexuality (it seldom does). Or marriage can be a "front" for homosexual activity, a disguise to appear straight while privately engaging in homosexual activities. The mates chosen for these marriages are sometimes individuals with low sex drives who make few heterosexual demands on their homosexual partners.

Viewpoints on Homosexuality

Many homosexual people view their orientation merely as an alternative life style that presents no more difficulties than the typical heterosexual person encounters. Homosexuality meets with varying responses from academics and professionals. Psychiatrists and other physicians, for example, disagree greatly on how to regard homosexuality. Is it an illness, a maladjustment, or an alternative life style? Should treatment be suggested for homosexuality in itself? Or should treatment be considered only if a patient's homosexuality is creating problems? In the latter case, should the goal of treatment be to reorient the patient sexually or merely to help the patient adjust to homosexuality and its associated problems? There are no clear answers to these questions, and controversy continues in many medical and psychiatric journals. In 1973 the American Psychiatric Association removed homosexuality from its *Diagnostic and Statistical Manual of Mental Disorders.*

An extremely important factor in the current professional view of homosexuality is the fact that almost all research conducted in the past twenty years has occurred in the clinical setting. That is, the people being studied as representative of the homosexual orientation were those who, for reasons of problems of adjustment, sought psychiatric care. Obviously, this kind of sampling could not give a balance and adequate view of the entire homosexual population. More recent studies, including a major one funded by the Kinsey Institute for Sex Research, have concentrated on the homosexual orientation of people living in various American communities (Bell and Weinberg, *Homosexualities: A Study of Human Diversity,* 1978).

The law's response to homosexuality is now being critically examined. The simplistic assumption that homosexuality is strictly a legal question, calling for arrest and imprisonment, is clearly being put aside.

TRANSSEXUALITY

Transsexuality is a psychological condition in which a person believes that he or she belongs to the opposite sex. "Trapped" in a body and social role of the wrong sex, the transsexual thinks, feels, and acts like a person of the opposite sex.

Transsexuality is often confused with *transvestism* (wearing clothing of the opposite sex) and with homosexuality. Although many transvestites and many transsexuals dress like the opposite sex, transvestites are not transsexuals. Similarly, pas-

Tennis player Renee Richards, a transsexual.

sive homosexuality in the male transsexual is common. The male homosexual on the contrary enjoys his penis as well as the interest other homosexuals have in it. He derives genital pleasure from his sexual contacts. The transsexual male, on the contrary, hates his penis and derives no pleasure from it at all. Usually he does not experience erection or ejaculation.

The transsexual is usually a biologically normal male or female. With rare exceptions, there are no anatomical, chromosomal, or hormonal abnormalities. Transsexual attitudes can sometimes be traced back to a childhood background of parents whose own sexual roles were unclear or who were disappointed that the child was not of the opposite sex. There also may be prenatal factors that influence the development of the transsexual.

Psychotherapy has not been highly successful in the treatment of adult transsexuals, though it is obviously preferable to more radical treatment. Many transsexuals have achieved happiness only after gender reassignment surgery and hormonal procedures. While this is undeniably an extreme measure, the alternatives for many transsexuals are continued suffering or even suicide. Several ethical sex-change clinics have been established within the United States. Many patients have expressed satisfaction with their sex changes. Those changing from male to female (the more common procedure) are provided with breasts and a vagina. They may experience sexual pleasure and even orgasm. In the reverse change, it is surgically possible to construct an artificial penis that can remain in a semierect

state. Of course it would transmit no semen. Some may attach an artificial penis just for intercourse. Of course, natural parenthood is impossible after either sex-change procedure.

VARIANT SEXUAL BEHAVIOR

Many patterns of sexual behavior are contrary to the standards of at least some members of our society. Some forms of sexual behavior are condemned by almost everyone. Other sexual practices are condemned by some and accepted by others. For example, if a truck driver eating lunch in a truck-stop café pinches the waitress, she may typically wink at him and promptly forget the incident. If a diner in a hotel dining room pinches the waitress, she will tell him to "watch it" or maybe even call the manager to talk to him. If a man pinches a woman on the street, she is apt to call a policeman and have him arrested. As another example, if a man peeks through a bedroom window at a partially dressed woman, he may be arrested and convicted of a sex offense. But looking at even more scantily clothed women in a nightclub act is perfectly acceptable behavior to many members of our society. In these two examples, the major factor seems to be the context rather than the details of the act itself.

Sexual variance also includes a psychological component. Some patterns of variant sexual behavior can be readily associated with particular emotional characteristics and poor social adjustment. The most common forms of variant sexuality can be discussed in terms of three types of psychopathology: (1) feelings of sexual inferiority, (2) developmental abnormalities, and (3) orientation toward violent behavior.

Sexual Insecurity

The fear of sexual inadequacy is associated with certain types of variance. The sexually insecure person feels ashamed or disgraced because of a real or imagined inability to function satisfactorily in sexual relationships. From this single foundation, a range of variant behavioral patterns can rise.

Exhibitionism Legally, an *exhibitionist* is an adult man who deliberately exposes his penis to an unsuspecting female, who is an involuntary observer, usually a complete stranger. The exposure must be intentional and not incidental as in the case of a drunk urinating. The penis may be flaccid or erect and the exposure may or may not include masturbation.

Although most exhibitionists cannot give a clear account of their feelings at the time of the exposure, the apparent intention of the act is to arouse an emotional expression in the victim. The most common intention of the exhibitionist seems to be to evoke fear and shock rather than pleasure from his victim. An amused reaction often sends the exhibitionist into a state of depression. It is important to note that the exhibitionist is not soliciting further contact with his victim. On the contrary, he is afraid of any closer contact. If a woman approaches him for sexual contact, he is likely to run away. The exhibitionist is one of the most harmless sex offenders.

The most significant psychological finding about exhibitionists is they experience deep feelings of inferiority and sexual inadequacy. Thus, through exposing themselves, they seek a feeling of power, dominance, and sexual adequacy. The reaction they are striving for is shock at the large size of their sex organs. This probably explains why they often expose themselves to children, who are more likely to be shocked than adults. In addition, this is probably the reason why an amused reaction can be so crushing. In most cases, the safest and best response to an exhibitionist is merely to ignore him.

Obscene phone calls *Obscene phone calls* are anonymous communications without intention of further sexual contact. While usually a male phenomenon, women occasionally are the callers. Like the exhibitionist, the caller finds sexual release in the act itself or in masturbation during the call. The hoped-for reaction is shock and consternation on the part of the victim. The fact that the victim is both inaccessible and unknown protects the caller from a possible sexual confrontation. The best way to deal with an obscene phone call is simply to hang up. The longer you stay on the phone, the more you are likely to encourage the caller. If you demonstrate fear or shock, you are playing the caller's game and inviting further calls.

Voyeurism A *voyeur* is a person who attains sexual gratification by looking at sexual objects or situations. One of the complications in studying voyeurism is the fact that almost all people have

● ASK YOURSELF

1. Why is it preferable to speak of homosexual orientation when describing homosexual behavior?
2. Why is there often a considerable difference between male and female homosexual behavior patterns?
3. You are a practicing heterosexual. You discover that a friend is a practicing homosexual (or lesbian). Is it possible for you to maintain or further that friendship and still be honest and open in your feelings toward him or her?
4. Transsexuals may feel trapped both biologically and socially. Why would they feel trapped? What steps might be taken to relieve them of this sense of entrapment?

some voyeuristic tendencies. Society accepts such forms of voyeurism as viewing "topless" shows at bars, reading sex magazines, and watching attractive people in brief bathing suits. The true voyeur is the peeper ("peeping Tom") and is almost always a male. Voyeurs look into private rooms or areas with the hope of seeing unsuspectedly nude or partially nude people. The peeper wants to see people behaving in presumed privacy. A few peepers call attention to themselves by such actions as tapping on a window, but most of them try to avoid detection. They often masturbate while or immediately after peeping.

Male peepers are generally shy with women and have strong feelings of inferiority. Their interests are heterosexual but their overwhelming fear of being rejected keeps them from seeking normal heterosexual activity. A few are mentally deficient. Peepers are rarely dangerous.

Bestiality *Bestiality* is engaging in sexual contact with animals. This is apparently a common occurrence among rural boys, yet it is one of the most taboo forms of sexual outlet. Even in rural areas, bestiality is the object of both condemnation and ridicule. Transient bestiality may occur for reasons of curiosity or novelty. It can become a fixed behavior out of the fear of failure with the opposite sex. Some men turn to bestiality to avoid the fear of incest; some perceive sexual relations with any woman as suggestive of incest with their mothers. Some men turn to bestiality in order to show hostil-

● ASK YOURSELF

1. *Deviancies* and *perversions* are labels that have been assigned to variant sexual behavior. Why might it be more constructive to speak of such behavior as variance?
2. What should a woman do when confronted by an exhibitionist?
3. What are the underlying causes of exhibitionism? Of obscene phone calls? Of voyeurism? Of bestiality? Of incest?

ity and contempt toward women in general by identifying them with animals.

Incest *Incest* is sexual intercourse between individuals too closely related to marry legally. The relationship can be father-daughter, father-stepdaughter, mother-son, mother-stepson, or brother-sister. Incest is one of the most ancient and widespread of the sexual taboos.

Most incidents of incest develop either within a subculture that takes a less strict attitude toward such behavior, or as a result of the mental incompetence of one of the partners. Even in the contemporary United States, certain subcultures consider incest to be an unfortunate but not a grave or unexpected situation. Incestuous fathers tend to be ineffectual, unaggressive, dependent, and preoccupied with sex. Many believe that incest damages children mostly because children perceive that the one in whom they have placed their greatest trust has violated that trust.

Developmental Abnormalities

Some sexual variations arise from failures in psychosexual development. Either because of an inability to attain normal sexual maturity or because of regression (return) to earlier, immature sexual habits, these individuals are likely to show one of the following behavior patterns.

Pedophilia *Pedophilia* is sexual involvement of an adult with a child. It may be either homosexual or heterosexual. Pedophilia is probably the least acceptable form of sexual behavior in our society. Since the variation lies in the sexual immaturity of the child, the natural break-off point for classifying an act as pedophilia would be the onset of puberty.

Pedophiles usually suffer from arrested psychosexual development (fixation). The offender had never grown beyond the immature prepubertal stage, or has regressed back to this stage of development. As a result, most pedophilic sexual acts involve childlike sex play, such as looking, showing, fondling, and being fondled. The nature of the sexual act usually corresponds to the maturity expected at the age of the victim rather than at the age of the offender.

In the vast majority of heterosexual pedophilia cases, the offender is part of the child's close environment and is usually known to the child and the child's family. Offenders are most commonly neighbors, family friends, or relatives. Less than one-fifth of the offenders are strangers or only casual acquaintances. In homosexual pedophilic offenses, the offender is more often a stranger.

Parents, police, and the courts can minimize the harmful effects on victims of pedophilic offenses by skillful handling of these cases. It has been found that the child is more often damaged by the events following the offense than by the offense itself. The effect on the child depends greatly on the reaction of parents and other adults on discovery of the offense. If the parents react with obvious fear, anger, disgust, or hysteria, the child is likely to suffer lasting effects. An additional problem is the appearance of the child as a witness in court. Interrogation and cross-examination can be far more damaging than the offense itself.

Fetishism *Fetishism* is deriving sexual gratification from contact with either a part of the human body or some inanimate object. Commonly used body parts include hair, hands, thighs, feet, ears, and eyes. Typical inanimate objects might include underclothing, hats, shoes, gloves, rubber, silk, or fur. Most people have fetishistic preferences for breasts, buttocks, legs, or hairy chests. Fetishism is viewed as deviant behavior when the person *must* have the fetish item to function sexually, or when the item serves as a substitute for a sexual partner.

Fetishism may be rooted in childhood conditioning in which the fetish object is associated with some moment of sexual excitement. Unsatisfactory interpersonal relations in adulthood may cause some people to seek comfort and sexual pleasure with an item they once associated with sexual pleasure.

Rape counseling.

Transvestism *Transvestism* is the practice of wearing the clothing of the opposite sex. The practice exists among both men and women. Transvestism usually indicates a distorted and confused sociosexual life that often, but not always, includes homosexuality. There are several possible motivations behind the transvestism.

First, some transvestism may be motivated by homosexuality. A few homosexuals dress in garments of the opposite sex as an outward sign of their homosexuality (to attract persons of the same sex) and as a symbol of their preferred role in homosexual acts. In this case, the clothing has no emotional or sexual value to the wearer. It is a means to an end, not an end in itself.

Second is the true transvestite, who wears the clothing of the opposite sex for the emotional or sexual gratification it provides. This type of transvestism is an end in itself and often involves fetishism. Most of these are heterosexual; many of them are married and have children.

Violent Behavior

A disposition toward unnecessary violence and the use of force to deal with emotional needs and desires is the basis of certain types of variant sexual patterns. For this reason, these variations pose the very real danger of physical and psychological harm.

Forcible rape A detailed study of men convicted of *forcible rape* (Gebhard et al., 1965) indicated that they fall into several distinct groups.

Violent life style The most common type, this rapist's life style includes the use of unnecessary violence. He does not commit rape because of a lack of willing sex partners. Instead, he prefers rape to conventional sex. For him, sexual intercourse is most gratifying if it is accompanied by physical violence or the serious threat of violence. There is a strong sadistic element in his personality. He dislikes women and gains satisfaction from punishing them. Often more violence is used than would be necessary to complete the rape. In some cases, the violence seems to substitute for sexual release or at least diminish the need for it. In fact, these rapists sometimes become impotent and are unable to complete the sex attack.

Amoral delinquents Such persons pay little attention to normal social controls and operate purely for their own gratification. They are not sadistic. They simply want to have intercourse, and the contrary wishes of the female are of no importance.

● ASK YOURSELF

1. What are the psychological characteristics of pedophiles?
2. Discuss some of the different types of rapists.
3. Fetishists may use both living and nonliving objects to obtain sexual gratification. What objects are commonly used?
4. You are confronted by a rapist in the semi-darkened parking lot of a shopping center. What steps ought you to take to defend yourself?

They are not hostile toward women, but look upon them solely as sexual objects whose role in life is to provide sexual pleasure to men.

Drunks   The drunk rapist's aggression ranges from uncoordinated efforts at seduction to hostile and truly vicious attacks. Such behavior occurs when intoxication lowers their inhibitions.

Explosive types   These are previously normal individuals who suddenly snap into a psychotic state as a result of emotional stress. An example might be a mild-mannered college student who suddenly and without warning rapes and kills.

The "innocent" male   Hoping to enter into a voluntary mutual relationship with a woman, this type of rapist often misunderstands her true wishes. He may be accustomed to a pattern of behavior in which a woman who says "No, no" actually means "Yes, yes." He does not realize that many women really mean No.

A woman can best reduce her chances of being a victim by avoiding those situations that most often lead to forcible rape. Of these, touring bars alone is the one most frequently mentioned by women who report forcible rape. Hitching rides is another dangerous practice.

The advice on what to do when faced with a rapist varies according to the circumstances. Sometimes women can calmly talk the rapist into changing his mind; others scream for help or trick him in one way or another. But even the best-planned defense may fall apart in the suddenness and fear of the actual attack. Rapists can be extremely dangerous. Some authorities advise women to fight back since some rapists are encouraged by submissiveness. Others recommend yielding to his demands and cooperating.

Historically, rape has been one of the most difficult charges to prosecute successfully. Most rapes go unreported by their victims. This failure to report is caused by the embarrassment and humiliation inflicted by unsympathetic authorities and by the fact that women were long considered to be wholly responsible for being raped.

Fortunately, public health officials and many police departments are realizing that rape is one of the most traumatic crimes. A victim of rape requires understanding and professional care to avoid permanent emotional damage. Women's groups have been establishing rape treatment centers that help victims through the experience and encourage them to assert their legal rights.

Gradually, through recent court decisions and enactment of restructured state laws, rape trials are beginning to favor the victims. Defendants are facing increased chances of conviction through new legal provisions that limit the amount of probing into the victims' previous sex lives. By dredging up a woman's past sex life, defense lawyers have often attempted to make her appear as the community whore. The defendant may be portrayed as someone who has been enticed into an unfortunate liaison. Such tactics have been used successfully to discredit victims' complaints. However, much still remains to be done to fully protect women from rapists, and to ensure successful convictions.

Sadism   *Sadism* is the infliction of pain upon someone in order to obtain sexual release. The sadist is often unable to achieve orgasm without resorting to some form of violence. There are both male and female and heterosexual and homosexual sadists. As was mentioned, sadism is often a motivating factor in rape. The rapist frequently uses more force than is necessary to complete the rape, and elements of torture are sometimes involved. Many men who are unable to obtain satisfaction otherwise resort to sadistic cruelty in their relationships with prostitutes or even their wives. Some forms of sadism seem to have no relationship to its sexual basis.

Masochism   *Masochism* is the attainment of sexual gratification from suffering physical pain. There are men and women who must be physically punished in order to achieve sexual arousal or orgasm. The punishment often involves beating, whipping, biting, pinching, scratching, burning, and similar painful treatment. Various psychological explanations have been offered for masochism. It has been interpreted in terms of destructive impulses of the unconscious mind. It has also been related to subconscious guilt feelings from which the masochistic punishment gives temporary release.

IN REVIEW

1. Our sexuality is the sum of all our sexual feelings and behavior. Gender identity is our own recognition of our sex; gender roles are the activities viewed as appropriate to our sex. Gender identity is determined by:
   a. Chromosomal factors
   b. Hormonal factors
   c. Social factors
2. Appropriate sexual behavior follows life-cycle periods of:
   a. Childhood
   b. Adolescence
   c. Adulthood
   d. Old age
3. Unmarried persons may fulfill their sexual needs through petting, masturbation, prostitution, or other forms of sexual intercourse.
4. Nonmarital sexual activity carries with it the responsibilities for:
   a. Nonmarital pregnancy, which may be resolved through single parenthood, abortion, adoption, or marriage
   b. Sexually transmitted diseases
   c. Meeting behavioral needs
   d. Effects on future marriage
   e. Perpetuating the double standard
   f. Legal considerations
5. A person's sexual orientation may range anywhere from the exclusively homosexual to the exclusively heterosexual. Male and female homosexual behavior patterns may differ considerably.
6. Transsexuals believe they are "trapped" in the body of someone of the opposite sex. Transvestites like their original bodies, but gain pleasure from cross-dressing.
7. Variant sexual behavior may have underlying causes in:
   a. Sexual insecurity: exhibitionism, obscene phone calls, voyeurism, bestiality, and incest
   b. Developmental abnormalities: pedophilia, fetishism, and transvestism
   c. Violent behavior: forcible rape, sadism, and masochism.
8. Forcible rapists may be classified into those who are:
   a. Following a violent life style
   b. Amoral delinquents
   c. Drunks
   d. Explosive personalities
   e. "Innocent" males

CHAPTER 14
SEXUAL PHYSIOLOGY AND RESPONSE

Until recent years the bulk of our understanding of human sexuality was largely medical. Physicians dealt with pregnancy and birth, and with female and male sex problems. Interestingly, within our lifetime the first detailed knowledge of the human sexual response was uncovered. Because of the work of Masters and Johnson (1966, 1970, 1979), and many others, our perspective on sexuality has become more physiologically accurate. Human sexuality is now viewed as a normal physical response.

Knowing what our bodies look like in the nude, and how they respond when aroused helps us feel at home with the richness of our sexual selves. No less important, this knowledge enhances the interaction of lovemaking. Artful stimulation increases sexual response and delight, while clumsy efforts inhibit and "turn off" a partner's interest. Knowing the sensory richness of each other's bodies enables loving partners to interact with true intimacy and caring.

This chapter is devoted to helping us discover our sexual bodies and the manner in which they respond.

SEXUAL ANATOMY AND PHYSIOLOGY

While differing somewhat in appearance, the female and male reproductive structures are basically alike. Both are located partly inside the body cavity and partly outside it. The internal sex organs are primarily for reproduction, while the external ones are more related to sexual pleasure.

The Female Sex Organs
The female sex organs are less conspicuous than the male. Since the external organs are more related to intercourse, they are known as the *genitals*.

The genitals Genitals are collectively called the *vulva* or the *pudenda*. They consist of the mons pubis, outer lips (labia majora), inner lips (labia minora), clitoris, vaginal opening, and hymen.

Mons pubis The *mons pubis* is a fatty pad lying over the pubic bone. The most conspicuous part of the female genitals, it becomes covered with curly hair during puberty.

Lips The *outer lips,* or labia majora, are two long folds of skin that extend down and back from the mons pubis. Located between the outer lips are the *inner lips,* or labia minora. Consisting of two smaller elongated folds of skin, the inner lips surround the *vestibule*. Located in the vestibule from front to back are the clitoris, the urethral opening, and the vaginal introitus.

Clitoris The *clitoris* is a small erectile organ that has a rich supply of nerve endings and is the chief site of sexual excitement in the female. The clitoris is similar to the penis in structure, yet is rarely longer than one inch. Unlike the penis, it does not contain the urethra.

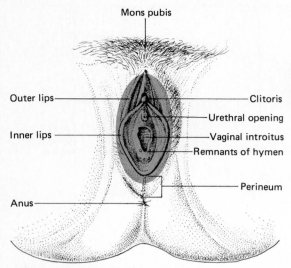

Female genitals, front view.

Vaginal introitus The *vaginal opening,* or *introitus,* may be surrounded by a partial membrane, or *hymen.* Varying in size and thickness, it may remain intact until first intercourse. It may, however, be reduced in size by the use of tampons, finger manipulation, participation in field sports, or by a physician. Contrary to popular belief, its reduction in size may not involve bleeding. Its absence should not be taken as a sign of prior sexual intercourse.

Internal sex organs The internal organs include the two ovaries, the fallopian tubes, the uterus, and the vagina.

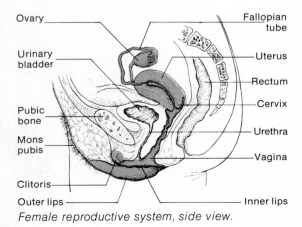

Female reproductive system, side view.

Internal female reproductive organs.

Ovaries The production of female sex cells, eggs (ova), is accomplished by the *ovaries* (the female *gonads*). They are situated deep in the pelvic cavity, one on either side of the uterus. The ovaries serve a dual function, producing both eggs and hormones. Within each ovary are many vesicles called *ovarian follicles.* Beginning with puberty these follicles mature at the rate of 1 about every 28 days and develop into *graafian follicles.* Each month, usually midway between menstrual periods, a follicle ruptures and releases a mature egg. Since the reproductive life of the female extends about 35 years (ages 12 to 47) and about 1 egg is produced every 28 days (13 a year), about 450 eggs mature.

As the follicle enlarges, it moves toward the surface until it finally appears like a little blister on the surface. Near the midpoint between menstrual periods, it ruptures and releases the egg enclosed within it, a process called *ovulation.* Actually ovulation may occur as early as the eighth day and as late as the twentieth day. After ovulation, the blood clot is soon replaced by yellow-colored cells called the *corpus luteum.*

Fallopian tubes The *fallopian tubes,* or *oviducts,* are about 4 inches long and extend from the uterus out to the ovaries. When the egg ruptures through the wall of the follicle, the egg passes into the tube. Once inside the fallopian tube, the egg is propelled toward the uterus by the movement of the cilia and by contractions in the walls of the tube.

Once the egg is released from the ovary, it can be fertilized by any sperm that is present. Three to

Phases of the Menstrual Cycle

The menstrual cycle can be divided into four distinct phases:

1. *Destructive, or menstrual, phase.*   This phase occurs because of the disintegration of blood-rich endometrial tissue. This results in the flow of both blood and cell fragments. The menstrual phase usually lasts four to six days.
2. *Proliferative, or follicular, phase.*   After menstruation has stopped, the uterine lining is thin. Meanwhile, the next follicle in the ovary is maturing. During this phase, which lasts about ten days, the follicle is producing increasing amounts of estradiol. Estradiol leads to reconstruction of the endometrium.
3. *Ovulatory phase.*   Ovulation usually occurs between days 12 and 16, but most commonly about day 14. As soon as the egg breaks through the follicle wall, the remains of the follicle become a corpus luteum.
4. *Secretory, or luteal, phase.*   The endometrium becomes thick, spongy, and very glandular. This phase of the cycle lasts about 14 days. In the event the egg is not fertilized, the corpus luteum disintegrates, stops producing its hormones, and the endometrium destructs. The next menstrual period begins.

four days is normally required for the transport of the egg from the ovary through the fallopian tube to the uterus. An egg is believed to remain viable for about 24 hours, after which it begins to degenerate. Fertilization, if it is to occur, must take place within this time. Thus, fertilization usually occurs within the outer third of the fallopian tube.

Uterus   The *uterus* (womb) is a hollow, pear-shaped organ located in the pelvis. It is slightly above and behind the bladder, but in front of the rectum. It is loosely suspended in position by several ligaments. Its normal position is a forward tilt.

In the adult it may be about 3 inches long and 2 inches wide. Its walls are thick and very muscular. In pregnancy it stretches to over 12 inches in length as it expands to accommodate the growing baby. The upper half of the uterus is the *corpus* (body), the lower half is the *cervix,* and the lower opening is the *os.*

The inner layer of the uterus, the *endometrium,* is richly supplied with blood vessels and glands. Following ovulation, the egg descends through the tube into the uterus. If the egg has been fertilized, it becomes implanted in the endometrium within three to four days.

Vagina   The *vagina* is a tube extending from the vestibule to the uterus. This muscular tube is 4 to 6 inches long and lies between the bladder and the rectum. It serves as the excretory duct for the uterus, the female organ for sexual intercourse, and the birth canal. The lining membrane gives off a mucuslike secretion during sexual arousal. At the peak, or climax, of sexual arousal, its muscular walls contract strongly and rhythmically, producing an intensely pleasurable sensation called *orgasm.*

The Menstrual Cycle

The sexual and reproductive lives of women go through a rhythm of changes somewhat like the seasons. Repeating about once a month, these changes are called the *menstrual cycle.* Marked by both biological and psychological events, the most noticeable sign is a monthly flow of menstrual products, the *menstrual period.*

The purposes of the menstrual cycle are twofold. One is the production by the ovary of an egg. The other is the development of the endometrium to receive and nourish the egg in the event it is fertilized. The preparation of the endometrium is very elaborate and is only temporary. If an egg is not fertilized immediately after it is released, the

TABLE 14.1 MENSTRUAL HORMONES

Gland	Hormone
1. Hypothalamus (brain)	LRF (luteinizing-releasing factor) PIF (prolactin-inhibiting factor)
2. Anterior pituitary (gonadotropins)	FSH (follicle-stimulating hormone) LH (luteinizing hormone) PRL (prolactin)
3. Ovary	Estradiol Progesterone

Hormone	Source	Functions
Menstrual cycle		
LRF	Hypothalamus	Stimulates ant. pituitary to release much FSH and some LH.
FSH	Ant. pituitary	Stimulates ovary to develop mature follicle (with egg). Follicle produces increasingly high levels of estradiol.
Estradiol	Ovary (follicle)	Causes rapid growth of endometrium. Rising levels of estradiol have *negative* feedback (inhibiting) effect on hypothalamus and LRF. LRF output reduced. Estradiol causes ant. pituitary to *inhibit* FSH production and *increase* LH production. Estradiol inhibits *release* of LH, which is stored in ant. pituitary. Very high levels of estradiol *reverse* effect on hypopthalamus, stimulating it to suddenly release large dose of LRF. LRF causes ant. pituitary to release sudden enormous *surge* of stored LH.
LH	Ant. pituitary	Stimulates follicle to break open and discharge ovum and follicular fluid.
Ovulation		
LH	Ant. pituitary	Follicle converted into corpus luteum, which secretes estradiol and increasing amounts of progesterone.
Progesterone	Ovary (corpus luteum)	Causes endometrium to become thick, spongy, and glandular; makes it receptive to a fertilized egg. Causes breast engorgement (may be sensitive or painful). Affects *negative* feedback on ant. pituitary, causing drop in LH production.
LH		As LH supply drops, the corpus luteum degenerates and can no longer produce progesterone and estradiol. In absence of progesterone, endometrium cannot be maintained. *Menstrual flow begins.*

MENSTRUAL CYCLE   OVULATION AND FERTILIZATION

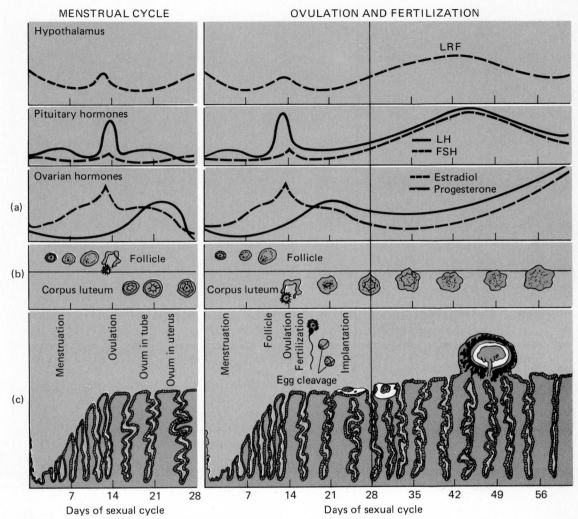

Menstrual cycles: Normal and during pregnancy.

endometrium is soon shed and the uterus reconstructs a new lining. This cycle is repeated about once a month. The two striking events of each menstrual cycle are *ovulation* (release of the egg) and the *menstrual period* (flow).

Menstrual periods usually occur once every 28 days. But the length of cycles may vary. Cycles as short as 21 days or as long as 38 days are known. The periods themselves are marked by a blood flow through the vagina. The average amount of blood and tissues lost ranges from 60 to 180 ml (about 4–12 Tbsp) each menstruation.

Hormones of the menstrual cycle The female menstrual cycle is under the control of three sets of hormones. One set arises in the *hypothalamus* of the brain. A second set, the *gonadotropins,* arises in the anterior pituitary gland. Located directly beneath the hypothalamus, the anterior pituitary is under the strict control of the hypothalamus. A third set of hormones is produced by the ovaries. Table 14.1 lists each of these menstrual hormones, tells where they are produced, and describes their effects.

Changes in hormone levels during the menstrual

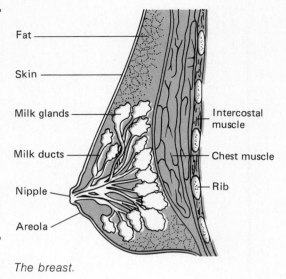

The breast.

● ASK YOURSELF

1. If ovulation occurs on the fourteenth day of a woman's menstrual cycle, how soon afterwards must fertilization occur? Where would it most likely occur?
2. What features in its structure make the uterus so well adapted to carry a growing baby during pregnancy?
3. Review the female sex organs. Distinguish between the external organs (genitals) and the internal organs.
4. As a woman, have you ever taken a mirror and examined your own genitals? If not, look at yourself and locate each of the structures mentioned.

cycle are seen in the accompanying figure. Peaks in estradiol and LRF provide the LH surge, which results in ovulation.

Other Special Events: Puberty Estradiol brings about the maturation of the secondary sex characteristics. Occurring during puberty, these include the "budding" of the breasts, deposition of fat around the hips, growth of pubic hair, growth of the genitals and reproductive tract, and the first menstrual cycle—*menarche*. Ready to operate much earlier than they actually do, a girl's ovaries await the necessary hormonal signals from the anterior pituitary, the FSH and LH. These must wait for the start-up signal of LRF from the hypothalamus. Thus the start-up system at puberty awaits the maturing of this part of the brain.

The onset of puberty is believed to be linked to nutrition. Around 1850, the average age for puberty in the United States was over 16. It now occurs at an average age of less than 13.

Pregnancy In the absence of pregnancy, the output of progesterone drops, the endometrium breaks down, and a menstrual period begins. If pregnancy intervenes, circumstances are altered.

The ovum, fertilized by sperm in the fallopian tube, finds its way into the uterus, and within several days embeds itself deep into the endometrium. Soon an embryonic membrane, the chorionic membrane, is formed and begins to produce hormones. One hormone, *human chorionic gonadotropin*

(HCG), is soon released. Acting almost exactly like LH, HCG rescues the corpus luteum from disintegration and causes it to maintain its production of progesterone and estradiol. The endometrium and its implanted embryo are thus preserved.

Since HCG is in the blood, it passes readily into the urine. Many tests for determining early pregnancy are based on the presence of HCG in urine.

High HCG production continues for about three months, then drops. By this time the placenta is producing large quantities of progesterone and estradiol, which continues to the end of pregnancy. Thus a pregnant woman has no menstrual periods until after the child and placenta are delivered. After delivery, menstrual periods may be delayed or irregular if the mother nurses her baby. This delay may be due to high output of PRL (prolactin) from the anterior pituitary.

Lactation Toward the end of pregnancy, levels of estradiol, progesterone, and prolactin are very high in the mother. These high levels of estradiol and progesterone stimulate the development of the nipples, the milk ducts within the breasts, and the glands where milk is formed. But high levels of progesterone *inhibit* actual milk production. After the placenta is expelled and progesterone levels drop, prolactin causes the breasts to fill with milk.

TABLE 14.2   MILK PRODUCTION

Hormones	Functions
Estradiol	Responsible for development of nipples, areolae, and ducts within breast that bring milk out to nipple.
Progesterone	Involved in the development of the breast's milk-producing of glands.
PRL (prolactin)	Stimulates breasts to produce milk only if breasts are first exposed to sufficient amounts of estradiol and progesterone.
Oxytocin	Causes contraction of cells around milk-producing glands. Cells squeeze milk into ducts and out nipple (milk "let down").
PIF (prolactin-inhibiting factor)	Inhibits anterior pituitary release of PRL. Normally present.

Nipple stimulation during breast feeding sends signals via sensory nerves to the hypothalamus, which transmits a signal to the posterior pituitary gland to release the hormone *oxytocin* (*see* Table 14.2). The breasts then "let down" the milk. Suckling also reduces the usual output of the hormone PIF, thus further enhancing milk production.

Various drugs may reduce a woman's milk production. Nicotine reduces lactation by inhibiting prolactin release by the anterior pituitary. Alcohol lowers milk production by inhibiting oxytocin release by the posterior pituitary gland.

Menopause   Eventually, the blood vessels of the ovaries begin to deteriorate. Over the course of several years, the ovaries become unresponsive to the stimulation of either FSH or LH. For most women, *menopause* (the cessation of menstrual cycles) occurs between the ages of 47 and 52.

No longer responsive to FSH and LH, the ovaries' output of progesterone and estradiol drops. The negative feedback of these hormones on the pituitary gland is thus removed, and LRF and FSH output goes on unabated.

Controlling blood circulation is another function of the hypothalamus. It does this by controlling the size of many small blood vessels. During menstrual life, the hypothalamus becomes accustomed to much estradiol (an estradiol addiction, of sorts). When the ovaries stop producing estradiol, the hypothalamus goes through an estradiol withdrawal, and *vasomotor* (blood-vessel control) instability occurs. Blood vessels of the face and neck, in par-

ticular, dilate and produce skin flushing and sensations of heat ("hot flashes") from time to time. Some women become depressed, nervous, or show other emotional symptoms around this time. It is believed that social and cultural factors, as well as hormones, play a role in a woman's perception of menopause.

The Male Sex Organs

The male sex organs are less complex than those of the female in that the male does not need to provide for the growth of a fertilized egg. The male's external organs, or genitals, are more conspicuous than those of the female.

● ASK YOURSELF

1. In the menstrual cycle, what is the relationship of the pituitary gland to the ovary and uterus?
2. What role does the brain play in overseeing the events of the menstrual cycle, lactation, and menopause?
3. If the flow of progesterone from the ovary is sufficient to inhibit LH production and bring on menstruation, what prevents this from happening when a woman becomes pregnant?
4. Pregnant women are advised not to smoke. How could smoking have an affect upon her production of milk?
5. Women talk about the physical changes they go through during menopause. What is the physical explanation for the "hot flashes" they sometimes report?

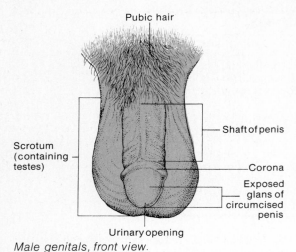

Male genitals, front view.

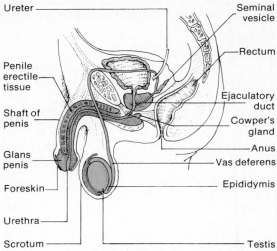

Male reproductive system, side view.

Genitals The male genitals, shown in the illustration, consist of the penis and the scrotum.

Penis The *penis* is the male organ of urination and sexual intercourse. It contains three columns of spongy erectile tissue. When the male becomes sexually aroused, blood fills this spongy tissue, causing the penis to become firm and erect. The erection is lost when blood leaves the penis faster than it enters. One of the columns of spongy tissue forms the head, or *glans,* of the penis. The glans, particularly its *corona* (the ridge at the back edge of the glans), is the most sexually sensitive area of the genitals. The body, or *shaft,* of the penis is covered by loose skin, which near the glans is known as the *prepuce* or *foreskin.* The surgical removal of this foreskin is known as *circumcision.*

The inability of the male to attain erection is called *impotence;* it may be the result of either physical or psychological factors. Impotence must not be confused with *sterility,* which is an inability to produce offspring. A man can be sterile yet fully potent.

Scrotum The *scrotum* is a loose pouch of skin suspended behind and beneath the penis. Each of the two compartments of the scrotum contains a testis and its spermatic cord, consisting of the vas deferens (pl. vasa deferentia), blood vessels, nerves, and muscles. Since sperm production cannot occur at normal body temperatures, the scrotum suspends the testes away from the body. Thus, their tempera-ture is 3 to 4 degrees F lower than normal body temperature. In cold temperatures, the thin scrotal muscles contract and pull the testes closer to the body wall. In hot temperatures the muscles relax, allowing the testes to be suspended further away from the body.

Internal sex organs The internal organs include the two testes, epididymis, vasa deferentia, and several fluid-producing glands.

Testes The *testes* are a pair of reproductive glands, the *male gonads.* They serve the dual function of producing both *sperm* (male gametes, sex cells) and the male sex hormone *testosterone.* Within each testis are many very small tubes, or *seminiferous tubules.* Sperm cells form inside these tubules. Beginning during puberty, these sperm cells are produced without letup for the lifetime of the male. Initial production starts slowly, then increases, until, in the sexually mature male, approximately 250 million sperm are produced daily.

Each sperm is microscopic in size. The length of each cell is about 60 micrometers (it would take over 400 of them end to end to cover an inch). Each sperm consists of a head, neck, body, and tail.

Human sperm cell.

TABLE 14.3 MALE REPRODUCTIVE HORMONES

Hormone	Source	Functions
Gonadotropins		
1. FSH (follicle-stimulating hormone)	Anterior pituitary	Stimulates seminiferous tubules of testes to produce sperm. In the female, this hormone stimulates the ovary.
2. ICSH (interstitial-cell-stimulating hormone)	Anterior pituitary	Stimulates interstitial cells around the seminiferous tubules to produce testosterone.
Testosterone	Testes (seminiferous tubules)	Causes development of sexual maturity in the male. Rising levels of testosterone have negative feedback (inhibiting) effect on the anterior pituitary.

The testes of a male fetus are formed in the abdominal cavity. About the eighth month of development, the testes migrate from the abdominal cavity into the scrotum. When the testes fail to descend, the male becomes sterile.

The testes serve also as endocrine glands, producing the hormone *testosterone*. Beginning during puberty (at age 12 or 13), testosterone production initiates the physical changes leading to sexual maturity.

Epididymis As the sperm mature, they move out of the seminiferous tubules and collect in a coiled tube called the *epididymis*. It lies on the upper side of each testis and would be about 20 feet long if uncoiled. Here sperm are stored until released from the body by ejaculation or until they disintegrate and are reabsorbed by the tubules.

Vasa deferentia The *vasa deferentia* are small ducts carrying sperm upward from each epididymis to the ejaculatory duct. Surgical sterilization of the male, or *vasectomy,* usually consists of cutting and tying off or cauterizing each vas deferens.

Fluid-producing glands Sperm cells make up only a small portion of the seminal fluid. The most important fluid-producing glands are the *seminal vesicles,* which secrete a fluid that initiates the motility of the sperm. The vesicles empty into each vas deferens at the ampulla. The second source of fluid is the single *prostate gland.* The vasa deferentia and the duct carrying urine from the urinary bladder, the *urethra,* unite within the prostate gland. The urethra then serves as a common duct for both urine and semen from the prostate to the tip of the penis. Emptying by ducts into the urethra just below the prostate are a pair of small glands called *Cowper's glands.* Secreted just prior to ejaculation, their clear, sticky fluid helps clear out any urine from the urethra in preparation for the passage of semen.

Semen, or **seminal fluid,** is the fluid ejaculated by the male. It consists of fluids from the testes, seminal vesicles, prostate gland, and Cowper's glands. The semen is a grayish-white sticky fluid, containing 60 million to 120 million sperm per ml.

Ejaculation Physical stimulation of the penis not only causes it to become erect but finally results in *ejaculation,* the forcible expulsion of semen. Ejaculation usually results in the discharge of about 2.5 ml of semen. Occurring at the *climax* of response to sexual arousal, it is usually accompanied by a feeling of intense sexual excitement and emotional release called *orgasm.* Shortly after ejaculation, the orgasm subsides, the penis becomes limp, and the male feels sexually satisfied.

Nocturnal emissions Commonly called "wet dreams," *nocturnal emissions* are involuntary discharges of semen with orgasm during sleep, often accompanied by sexual dreams. Almost all males experience these emissions at some time. They are perfectly harmless and serve to relieve the pressure of fluid in the seminal vesicles if no other sexual outlet is available.

Male reproductive hormones The male hormones are collectively called *androgens.* These hormones and their sources and actions are summarized in Table 14.3.

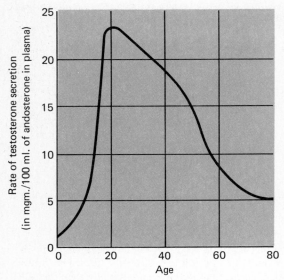

The rate of testosterone secretion by age in males. Note the lifetime peak at ages 17 to 21.

The absence or removal of the testes may cause hormonal deficiencies in the male. Since the production of testosterone (the most important androgen) becomes increasingly important to male sexual traits after puberty, the effects of insufficient testosterone depend on when the deficiency occurs. If it occurs before puberty, the male fails to develop secondary sex characteristics. A boy who loses his testes prior to puberty becomes a *eunuch.* A male who loses his testes after the onset of puberty will

---

● ASK YOURSELF

1. Men who work near hot objects (ovens, foundries, blast furnaces, hot engines) are sometimes sterile. What would cause this?

2. In order for a man to be sterilized against producing more children, some "amateurs" recommend removing the testes. For what reasons isn't this a good idea?

3. Older men suffering from swollen prostate glands may have trouble urinating. Why would there be such a relationship?

4. In an ejaculation, males report feelings of contraction deep in the pelvic area. Where might such feelings be coming from?

5. Review the male sex organs. Distinguish between the external organs (genitals) and the internal organs.

---

retain some male secondary sex traits. The removal of the testes is called *castration.*

SEXUAL STIMULI

Sexual responses are a basic element of human physiology and psychology. Not surprisingly, they can be induced by a wide range of stimuli. The sounds, smells, touch, sight, and indeed even the thought of sexual situations can sexually arouse most people. The sexual responses of both men and women can be divided into identifiable phases, each with its physical and emotional characteristics.

The *excitement phase* is the first, accompanied by the initial signs of physical arousal. The *plateau phase* follows, in which the sex organs further change in size and shape. Breathing and cardiac rates increase. The *orgasm phase* brings the intensely satisfying sensations of sexual climax. In the *resolution phase,* the organs return to their normal sizes and shapes, and the intense feelings of climax subside.

The Female Sexual Response

The response of a woman's body to sexual stimulation is widespread, involving organs besides the genitals. Both the genital and extragential events of female sexual arousal are summarized in Table 14.4.

As the female becomes excited the clitoris swells and elongates, the vagina lubricates (faster with greater arousal), and the breasts enlarge and the nipples stiffen. Skeletal muscles tighten throughout the body. Pulse and breathing rates increase.

As response heightens, the clitoris moves under its hood and becomes so tender that efforts to touch it directly may be painful. The inner lips increase in size and turn bright red. The outer third of the vagina contracts, while the inner two-thirds expands.

In the orgasm phase, strong contractions occur in the outer half of the vagina (3 to 5 in a mild orgasm, 8 to 12 in an intense one). The uterus contracts as it does in labor, and the anus contracts tightly. The intensity and duration of orgasm in the female varies more than in the male. As do males, females may come to orgasm involuntarily during sleep, accompanied by sexual dreams.

The period of sexual stimulation preceding orgasm varies from woman to woman and from time to time for the same woman. The experience of many couples is that the woman needs a longer period of stimulation in intercourse to reach orgasm than does the man. But Alfred Kinsey et al. (1953) reported that the average woman can masturbate to orgasm almost as fast as the average man. They attributed the difference to the fact that the masturbating woman can manipulate her sensitive areas more specifically than is possible in intercourse. This conclusion seems to be confirmed by Masters and Johnson (1966), who found the measurable physiologic intensity of female orgasm greatest in masturbation, moderate in partner manipulation, and lowest in intercourse.

This conclusion is further endorsed by *The Hite Report* (1976). Based on her survey, Hite concluded that clitoral stimulation is central to women's sexuality. She found that most women did not regularly reach orgasm during intercourse without direct massaging of the clitoris. She believes that female orgasms during intercourse may be emotional orgasms—a desirable form of release as long as women are not pressured into using emotional orgasm as a substitute for that induced by clitoral massage.

The subjective experiences of a woman during orgasm, as reported by Masters and Johnson, begin with a feeling that orgasm is imminent, followed by an intense sensual awareness of the pelvic region. The woman becomes almost completely oblivious to the surrounding environment, focusing on her own sensations. The next feeling, reported by almost every woman, is that of warmth, starting in the pelvic region and spreading throughout the body. A final feeling, reported consistently, is that of a pelvic throbbing.

During and directly following orgasm, there is a marked coolness of the lips of the mouth and surrounding areas. This is due to sudden release of blood vessels. Muscular and psychological relaxation accompanies orgasm.

Many women are multiorgasmic (experiencing three or more orgasms within a few minutes) if stimulation is repeatedly resumed before sexual arousal drops below plateau-phase levels. Masturbation often best contributes to multiorgasmic ability, because it frees the women from dependence on her partner's abilities. In lovemaking, the woman's ability to be multiorgasmic depends to a greater extent on foreplay and other forms of stimulation than on prolonged intercourse. In intercourse, the possi-

William Masters and Virginia Johnson.

bilities for multiple orgasm depend somewhat on the partner's ejaculatory control—his ability to engage in prolonged, vigorous intercourse without ejaculation and subsequent loss of erection. Many women also enjoy reaching multiple orgasms through oral or manual stimulation by their partners.

The Male Sexual Response

As with the female, the male's body response to sexual stimulation is widespread and involves organs other than the genitals. Both the genital and extragenital events of male sexual arousal are summarized in Table 14.5

● ASK YOURSELF

1. The vagina is the site of penile insertion and insemination during intercourse. Trace the changes that occur within the vagina during each of the four phases of the sexual response cycle.
2. The clitoris is similar to the penis in structure and sensitivity. The clitoris differs in that it is much smaller and contains no urethra. But if the vagina is where penile insertion occurs, what is the function of the clitoris?
3. Much has been written about female orgasm. If women do not perceive all orgasms as being the same, how and why might orgasms differ in nature?
4. If a woman would like to develop her capacity for multiple orgasms, how might she enhance her chances?

**TABLE 14.4a   SEXUAL RESPONSE CYCLE
OF THE HUMAN FEMALE—GENITAL REACTIONS**

	I. Excitement phase	II. Plateau phase	III. Orgasmic phase	IV. Resolution phase
Outer lips	Lift up and out—away from vagina—in woman who has never had a child; move slightly away from midline in woman who has given birth	May become engorged with blood during prolonged phase	No observed reaction	Returns to normal slowly
Inner lips	Slight thickening and expansion that extends vagina outward by 1 cm	Vivid color changes ranging from bright red to deep wine;	No observed reaction	Color subsides to normal light pink in 10–15 sec; return to normal size slowly
Clitoris	Tip of clitoris (glans) swells. Shaft elongates and its diameter increases; clitoral color deepens due to extra blood	Clitoris withdraws and retracts against pubic bone and under the hood	No observed changes	Returns to normal position 5–10 sec after orgasm; slower return to normal size color
Vagina	Vaginal lubrication appears 10–30 sec after initiation of any form of sexual stimulation; vaginal barrel expands and distends; vaginal wall color deepens markedly	Orgasmic platform develops in outer third of vagina; gross distension by blood reduces size of opening by 1/3; inner 2/3 of vaginal barrel widens and deepens	Orgasmic platform contracts at 0.8 sec intervals for 5–12 times; after first 3–6 times, contractions slow down and are less intense	Swelling and engorgement of orgasmic platform rapidly subsides; vaginal wall relaxes and slowly (15–20 min) returns to normal color
Uterus	Uterus begins to elevate to upright position from anterior one; uterine muscles show response to stimulation	Full uterine elevation producing tenting effect over vagina; increased reaction of uterus to stimulation	Contraction of uterus starts at top and spreads down to cervix; parallels the intensity of orgasm; usually begins 2–4 sec after woman is aware of orgasm	Widening of cervical opening continues for 20–30 min; cervix dips into seminal pool; uterus returns to normal position

Adapted from Masters and Johnson, *Human Sexual Response* (Boston: Little, Brown, 1966).

Sexual arousal in the male is signaled by erection of the penis. The scrotum tightens, and the testes draw up close to the body. During prolonged sexual arousal, the penis may erect and relax and then erect again several times.

As arousal heightens, a few drops of preejaculatory fluid from the Cowper's glands are gradually emitted from the penis. This fluid contains active sperm. The testes become further elevated (the more complete the testicular elevation, the greater the ejaculatory pressure), and they increase in size. The increase in sexual arousal to the point of orgasm (ejaculation) requires, in all but a very few males, tactile stimulation of the penis.

**TABLE 14.4b SEXUAL RESPONSE CYCLE
OF THE HUMAN FEMALE—EXTRAGENITAL REACTIONS**

	I. Excitement phase	II. Plateau phase	III. Orgasmic phase	IV. Resolution phase
Breasts	Nipples erect; breast size increases; veins become visible; areolae swell	Nipples turgid; further increase in breast size; areolae engorge	No observed changes	Nipples and areolae rapidly return to normal; slow decrease of breast volume and normal vein pattern
Sex flush (not present in all women)	Raised reddish rash appears first on upper abdomen, spreads rapidly over breasts	Rash spreads to rest of body	Degree of flush parallels orgasmic experience (estimated at 75% incidence)	Rapid disappearance of flush in reverse order of its appearance
Muscle tensions	Voluntary muscles tense; involuntary: vaginal wall, abdominal, and rib muscles tense	Further increase in both voluntary and involuntary tension; may show spastic contractions of facial, abdominal, and rib muscles	Loss of voluntary control; involuntary contractions and spasms of muscle groups	Spasms and involuntary contractions may last up to 5 min; slower return to normal
Rectum	No observed reaction	Voluntary contractions of anus (sometimes)	Involuntary contractions of anus simultaneously with contractions of orgasmic platform	No observed changes
Respiratory rate	No observed reaction	Appearance of reaction occurs late in phase	Respiratory rates as high as 40/min; intensity and duration indicative of sexual tension	Resolves early in phase
Heart rate	Rises parallel to tension regardless of technique of stimulation	Rises to average 100–175/min	Ranges from 110–180 + beats/min; rates indicate variation in orgasmic intensity	Return to normal
Blood pressure	Elevation parallels rising tension	Elevation rise above normal by: Systolic: 20–60 mm Hg Diastolic: 10–20 mm Hg	Elevations rise above normal by: Systolic: 30–80 mm Hg Diastolic: 20–40 mm Hg	Returns to normal

Adapted from Masters and Johnson, *Human Sexual Response* (Boston: Little, Brown, 1966).

Masters and Johnson divide the male orgasm into two stages. The first is a period of two or three seconds before ejaculation during which a male can feel an ejaculation coming and can in no way restrain or control it. The second is the actual ejaculation of semen from the penis by contractions of the urethra and related muscles. The first three or four contractions expel the semen under great pressure.

**TABLE 14.5a SEXUAL RESPONSE CYCLE
OF THE HUMAN MALE — GENITAL REACTIONS**

	I. Excitement phase	II. Plateau phase	III. Orgasmic phase	IV. Resolution phase
Penis	Rapidly becomes erect; may be partially lost and regained during prolonged phase or impaired by introduction of nonsexual stimulation	Increase of circumference at coronal ridge; color deepens at coronal ridge (in estimated 20% of men)	Expulsive contraction entire length of penis at 0.8 sec intervals; after first 3–4, frequency and force lessen; minor contractions continue for several seconds	Two-stage return to normal: 1. Rapid loss of congestion to 1–1½ times enlargement 2. slowly subsides to pre-arousal size
Scrotum	Loses its folds and free movement; sac constricts, becomes congested and appears tense	No specific reactions	No specific reactions	Rapid loss of tense appearance; return of folds and free movement sometimes delayed
Testes	Spermatic cord shortens, causing testes to partially lift toward body	Testes enlarge to 50% of original size; the longer this phase lasts the greater the increase in size due to congestion; may be 100% full elevation of testes, particularly the left or lower testicle; signals impending ejaculation	No recorded reaction	Return to normal size and position in relaxed scrotum may occur rapidly or slowly, depending on length of plateau phase
Secondary organs (vas deferens, prostate, seminal vesicle, and ejaculatory duct)	No observed changes	No observed changes	Secondary organs contract, creating sense of pending ejaculation; initiates ejaculatory process	No observed changes
Cowper's glands	No observed changes	Emits 2–3 drops of mucoid fluid containing active sperm	No observed changes	No observed changes

Adapted from Masters and Johnson, *Human Sexual Response* (Boston: Little, Brown, 1966).

As in a female, a male's breathing rate may increase from about 17 to 40 times per minute, and the heart rate from about 70 to 180 beats per minute.

Most men are incapable of another erection for a period of time after ejaculation. This is called the *refractory period*. However, the minimum time interval required before repeated male erection and

**TABLE 14.5b   SEXUAL RESPONSE CYCLE
OF THE HUMAN MALE—EXTRAGENITAL REACTIONS**

	I. Excitement phase	II. Plateau phase	III. Orgasmic phase	IV. Resolution phase
Breasts	Nipples erect (inconsistent—may be delayed till second phase)	Nipples erect and turgid (inconsistent)	No observed change	Slow return to normal
Sex flush (not always present)	No observed reaction	Raised, red rash late in phase (inconsistent; starts on upper abdomen, spreads to chest, neck, face, and maybe arms)	Well-developed rash, parallels intensity of orgasm (estimated at 25% incidence)	Rapidly disappears in reverse order of appearance
Muscle tension	Voluntary muscles tense; involuntary show some signs; testes begin to elevate; abdomen and rib spaces tense	Increase of both voluntary and involuntary tension; semispastic contractions of face, abdomen, and rib musculature	Loss of voluntary control; involuntary contractions and spasm of muscle groups	May take up to 5 min to subside
Rectum	No observed reaction	Voluntary contraction of rectal sphincter (inconsistent)	Involuntary contraction of rectal sphincter at 0.8 sec intervals	No observed changes
Respiratory rate	No observed reaction	Reaction appears late in phase	Rate rises to as much as 40/min; intensity and duration indicate sexual tension	Resolves during refractory period
Heart rate	Increases parallel to rising tension regardless of technique	Rates range from 100 to 115 beats/min	Rates range from 110 to 180 beats/min	Returns to normal
Blood pressure	Elevates directly parallel to rising tension regardless of technique of stimulation	Elevates from normal by: Systolic: 20–80 mm Hg Diastolic: 10–40 mm Hg	Elevates from normal by: Systolic: 40–100 mm Hg Diastolic: 20–50 mm Hg	Returns to normal

Adapted from Masters and Johnson, *Human Sexual Response* (Boston: Little, Brown, 1966).

orgasm varies greatly, both among different men and for the same man at different times. It may range from minutes to hours. This time interval increases with the age of the man and with general physical or emotional fatigue. It decreases with a high degree of original sexual arousal, a high degree of sexual restimulation after ejaculation, and a long period of sexual restraint prior to the first ejaculation. A young man who has had no sexual release for some time, or has been restimulated after ejaculation, may be ready for further intercourse within just a few minutes.

Lovemaking involves much more than good technique. Sexual enjoyment depends on a willingness to give and receive pleasure, open communication, and sensitivity to one's partner.

Techniques of Intercourse

It is not the intention of this book to give detailed instructions for achieving sexual satisfaction. We will, instead, give a few general comments that may or may not be useful to a particular couple. A good sexual relationship can be achieved if each partner has an adequate knowledge of sexual psychology and physiology, a positive attitude toward sex, and a concern for the sexual satisfaction of the partner. Open and uninhibited discussion of sexual desires and responses is important. Each partner should feel free to tell the other which practices increase

● ASK YOURSELF

1. Since the penis is the main indicator of sexual arousal in the male, trace the changes that occur with the penis during each of the four phases of the sexual response cycle.
2. What might be some advantages for the propulsive force of the semen during ejaculation?

sexual enjoyment and which decrease it. If a couple is satisfied with their sexual patterns, they may not feel the need to experiment with varied positions and practices. As long as open, honest communication exists, and the couple is aware that their own desires and interests may change with time, then they should practice whatever is comfortable for both of them.

Preparing for intercourse There is no set routine that is universally necessary or useful in preparing for intercourse. The sexual responses of men and women are highly variable and individual, so a technique that is ideal for one couple might have no value for another.

It is common, however, for the women to need a longer period of stimulation before intercourse than the man in order to prevent pain upon first penetration and increase her chance of reaching orgasm. While a man may attain erection in a few seconds, a woman may need several minutes to become fully aroused and produce adequate vaginal lubrication.

Most women enjoy a period of sex play before actual penetration of the vagina as it prepares them physically and psychologically for successful intercourse. Every woman will discover particular types of caresses and stimuli that provide her with the most intensive sensations, and she should freely communicate this information to her partner.

Among the types of sex play that stimulate many women are kissing various parts of the body; gently fondling or tightly squeezing the breasts; lightly rubbing, pinching, pulling, sucking, or lip-biting the nipples; squeezing or lightly rubbing the buttocks; lightly stroking the insides of the thighs; and manual or oral manipulation of the genital organs. Genital manipulation is mentioned last because many women prefer to become somewhat aroused before this begins. Masters and Johnson (1966) suggest that indirect clitoral manipulation by stimulating the mons area is preferred by most women as direct clitoral stimulation can be painful. Women who are fatigued or under emotional stress may need more than usual stimulation prior to intercourse or may prefer to postpone intercourse.

The test of whether a woman is ready for penetration is the extent of lubrication of the vaginal walls. The man can readily determine this with his fingers.

Men also respond to a great variety of stimuli,

such as kissing; licking and kissing the ears; body kisses; and oral and manual stimulation of the nipples, penis, scrotum, and anus. In sex play, any practice accepted and enjoyed by both partners should be used.

In all types and phases of sexual relationships, open and honest communication is important. Many people are very reluctant to talk about sex, even to the extent that they fail to tell their partners just what pleases or displeases them sexually. Therapists report many cases where the primary cause of sexual problems is lack of communication between partners. It is of utmost importance that couples develop the ability to openly and honestly communicate their sexual feelings.

Sexual intercourse Intercourse proper begins with the insertion of the penis into the vagina. This can be achieved from a variety of positions. There has been a tendency in the sex manuals to stress positions that place the penis in direct contact with the clitoris, based on the assumption that if the clitoris is the most sexually responsive part of the female body then it should receive direct stimulation. However, because the clitoris reacts beneath the clitoral hood when aroused, it becomes unimportant whether the clitoris is stimulated by the hood or by the penis (which may be difficult).

Coital positions Any position that allows a full penetration of the vagina by the erect penis will often provide an adequate amount of indirect clitoral stimulation. Most important is that intercourse be a true act of concern and participation between two people. Intercourse should not be something that is done *to* someone else, but the enjoyable sharing of a love act by two people.

A couple who try many different positions for intercourse will probably find several that seem particularly good for them and will probably use all of these positions from time to time. Some of the more common basic positions are discussed briefly below. Each of these positions has many variations.

1. *The man-above position.* This position is the most natural and comfortable one for many couples. The woman lies on her back with her legs spread apart, either drawn up or straight out. The man lies facing her, between her thighs. If guidance of the penis into the vagina is necessary, either the man or woman can give this assistance. This position allows most couples to kiss freely, but is more restrictive of breast manipulation than some of the other positions. Masters and Johnson found it to be the best position in which to achieve pregnancy.

2. *The woman-above position.* In this common position, the man lies on his back while the woman lies above and facing him. Some couples roll from the man-above position to the woman-above position, or vice versa. Some women can achieve orgasm more easily in this position by controlling the pelvic thrusts while the man lies more or less passively.

3. *Face-to-face positions.* There are many lateral positions. They offer several advantages. Neither partner must support the weight of the other, and thus are good for prolonged sexual connections. Kissing is easy, as is breast manipulation or any other type of caressing desired.

4. *Rear-entry positions.* There are several different rear-entry positions. The woman may lie on her side or face down or may kneel on her hands and knees. In any case, the man approaches from behind, passing his erect penis between her legs and into her vagina. These positions facilitate breast manipulation, but, of course, kissing is more difficult.

5. *Other positions.* The variety of sexual positions is limited only by the imagination and agility of the couple. For variety in their lovemaking, many couples occasionally enjoy more unusual positions. Sitting positions are enjoyed by some couples. In one of these, the man lies on his back while the woman sits astride him, feet forward. This position allows very deep vaginal penetration. In other sitting positions, the man may sit on a chair or the edge of a bed while the woman sits astride him. Some couples even enjoy intercourse while standing up, using front or rear entry. Any position that affords mutual pleasure should be used.

After the penis has been fully inserted into the vagina, a man is often at the very peak of his sexual

● ASK YOURSELF

1. Some sexual partners feel their real wishes as to kinds of sexual stimulation are misunderstood. How might a couple improve their communication of feelings and desires?
2. During sexual play, inserting a penis into an unaroused vagina can be painful for the woman. Aside from verbal instruction, how might a man determine when the vagina is prepared for entry?
3. There is much interest in oral sex. What might be some of the pros and cons of cunnilingus? Some pros and cons of fellatio?
4. Of all the coital positions, which one is best for achieving conception? Why?

arousal and near orgasm. In order to prevent premature ejaculation, many couples find it desirable to lie together quietly for a short time before beginning the pelvic thrusts of intercourse. By lying quietly at this time and any subsequent time that orgasm seems near, many men can delay ejaculation.

One criterion in the choice of positions is whether they allow for greater or lesser penetration. Depending on differences in length of penis and vagina, the level of sex drive, and the absence or presence of pain during a given lovemaking, the woman may or may not desire greater penetration. Subtle changes in position can allow for this. For instance when the woman is on her back she may, after initial penetration, bring her legs together to prevent too deep a penetration. A creative couple can usually find the preferable position for a given lovemaking technique.

Orgasm Some couples place great importance on *simultaneous orgasm*. Simultaneous orgasm is not an important goal in sexual intercourse and concentration toward this goal may decrease the pleasure received by each partner, rather than increase it. Since the orgasm of the young male is usually assured (except in impotence, as discussed later), he should be concerned with helping the woman to orgasm. However, many men maintain prolonged erection in lovemaking without reaching orgasm. As with the female, intercourse without orgasm can, on occasion, be satisfying for the man.

Women vary in the ease with which they reach orgasm. The difference may be either physical or psychological, but it is very real. Every study made has shown that significant numbers of women reach orgasm only with difficulty, if at all. Kaplan (1974) believes that 30 percent of women never reach orgasm, or if so, only by masturbation. Hite (1976) found that only 30 percent of women reach orgasm regularly from intercourse, 22 percent reach orgasm rarely from intercourse, 19 percent reach orgasm during intercourse with the addition of simultaneous clitoral stimulation by hand, and that 29 percent do *not* reach orgasm during intercourse.

A woman may not expect to experience orgasm with every intercourse. But the woman who reaches orgasm too infrequently is likely to lose her enthusiasm for sex. The man who wants to keep his sexual relationship happy and vigorous should try to satisfy his partner as often as possible.

No single pattern of orgasm is best for every couple, though many couples find, through experimentation, a pattern that seems best for them. Many couples develop a pattern in which the woman reaches one or more orgasms first, followed by the ejaculation of the man. In other couples, the woman finds that the stimulus of the penis throbbing in ejaculation is just what she needs to push her to orgasm.

Oral sex Oral sex involves stimulation of the genitals with the mouth. It may be used as the sole means of reaching orgasm or it may be used to heighten sexual arousal. *Cunnilingus* is stimulation using the tongue on any part of the female genitals, often with special attention to the clitoral area. It may be part of either heterosexual or lesbian sex plan. The fact that some women prefer orgasm through oral or manual stimulation rather than by penile-vaginal intercourse does not imply that they are sexual failures or immature or abnormal. It is simply one variation within the range of natural sexual responses.

Oral stimulation of the male's genitals by a partner is *fellatio*. It may include sucking or licking the glans penis, nibbling along the shaft of the penis or on the testicles. While it may be part of either heterosexual or homosexual sex play, a man's preference for it does not indicate a lack of masculinity or homosexual tendencies.

Some partners may practice mutual oral-genital

stimulation. Sometimes called "69," partners may lie abdomen to abdomen, but facing in opposite directions.

Oral stimulation may be especially useful whenever other forms of arousal are not satisfactory. Some women find it difficult or impossible to achieve orgasm through intercourse. Some men, especially those who are older, need manual or oral stimulation of the genitals in order to achieve erection.

Oral sex should be engaged in *only* if mutually acceptable to both partners. Other forms of pleasurable genital stimulation may be used if oral sex is not enjoyed by both. Since several genital diseases may be transmitted to the mouth, the genitals should be thoroughly cleaned with soap and water *before* oral stimulation.

Anal sex Anal practices include insertion of a finger into the anus of either partner, prior to or during coitus, and anal intercourse in which the penis is inserted into the anus. About half of all men and women report that they experience erotic reactions to some form of anal stimulation. Some women are able to reach orgasm through rhythmic contractions of the anus alone. In anal intercourse certain precautions are necessary. It is usually advisable to lubricate the penis to prevent tearing the delicate anal tissues. A further precaution is necessary: if the anus is tense (because the person fears pain), it must be slowly relaxed, or penetration may in fact be painful. Many physicians also recommend washing the penis after anal penetration if vaginal intercourse is to follow. This should be done in order to avoid possible vaginal injection with rectal bacteria. Anal sex may progress to orgasm or may lead to vaginal coitus.

Frequency of Intercourse

The frequency of sexual intercourse is a source of conflict in some relationships. In almost any relationship, there will be times when one partner would like sexual intercourse and the other partner is either not interested or incapable of responding. Such situations may place a great burden on the relationship unless both partners are understanding and tolerant. Sex is only a part of any relationship, and the success of a couple's sexual relationship is often a reflection of their total ability to communicate and relate to each other.

It is very difficult to compare the sex drives of men and women, because the drives themselves are fundamentally different. To be sexually arousable, a woman must be sensitive to tactile stimuli and able to enjoy penile penetration. Most women experience compelling sexual drives that demand release, particularly during sexual arousal when blood engorges their genital organs.

Males possess an internal physical sexual stimulus from the accumulation of fluids within the seminal vesicles and prostate gland. The longer these fluids build up, the greater becomes the need for release. Males who have not ejaculated for an extended period of time may experience an extraordinary level of sexual tension.

There is no particular "normal" frequency of intercourse that is most desirable. Among happy and physically healthy couples, the frequency of intercourse ranges from once a month or less to several times a day. Statistically, the average frequency, though this should not be interpreted as a goal for an individual couple, is between two and three times a week. A young couple is likely to exceed this frequency, whereas an older couple may not engage in intercourse that often.

Disagreement over frequency Let us first consider the more likely situation—a couple wherein the man desires intercourse more often than does the woman. First, there are several things the man should ask himself: Am I keeping myself physically attractive? Am I pleasant and loving with my partner at all times, or just in bed? Do I spend enough time in precoital sex play? Do I seriously try to help her reach orgasm as often as possible? (The woman who is too often disappointed soon learns to avoid frustration by avoiding sex.) Am I expecting too much of her? Few men appreciate the amount of energy women spend on careers, housework, child care, and community activities.

The woman might ask a few questions of herself: If I have ever refused him, have I honestly explained to him why? Would I be more confident with another contraceptive method? Am I trying to punish him for something that is beyond his control?

The situation in which the woman wants intercourse more often than the man is more difficult. The man must attain erection and generally take a more active role in order to satisfy her. Again the

● ASK YOURSELF

1. What steps might a couple take to overcome disagreement over the frequency of having sexual intercourse?
2. Why should a man with a short penis anticipate as much sexual fulfillment as a man with a long penis?
3. Is there any basis for the common belief that large-breasted women are more responsive sexually than those with smaller breasts?

woman might consider possible reasons for his reluctance: Do I try to be sexually appealing? Are my demands excessive?

As men grow older, their biological capacity for sex diminishes. Older men are thus in a particularly bad emotional position. They are faced with the demands and perhaps derision of their partners in addition to their own feelings of inadequacy. The man who feels sexually inadequate is often reluctant to make any attempt at intercourse, for fear of impotence. He avoids the risk of ego-damaging impotence by avoiding sex. Because this situation often becomes worse with the passage of time, prompt efforts should be made to remedy it.

Anxieties

Many people, more sexually inexperienced ones in particular, suffer needless anxieties over the appearance of certain parts of their body. They are concerned lest the shape or size of their breasts, vagina, or penis, or fears of pain during intercourse, interfere with their sexual attractiveness or in some way limit their sexual expectations.

Male anxieties Many young men suffer needless anxiety regarding the size of their penis. Some worry that it may be too small to satisfy a woman. Some fear that it may be too large to be accommodated by the vagina. The size of the penis and whether or not it is circumcised has little if anything to do with the sexual satisfaction of the man or of the woman. Sexual technique, experience, and care about the partner's needs are the important things.

The vagina can accommodate a penis of virtually any size. Masters and Johnson found that even in very small women, the vagina, when sexually aroused, lengthened sufficiently to accept the largest fully erect penis. Thus there need be no

apprehensions over sexual intercourse between a man with a large penis and a woman with a small vagina.

The man with a short penis should realize that there are very few sensory receptors in the deeper part of the vagina. These receptors are concentrated in the clitoris and labia and outer vagina. As a result, the sensory satisfaction a woman receives will be as great with a short as with a long penis.

Female anxieties Because of the erotic significance our culture has attached to a woman's breast, some women are unduly concerned about the development of their breasts. Breast size has no bearing on a woman's responsiveness. Flat-chested women report as much response to sexual stimulation as big-bosomed ones. The size of the breast has no relation to its sensitivity. Nor is there any relationship between breast size and the consistency of attaining orgasm, or to the frequency of intercourse. Breast size has no relation to the woman's ability to achieve orgasm.

Some women with no coital experience hold unconscious, irrational fears of mutilation from penetration of the penis. With sufficient partner tenderness and sex play, the entry of the penis brings great pleasure to the woman in the absence of any pathological condition or psychological problem. In fact, during sexual arousal a woman is less sensitive to pain than she is when not sexually aroused.

PROBLEMS IN SEXUAL RESPONSE

Few people go through life without experiencing at least some instances of difficulty in their sexual responses. The difficulty may be as minor as that of the man who once in a great while is unable to achieve erection or to delay ejaculation until his partner is satisfied. The woman on rare occasions may not respond in the least to stimulation from her partner. Or if aroused, she is unable to achieve orgasm. But some problems in sexual response are not always so temporary. Some persist and become so regular that they prevent almost all sexual satisfaction.

The term *dysfunction* means painful or difficult functioning; something has gone wrong in what should be a normal response. The term underscores the functional nature of most sexual problems. Some of them arise from definite physical or

Sexual estrangement in marriage often indicates emotional alienation.

organic causes. Some anatomical structure is malformed, is infected, or may be affected by some drug or chemical. Other causes may be psychological. These may arise out of a troubled marriage which ends up focusing on sexual relations. Underlying feelings of rejection, mistrust, or resentment set off a chain of physical-emotional events. But whatever the cause, the result is sexual dysfunction.

Female Sexual Dysfunction

A variety of problems detract from women's sexual satisfaction. Descriptions vary and symptoms are diverse, yet the underlying causes are often quite similar. It is important to identify each problem accurately.

The causes of sexual dysfunctions are either physical or psychic. Tenderness from surgery that has not healed sufficiently, genital malformations, poor genital muscle tone, damage from sexual abuse, infections, sensitivity reactions to contraceptive materials, and tumors are physical causes that may respond to medical treatment.

The psychological causes may be somewhat more subtle. Some women carry fears—of men, of pregnancy, of loss of identity, of physical or emotional hurt, or of sexually transmitted diseases. Some are "turned off" because they are hostile toward men in general or toward a specific man; because they have conscious or unconscious love conflicts with another man; because certain acts violate their moral standards; because their partner is unloving; or because of a past emotional trauma. Such causes may require psychotherapy or counseling. None of them should be labeled *frigidity*.

This is an obsolete, misused, "wastebasket" term that has often been used to ascribe antierotic attitudes to some women; it is negative and has no place in discussions such as this.

Hyporesponsiveness Identified by some therapists as general sexual dysfunction, this is an impairment of the vasocongestive or erotic arousal phase of sexual response. The woman lacks erotic feelings, her vagina does not lubricate, and her genitals do not enlarge. She may report little or no enjoyment of sexual activity, find penetration painful, and avoid sexual relations whenever possible. While many women experience such symptoms very infrequently, some experience them in every sexual encounter.

Orgasmic dysfunction This is a specific impairment of the orgasmic reflex. The woman has full erotic feelings, adequate lubrication, and genital swelling. Her sexual response rises to the plateau level, but she is unable to reach orgasm in spite of adequate stimulation. Such a woman enjoys all the aspects of lovemaking and looks forward to it.

Aside from some of the causes mentioned above, another may be the new cultural expectations for women's orgasmic performance. An occasional inability to achieve orgasm may cause a woman to worry about being *anorgasmic,* or unable to achieve orgasm. The more she worries the worse it becomes—a kind of self-fulfilling prophecy. No woman can *will* or force an orgasm. By relaxing, letting themselves go, enjoying all of the aspects of lovemaking, and accepting erotic stimuli, most women should be able to become orgasmic.

Other problems *Vaginismus* is a condition in which involuntary contractions in the vaginal muscles make entrance of the penis difficult if not impossible. An imperforate hymen, recollections of severe pain on first intercourse or of some other trauma, or various forms of psychological conditioning may be causes. The gentle use of a finger or of instrument dilators may help reassure a woman and help her accept erotic stimuli.

Pain during intercourse is termed *dyspareunia*. There may be any number of physical causes: malformations, growths, sensitivity reactions, infections, damage from rape, or uterine problems. Psychological barriers may prevent vaginal lubrication, or fears may bring on perceptions of pain. When such a problem persists, it should be addressed medically or through counseling.

Male Sexual Dysfunction

The male genitals seem quite unlike the female genitals. Yet in their response to erotic stimulation men face the same kinds of problems. As with women, some male problems have physical causes, many of which can be successfully treated medically. Others have conscious or unconscious psychological causes that may require psychotherapy.

Men have an important ego need to feel sexually adequate, just as do women. Some men measure their self-worth in terms of their penile abilities. The slightest experience of failure, along with any doubts of masculinity, may set off a self-perpetuating chain of anxieties and failures. Some men avoid situations that call for closeness and intimacy because they fear of ego-damaging failure. While the symptoms of such masculine sex problems may vary, the causes may be the same.

Erectile dysfunction (impotence) Impotence is the inability to attain or maintain an erection of sufficient strength to perform intercourse; it is a problem that affects most men at some time. *Erectile dysfunction* is a preferable name instead of impotence, with its implications of powerlessness. In some cases the cause may be organic, or due to an anatomical defect. Functional causes may include hormonal imbalances, circulatory problems that bring too little blood to the penis, nervous disorders, drugs, aging, or physical exhaustion.

Most frequently the cause is *psychogenic* — that is, caused by the emotions. One example is *primary erectile dysfunction*, or never having achieved or

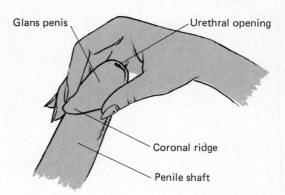

Glans penis · Urethral opening · Coronal ridge · Penile shaft

To delay ejaculation, a woman may apply the squeeze technique to her partner's penis.

maintained an erection of sufficient firmness to engage in coitus. Men who have had successful intercourse in the past but are now incapable of it suffer from *secondary erectile dysfunction*. Men with psychogenic erectile problems have nighttime erections during REM sleep or on awakening from REM sleep.

Most men experience temporary erection problems. Persistent problems are of greater concern. The causes may be deep-seated relational problems, inhibiting religious teachings, psychological problems rooted in adolescence, or homosexual conflicts. Psychotherapy often uncovers the nature of the cause and then reeducates a man to learn to relax and enjoy the physical pleasures of bodily contact without pressure to perform.

Premature ejaculation Masters and Johnson define *premature ejaculation* as the inability to delay ejaculation long enough to satisfy a normally responsive female in at least 50 percent of coital experiences. Such a definition excludes those times when it is transient or due to an exciting new relationship.

Psychological causes of ejaculatory dysfunction are similar to those of other sex problems. Hostility toward a partner, guilt about sex, feelings of sexual inadequacy, fear, and love conflicts may be typical causes. Minimizing the problem is of prime importance; a man needs to be assured unequivocally that the problem can be reversed. With the young, highly aroused male, repeated intercourse may bring on a more controlled response. Regular sex partners can learn to delay orgasm by holding very still when premature ejaculation threatens. Wearing a condom or applying a topical anesthetic ointment to the glans of the penis may reduce excitement. The female may use the "squeeze technique," in which she tightly pinches the glans when ejaculation is about to occur. This conditions the male's body to delay ejaculation. Whatever the cause, the ejaculatory dysfunctions can be successfully treated.

Other problems *Ejaculatory incompetence* is failure of a male to reach a climax. He can become sexually aroused to the plateau level, but is unable to have an orgasm in spite of adequate stimulation. This condition is similar to orgasmic impairment in a female. As with women, relaxation, learning to accept tactile stimulation, and removing the performance expectations from intercourse may enable a male to become orgasmic.

Dyspareunia, or painful intercourse, in the male may be due to a number of physical problems. These may include infection in the genitals, prostate problems, a very sensitive glans penis, or penile structural malformations. Prompt medical treatment should be sought if the problem persists.

Hyperactive Sexual Desire

Common among both sexes are people whose sex lives consist of an endless series of brief encounters with ever-changing partners. They seldom establish any level of emotional intimacy or gain any emotional satisfaction.

The term "satyriasis" or "Don Juanism" has been applied to men who have unusually high numbers of heterosexual encounters. In women, the problem has been called "nymphomania."

Several personality traits are commonly associated with hyperactive sexual desire. The foremost is a feeling (which can be unconscious) of sexual or personal inadequacy. The person needs to constantly "prove" his or her ability to attract lovers and function sexually. But regardless of the number

● ASK YOURSELF

1. What meaning might there be in the word *dysfunction* to show that sexual problems are not the desire, or wish, of the affected partner?
2. Two common female sexual dysfunctions concern the inability to respond sufficiently sexually. How would *hyporesponsiveness* differ from *orgasmic dysfunction?*
3. Earlier, the chapter discussed male sterility and inferility. Contrast these two conditions with impotence.
4. In what respects might sexual dysfunctions in the female be similar in cause to sexual dysfunctions in the male?
5. Some people fantasize about the pleasures of making love to a person (male or female) who has an insatiable appetite for sex. Why might such situations not be everything our fantasies build them up to be?

of lovers attracted or orgasms reached, the fear of inadequacy remains.

Fear of intimacy is also common. Many people, perhaps as a result of past emotional hurts or their feelings of inadequacy, lack the confidence required to establish a close relationship with another person. Lacking intimacy, their sexual relationships always seem incomplete, so the search for satisfaction must continue.

Many people have been raised in sexually repressive atmospheres. Sexual guilt or inhibition, perhaps as a result of childhood conditioning, may act to deny sexual satisfaction, even though full physiological orgasm may occur.

Like other forms of sexual dysfunction, hyperactive sexual desire can often be treated successfully by a qualified therapist. Treatment may involve building a sense of personal adequacy, developing the ability to establish intimacy with another person, or extinguishing feelings of guilt or inhibition.

IN REVIEW

1. While differing somewhat in appearance, the female and male reproductive structures are basically alike.
2. The female sex organs consist of:
   a. The external organs (genitals, vulva, pudenda): mons pubis, lips (both inner and outer), clitoris, and vaginal introitus

b. The internal sex organs: ovaries, fallopian tubes (oviducts), uterus (womb), and vagina

3. The monthly uterine-ovarian-hormonal events in the female constitute the menstrual cycle. The cycle can be divided into four phases:
   a. Destructive (menstrual) phase
   b. Proliferative (follicular) phase
   c. Ovulatory phase
   d. Secretory (luteal) phase

4. The hormone events of the menstrual cycle are under the control of three sets of hormones: hypothalamic (brain), gonadotropic (pituitary), and ovarian.

5. The female secondary sex characteristics develop during puberty. Included is menarche—the female's first menstrual cycle.

6. In pregnancy the fertilized egg embeds itself in the uterine lining (endometrium). Since the lining must be retained for the duration of the pregnancy, menstrual discharges do not occur during pregnancy.

7. Toward the end of pregnancy, the breasts become functional. Nipples enlarge, milk ducts and glands develop within the breast, milk is formed, and after childbirth the milk is "let-down."

8. Menopause marks the completion of a woman's reproductive capacity, and indicates the cession of menstrual cycles.

9. The male sex organs consist of:
   a. The external organs (genitals): penis and scrotum
   b. The internal sex organs: testes, epididymis, vasa deferentia, and fluid-producing glands (seminal vesicles, prostate, and Cowper's glands)

10. Ejaculation is the explosive discharge of semen from the penis. Nocturnal emissions (wet dreams) are ejaculations that occur during sleep.

11. The sexual response cycle includes excitement, plateau, orgasm, and resolution.

12. Female sexual arousal is best indicated by vaginal lubrication. Male sexual arousal is best indicated by erection of the penis.

13. The most important factor in a good sexual relationship is open and honest communication. Both partners need and enjoy a period of sex play before actual intercourse.

14. Some of the more basic coital positions include: man-above, woman-above, lateral face-to-face, and rear-entry.

15. While orgasm is pleasurable and should be an expected part of sexual intercourse, it should not be overemphasized. Simultaneous orgasm is *not* a significant goal.

16. Oral sex is known as cunnilingus when the female is stimulated, as fellatio when the male is stimulated.

17. Sexual intercourse should be engaged in as often as is desired by both partners.

18. Sexual anxieties over the appearance of body parts may affect both males and females.

19. Problems in sexual response are known as sexual dysfunctions. They may be both physical and psychological in cause.

20. Female dysfunctions include: hyporesponsiveness, orgasmic dysfunction, vaginismus, and dyspareunia.

21. Male dysfunctions include: erectile dysfunction (impotence), premature ejaculation, ejaculatory incompetence, and dyspareunia.

22. Compulsive sexuality in males is known as satyriasis; in females it has been called nymphomania.

CHAPTER15

SEXUAL PARTNERSHIP

This generation stands witness to the changing purposes of sexuality. With the increasing concern over excess population, the role of sex is slowly changing. Sex, with love, is more important now in the selection of sexual partners. It is used to express emotional intimacy; it indicates competency in interpersonal relations. It can even be a statement of rebellion.

In the selection of partners, sexual considerations walk hand in hand with romantic love. Arguments to the contrary, sexual intercourse for many is deeply bound up with affection and emotional commitment; it is a part of the attachment-choice process, and it intensifies a relationship.

Together, love and sex are tools for assuring ourselves that we are making the right choice. "Going all the way" is an act that confirms a set of choices. Thus for many, sex is a part of being in love; for them, love justifies sex, and sex justifies love.

The settings of the process vary; they may be heterosexual, homosexual, or bisexual. The conditions may be nonmarital, premarital, marital, or extramarital.

FORMING SEXUAL PARTNERSHIPS

The formation of sexual partnerships begins early. All studies indicate an increased incidence of non-marital sex. More than half of the 21 million young people aged 15–19 are estimated to be sexually active. More than this, about one-fifth of the 13–14

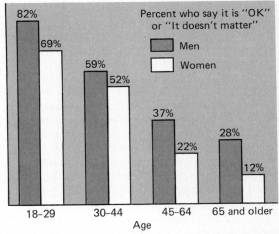

Public tolerance of unmarried couples living together

year olds have already begun their sexual experimentation. A growing proportion of sexually active teen-agers, having begun their sexual activity at an earlier age, is another indication of the changing purposes of sexuality.

Eventually, couples begin living together as another step in the attachment-choice process. Feeling good about their sexual attraction to each other, they want to begin testing a closer relationship. About 40 percent of couples who live together eventually marry each other. For some, living to-

TABLE 15.1   CHANGES IN U.S. FAMILIES SINCE 1970

	1970	Latest	Percent Change	
Marriages performed	2,159,000	2,413,000	Up	11.7%
Divorces granted	708,000	1,182,000	Up	66.9%
Married couples	44,728,000	47,662,000	Up	6.6%
Unmarried couples	523,000	1,346,000	Up	157.4%
Persons living alone	10,851,000	17,202,000	Up	58.5%
Married couples with children	25,541,000	24,625,000	Down	3.6%
Children living with two parents	58,926,000	48,295,000	Down	18.0%
Children living with one parent	8,230,000	11,528,000	Up	40.1%
Average size of household	3.3	2.8	Down	15.2%
Families with both husband and wife working	20,327,000	24,253,000	Up	19.3%
Total population	203,558,371	220,100,000	Up	8.1%

Source: U.S. Department of Commerce data.

gether does not fulfill their deeper emotional needs for security and intimacy. Others want to legitimize the arrangement or want their children to have the legal benefits of a marriage.

Marriage

Most Americans want to marry, and most do—at least once. Because our culture restricts the sexual pleasuring of adults more than many other cultures, most adult sexual intercourse takes place within marriage. One of the concerns of early adulthood is choosing a marriage partner and developing a sexual relationship with that person. This creates high interest in marriage.

Yet marriage is no guarantee of a long, happy relationship. A record number of marriages are ending in divorce. Many divorces are the result of marriages that should never have occurred. The seeds of failure are evident from the beginning. Marriage must be entered into rationally as well as emotionally.

BEING READY FOR MARRIAGE

Looking at marriage, partners ought to reflect on the nature of their relationship and their readiness to formalize it by marriage. There are no chronological, physical, or emotional criteria that qualify a person for matrimony. Yet, certain objective standards, taken together, may give clues of marital readiness.

Age at Marriage

The age at which a couple marries is a good indicator of the chances of that marriage succeeding. Two-thirds of all marriages in which the woman is between the ages of 14 and 17 will end in divorce. One-half of all marriages in which the woman is aged 18–19 will terminate in divorce.

Many studies have shown that the level of satisfaction and success in marriage increases with the age of the couple at the time of marriage. Emotional conflict, sexual adjustments, money problems, in-law trouble, and divorce are all much more common among couples who marry in their teens. Marriages where the husband was in his teens are particularly unhappy.

Some people feel pushed to marry at a fairly early age to avoid getting "left out." Such fear is not warranted because at any age, numerous individuals eligible for marriage will be found. In fact, some excellent marriage prospects delay marriage for several years in order to reach educational or career goals. The average age at first marriage is now over 24 years for men and over 21 years for women, and is rising. The urgency to marry at prescribed ages is not as strong as in the past.

Many people greatly change their value systems and life styles between the ages of 16 and 22. If people marry before this change, there is a strong possibility that they will no longer meet each other's emotional needs later.

A related problem is that early marriage often in-

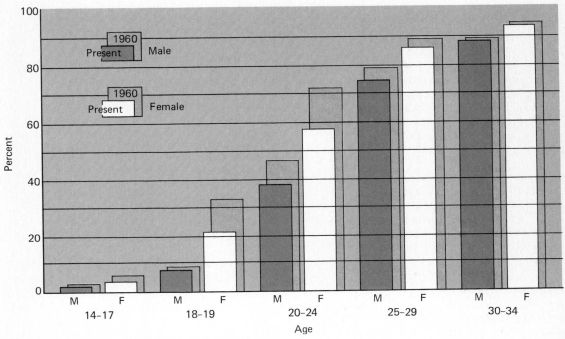

Young Americans: Percent married by sex and age.

terferes with the development of a mature philosophy. Individuals tend to stop growing intellectually when they marry. This can be prevented, of course, but many young people fall into a deep philosophical and intellectual rut from which they never escape. Or one partner grows while the other stagnates, a sure formula for unhappiness.

Emotional Maturity

The emotional demands of marriage are much greater than those a couple experiences during dating. Thus, an important element of marriage is emotional maturity. This generally increases with age, but some individuals remain emotionally adolescent even though they have legally become adults.

Before marriage, a person should be as free as possible of emotional maladjustments, such as moodiness, jealousy, anxiety, depression, and insecurity. The presence of such traits in a marriage can be destructive. A person who is subject to marked trouble in these areas should seek qualified professional counsel.

The truly mature person has skill in establishing

and maintaining good interpersonal relationships. He or she recognizes the needs of others and is willing to assume some responsibility for meeting these needs. Each partner in a happy marriage must have concern for the other.

Social Maturity

Social maturity develops through social interaction. Before marriage, social maturity may be built through dating different individuals. This gives a better basis for selecting a marriage partner and helps satisfy social curiosity. The person whose dating is more restricted may later, after marriage, feel he or she has missed something. Some try to compensate by having extramarital affairs.

It is important to experience a period of single, independent living before marriage, a time of freedom between the dependence of living with one's parents and the responsibilities of marriage. It is only by living away from parents that one can really come to know oneself, develop full social competency, and learn to manage one's own affairs. Anyone who is too immature, insecure, or otherwise in-

● ASK YOURSELF

1. What might be a "best" age for either a man or woman to marry? Support your point of view.
2. Suppose you enjoy the company of a boyfriend (or girlfriend) more than any other you've ever gone with. You've thought of marriage, but you would hate to make a mistake. A list of the most important considerations in helping you decide might include which items?

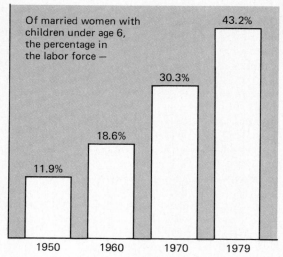

Of all married women with children ages 6 to 17, 59.1 percent now work or seek jobs — up from 28.3 percent in 1950.

The increase in U.S. working mothers, 1950–1979.

competent to live independently is, for the same reason, not ready for marriage. Many people, after enjoying their independence, feel a desire to "settle down" into marriage. Others find single life permanently satisfying and prefer not to marry. There is absolutely no reason why anyone should feel an obligation to marry.

Financial Resources

Although less important than the preceding personal characteristics, financial factors must be considered before marriage. The minimum amount of money a couple needs to live on is highly variable. Most young couples enter marriage without great amounts of money. But if either or both of the young married people are to be students, they must carefully evaluate the marriage situation to avoid ending an educational program for financial reasons.

Because few couples ever have what they consider to be enough money, a couple's attitude toward money and how to use it is likely to be more important than the size of the paycheck. A given amount of spendable income for one couple may be sufficient to meet common interests, while the next couple may be suing for divorce because the same amount of money is not enough. The relation between income and happiness depends on expectations. When a couple feels committed to an occupational field where income is lower, they can be happy if they are content to live at that level. If, however, their expectations run higher than income, their marriage may be troubled.

Increasingly common is the *two-career marriage*. More than half of the nation's mothers work outside the home. Six out of ten married women with school-age children work. Of married women with children under six years, 43 percent work.

Positive Personality Traits

By far the most important characteristic in a potential marriage partner is his or her personality. Some people have positive personality traits that enable them to enjoy life to its fullest and to bring joy to anyone in contact with them. Others are so burdened with negative reactions that it is impossible for them to be happy, as it would be for anyone who must live with them. Traits that help produce happiness in marriage include the ability to adjust easily to changes in conditions, optimism, a sense of humor, an honest concern for the needs of others, a sense of ethnics, and freedom from such negative traits as anxiety, depression, insecurity, and jealousy.

Mutual Need Satisfaction

The happy and lasting marriage is one in which the needs of each individual are adequately satisfied. While the idea may not appeal to romantics, the basic reason why people marry is to satisfy their needs. A good marriage satisfies needs for sex, love, companionship, and security, as well as many subtle psychological needs. Since everyone is unique in his or her psychological needs, the characterization of the imaginary ideal partner is a highly individual matter. Only by knowing each other very well over a long period of time and in a variety of situations (both pleasant and unpleasant), can two people learn whether they fulfill each other's needs.

Genuine Mutual Love

The distinction between genuine love and infatuation is not always clear. Infatuation is frequently a substitute for love until a person has the capacity to love someone fully and deeply. It tends to involve sexual attraction more than personality attraction. Infatuation is unrealistic, a fantasy. The object of the infatuation is seen as a "dream mate" who lacks any undesirble traits. Infatuation is often immediate, whereas love develops with time. Usually, infatuation wears off quickly. Yet it may, with time, develop into mature love.

A person truly in love is concerned with the loved one's happiness and well-being. He or she is tender, protecting, loyal, and is willing to sacrifice some pleasures in order to bring pleasure to the loved one. There is a desire to share ideas, emotions, goals, and experiences. Love continues to grow between partners committed to serving each other's needs and interests.

A strong sexual attraction should exist between prospective marriage partners. But it is important not to confuse sexual appeal with love. Love and sex appeal are not the same thing. Sexual attraction ought to confirm, or validate, one's feelings of love for the partner. A couple may have a good sexual relationship without loving each other, but such a relationship makes a poor basis for a happy marriage.

Hereditary Traits

Some individuals carry obvious hereditary defects. Others seem perfectly normal, but come from families in which such defects are known to occur. The latter individuals may or may not be carrying undesirable hidden genes. If there is any question regarding the possibility of transmitting defective genes, it is wise to seek genetic counseling, either from a physician or a genetic counselor recommended by a physician. For such people any decision to marry and have children, to marry and not have children, or not to marry should be based upon competent guidance. It should not be on the advice of well-meaning but uninformed friends and relatives.

Agreement on Parenthood

Any couple considering marriage should reveal their feelings about having children. Ideally, they should agree on whether they want children and, if so, how many. It is always unfortunate when a person who wants children marries one who would rather remain childless. Automatically, one or the other is destined to be unhappy. If there is serious disagreement on this matter, it would be a good idea for each individual to look for another mate.

If neither person wants children, there is no reason to feel guilty about a decision to remain childless. Studies have shown that children are not essential to happiness. In fact, they have been shown to place additional strain on already unhappy marriages. A couple need feel no obligation to themselves or to society to produce children. Many people now feel an obligation to limit the number of children produced.

Similarity in Background

Any couple considering marriage should take a critical, objective look at their differences in personality and family background. These differences may be minor and insignificant or major and may have a great bearing on the marriage. Many studies have shown that the more similarities between two individuals, the greater their chances of marital success. Significant differences may involve age, nationality or ethnic background, economic status, education, intelligence, religion, or previous marital status. Most marriages can be successful despite these differences if the couple is willing to work out the special problems involved.

Age Age differences at marriage are not uncommon. In four out of ten marriages, the husband is three to nine years older than the wife. But where there is a wide difference in age, the individuals must examine their motives. Is it the inability to find a partner close to one's own age? Is it the desire for immediate economic security? Is the older person seen as a "father" or "mother" figure? Does the older person need to dominate or the younger to be dominated? On the average, marriages are happiest when the man and woman are

An interracial couple.

within a few years of each other in age. Wide-age marriages can be successful if both partners assess their motives and discuss them honestly with each other.

Ethnic and racial differences Marriages between members of different ethnic or racial groups face the most difficult problems of any type of mixed marriage. Not only may there be problems within the marriage, but the couple may experience resentment and prejudice from family members.

The internal problems in these marriages may revolve around customs, standards, and points of view. For example, the attitudes toward women and their rights, duties, and status may be quite different. Family patterns of authority and the role expected of each member may conflict. Attitudes on

raising children and care of elderly relatives may be another area of disagreement. These problems do not appear in all mixed marriages, but such topics should be discussed thoroughly before marriage.

The problems caused by prejudices of family members and society are particularly frustrating, because they should not exist in an enlightened society. The source of many of these problems is the ethnocentric attitude of groups that guard their ethnic heritage to excess and often sincerely believe in the supremacy of their group over all others. The elders of some of these groups encourage their youth to maintain a distance from outsiders, to continue to respect the traditions and customs of the group, and to marry within the group. The young man or woman who marries outside the group may even be rejected by immediate family members.

Other problems may arise in finding housing and employment, especially in black-white marriages. There may be problems in finding friends who will honestly accept both partners. The amount of social prejudice felt by the interracial couple will vary from city to city and with the part of the country. Such couples may be most comfortable in college towns and large cities, where the general attitude is usually more open.

Since 1970 the percentage of black-white married couples has doubled, and the number of interracial marriages is likely to increase.

Economic status Even though our society has always claimed that one of its goals is social equality regardless of economic status, patterns of behavior do vary greatly with the socioeconomic level. Behavior that is "correct" at one level may meet with disapproval at another level. Attitudes toward authority, freedom, ethics, education, and other values may differ. Married individuals from widely different economic backgrounds must often adjust some of their attitudes.

One problem area for these couples is in-law relationships. The wealthier set of in-laws may not entirely accept the son-in-law or daughter-in-law who comes from a less affluent background. Other problems can arise when such a couple must live on a very limited income. They may need to accept financial aid from the wealthy in-laws (if it is offered). Or they can try to live within their income, which may put burdens on their relationship. While discrepancies in economic background may turn